Advance Praise for *Straw Bale Building Details*

Straw Bale Building Details is an incredibly comprehensive expl...
and why so many of us are utilizing this housing typology. I'm grateful to the authors of this book for coming together to publish the answers to our straw bale FAQs, and I am indebted to them for creating a text I plan to share with young, emerging professionals as well as experienced conventional builders. Sharing the details will surely lead to the growth of our industry.

—Emily Niehaus, founder, Community Rebuilds

This is the book I have been waiting for — the complete guide to straw bale construction. A remarkable group effort to bring lessons learned from architects, builders owner-builders together in one volume. Detailed illustrations and photos clearly show how it is done, and it's also excellent for working with building inspectors, banks, and insurance companies. For a more fire resistant, comfortable and efficient home — read this book.

—David Bainbridge, sustainability consultant, co-author, *The Straw Bale House*

Get this book, read it immediately. Co-created by some of the greatest eco-architects and builders who have ever lived, it is highly readable, comprehensive, and absolutely relevant to the immense challenges of our time. More, it will inspire every reader to see how we can engage small and large issues through the power of design, beauty, and inspired action.

—Mark Lakeman, founder, communitecture, architecture and planning;
and cofounder, City Repair and the Village Building Convergence

Straw Bale Building Details is a must-have for anyone interested in building with bales. A collaboration of many experts, the book provides multiple options for building a high quality, long lasting straw bale structure with enough flexibility in design practice to allow for differing approaches to that same end. Whether you're a professional or owner-builder, *Straw Bale Building Details* will prove invaluable in the development of your dream straw bale home.

—Andrew Morrison; Straw Bale Mentor/Consultant, StrawBale.com

To anyone contemplating building with bales: Start here! CASBA's *Straw Bale Building Details* is an unparalleled resource, representing decades of cumulative wisdom from dozens of top architects, engineers and contractors. Much more than the title suggests, included are structural options and finishing details for varying climates, plus instruction on stacking bales, plastering walls, and even to how to hang up a picture. Attractive photos and illustrations offer inspiring aesthetic possibilities. Thank you, CASBA!

—Catherine Wanek, author/photographer, *The New Strawbale Home* and *The Hybrid House*;
former editor, *The Last Straw Journal*; and co-editor, *The Art of Natural Building*

Years in the making and greatly anticipated, *Straw Bale Building Details*, was well worth the wait. The details are extensive, thoroughly and carefully described, illustrated, and explained—a tour de force. But what makes this book enormously more useful and beneficial is that those details are placed into their appropriate context, shifting wherever needed from the details to the bigger picture. This is a masterwork that belongs in the hands of anyone designing, building, or issuing permits for straw bale buildings.

—David Eisenberg, director, Development Center for Appropriate Technology, and co-author, *The Straw Bale House*

Straw Bale Building Details

An Illustrated Guide for Design and Construction

California Straw Building Association

(CASBA)

new society
PUBLISHERS

Cover by Diane McIntosh
Cover Photograph—Edward Caldwell
Cover Illustration—Devin Kinney
Black and white photography by Rebecca Tasker and
Jim Reiland unless otherwise noted

Printed in Canada. April 2019.

Inquiries regarding requests to reprint all or part of *Straw Bale Building Details* should be addressed to New Society Publishers at the address below.
To order directly from the publishers, please call toll-free (North America) 1-800-567-6772, or order online at www.newsociety.com

Any other inquiries can be directed by mail to:

New Society Publishers
P.O. Box 189, Gabriola Island, BC V0R 1X0, Canada
(250) 247-9737

Library and Archives Canada Cataloguing in Publication

Title: Straw bale building details : an illustrated guide for design and construction / California Straw Building Association (CASBA).

Names: California Straw Building Association, author.

Description: Includes index.

Identifiers: Canadiana (print) 2019005865X | Canadiana (ebook) 20190058684 | ISBN 9780865719033 (softcover) | ISBN 9781550926965 (PDF) | ISBN 9781771422925 (EPUB)

Subjects: LCSH: Straw bale houses—Design and construction. | LCSH: Straw bale houses—California.

Classification: LCC TH4818.S77 C35 2019 | DDC 693/.997—dc23

Funded by the Government of Canada | Financé par le gouvernement du Canada | Canada

New Society Publishers' mission is to publish books that contribute in fundamental ways to building an ecologically sustainable and just society, and to do so with the least possible impact upon the environment, in a manner that models that vision.

Contents

Acknowledgments

Our inspiration for this book came from Ken Haggard and Scott Clark and their editorial work on *Straw Bale Construction Details: A Sourcebook*, published through the California Straw Building Association (CASBA) in 2000.

Our original goal here was to refresh this work—to "update" it. But early in our research it became clear that large amounts of new information had emerged in straw bale construction since 2000, so we decided to reorganize the framework of the original book to incorporate this new material while honoring the straightforward presentation of the original book—including its easily accessible illustrations and clear text. We remain greatly indebted to Haggard and Clark for their trailblazing work and insights into an effective way to present this material.

This book has been made possible by the support, encouragement, and involvement of the CASBA Advisory Board, whose members have been a sounding board, giving feedback and sharing ideas. In particular, we acknowledge the support of Maurice and Joy Bennett, directors of CASBA from 2000–2013, and David Arkin, who has served as CASBA's director since 2014. These people have given countless hours to the promotion of straw building in California and beyond. Their leadership, enthusiasm, and encouragement saw this project to completion.

Finally, we thank the book's many contributors. This was a group effort, with most of CASBA's active members playing a role either contributing, reviewing, or patiently supporting this work between 2006 and 2018. They shared construction drawings and specifications, participated in breakout group sessions, and offered valuable insights and feedback through many meetings and revisions. They helped shape the ideas found in this book.

Soon after the project began, it became clear there was no single "best" way to design and build with straw bales. The concept of *good boots, a good hat, and a coat that breathes* was well understood, but the application of this principle varied with different design aesthetics, construction preferences, and each of California's coastal, desert, foothill, valley, and mountain climates. The strong earthquakes that can occur throughout the state factored into the conversation as well. This book needed to reflect this variety in order to illustrate the wide range of straw bale building solutions useful in different regions in California, but also in the United States and other countries.

CASBA's *open source* character made this project possible. Members volunteered their time and expertise and freely contributed straw bale design and building ideas and experiences to the melting pot that became this book. Because of our long history of collaboration, we had some difficulty knowing the exact lineage of some content and attributing credit for specific contributions. We can all claim a hand in this work. More importantly, what has emerged is far better than what any of us might have produced on our own.

We also acknowledge those who shouldered a larger load—contributors who spent additional time developing a chapter or illustrations. This smaller group of people periodically stepped away from their busy work and family lives to commit more time to this effort, and they deserve special recognition and heartfelt thanks.

To start at the beginning, we owe a great debt to the original Detail Update Committee (the DUCs), who steered the effort during the early years, beginning in 2006. Janet Johnston, Tim and Dadre Rudolph, Darcey Messner, Celine Pinet, and Cole Butler gave this book its initial vision, content, and form. In 2008, Jim Reiland, Rebecca Tasker, and Bob Theis picked up where the DUCs left off; they invited other contributors to share their expertise, and shaped the resulting material into the book you're holding today. Jim Reiland enlisted other contributors and managed the effort, editing for content, voice, and consistency, and, along with Martin Hammer and Bob Theis, he shaped the final draft. Martin Hammer also made the invaluable contribution of refining the text and aligning it with the *IRC Appendix S—Strawbale Construction*. Special thanks to Lesley Christiana for negotiating the publishing agreement, managing the production schedule, and coordinating the illustrations.

Each chapter had core contributors who created a first draft or significantly added to, rearranged, or shaped the work:

Chapter 1: Why Build with Straw Bales? Nehemiah Stone wrote most of the building science portions of this chapter; Dennis LaGrande gave us a farmer's perspective on grain crops and harvesting straw bales; and Jim Reiland, Janet Johnston, and Bob Theis supplied additional text.

Chapter 2: Designing with Straw Bales. Janet Johnston crafted the first draft, which Bob Theis and Jim Reiland revised and updated.

Chapter 3: Structural Design Considerations. Tim Rudolph worked on the initial draft, which he completed and published in pamphlet form in 2009. Jim Reiland, Anthony Dente, Eric Spletzer, Darcey Messner, and Martin Hammer developed and updated the material with new information and perspectives.

Chapter 4: Electrical, Plumbing, Ducts, and Flues in Straw Bale Walls. Janet Johnston, Bob Theis, and John Swearingen contributed to this chapter about running wires, plumbing, vents, and flues through straw bale walls.

Chapter 5: Stacking Straw Bale Walls. Jim Reiland developed this chapter about site-built straw bale building, with help from Rebecca Tasker, Janet Johnston, and John Swearingen.

Chapter 6: Plastering Straw Bale Walls. Special thanks to Tracy Thieriot for creating the framework and initial draft, and to Bob Theis, Rebecca Tasker, and Jim Reiland for filling in and rounding out the text and ideas presented.

Chapter 7: Straw Bale Construction and Building Codes. Martin Hammer wrote the introduction to this chapter, which is mostly comprised of Appendix S—Straw-bale Construction with commentary, from the 2018 International Residential Code and Commentary. Martin is also lead author of Appendix S and its commentary, with major contributions from David Eisenberg, Dan Smith, Mark Aschheim, Nehemiah Stone, Kevin Donahue, and over a dozen other straw bale building practitioners.

Illustrations. The quest for effective graphics took many years and several iterations—all of which inspired and informed the ultimate detail illustration. Dan Smith, John Swearingen, Bob Theis, and Anni Tilt had many lively discussions about not only which bale systems and details should be included but how best to illustrate them so that they clearly conveyed concepts to both a professional and lay audience, and were easy to update. Devin Kinney and John Koester provided invaluable skill and countless hours translating sketches into virtual 3-D and producing annotated drawings from them. Rebecca Tasker created supplemental illustrations, tables, and graphs and managed the effort to collect and provide photographs suitable for publication. Eric Spletzer created illustrations for Chapter 3: Structural Design Considerations.

In the earlier detail book, Haggard and Clark knew they were writing in the midst of rapid changes. The last sentence of the preface reads: "This source book is an attempt to document the rapid evolution of straw bale construction and clarify many of the exciting possibilities and practicalities of this method of sustainable building." True then, and true today. We recognize that as thorough and comprehensive as we have tried to be, change is the only constant. We fully expect this book to be updated as straw gains traction in the effort to build beautiful, healthy, and truly sustainable structures.

Finally, as this book goes into publication, we realize it owes its existence to the generosity and encouragement of many around us, especially our families, who at times wondered if we'd ever finish this project, which has been affectionately called "The Never-Ending, Ever-Evolving Book of Straw Bale Building Details." This effort is a testament to the community spirit and social consciousness that is the heart of the straw building community and the California Straw Building Association.

All proceeds from the royalties earned by this book will go to further CASBA's mission of outreach, research, and promoting the use of straw as a building material.

Book Contributors

Original Detail Update Committee Members

Darcey Messner, P.E.
Janet Johnston, Architect
Celine Pinet, PhD
Cole Butler
Tim Rudolph, P.E.
Dadre Rudolph

Jim Reiland, Managing Editor
Rebecca Tasker, Image Manager
Lesley Christiana, Project Manager

Disclaimer: Recommendations and drawings are for informational purposes only. Final details and construction applications must be developed by the designer for every project's specific conditions.

Contributors

Danielle Alvarez
David Arkin, Architect
Maurice and Joy Bennett
Bob Bolles, Consultant
Cole Butler, Architect
Scott Clark, Architect
Chris Church
Ken Haggard, Architect
Martin Hammer, Architect
Marcus Hardwick, Consultant
Drew Hubble, Architect
Michael Jacob, Contractor
Janet Johnston, Architect
Bruce King, P.E
Devin Kinney, Architect
James Kloote, Architectural Intern
John Koester, Owner-Builder, Illustrator
Dennis LaGrande, Farmer
Kelly Lerner, Architect
Mike Long, Contractor
Dietmar Lorenz, Architect
Chris Magwood, Designer-Builder
Henri Mannik, PE
Darcey Messner, P.E.
Tim Owen-Kennedy, Contractor
Celine Pinet, PhD
Jacob Deva Racusin, Designer-Builder
Jim Reiland, Contractor
Tim Rudolph, P.E.
Turko Semmes, Contractor
Daniel Silvernail, Architect
Dan Smith, Architect
Nehemiah Stone, Consultant
John Straube, PhD
John Swearingen, Contractor
Rebecca Tasker, Contractor
Tracy Vogel Theiriot, Contractor
Bob Theis, Architect
Mark Tighe, Consultant
Anni Tilt, Architect

Evaluators/Reviewers

Craig Dobbs, PE
Bill Donovan, Contractor
Jim Furness, Contractor
Kathy Gregor, Owner-Builder
Martin Hammer, Architect
Mike Long, Contractor
Henri Mannik, PE
Charan Masters, Contractor
Greg McMillan, Contractor
Jim Reiland, Contractor
Joy Rogalla, Owner-Builder
Alan Schmidt, Contractor
Turko Semmes, Contractor
Bob Theis, Architect
Mark Tighe, Consultant

Photo Contributors

David Arkin
Emily Baranek
Erica Bush
Edward Caldwell
Claudine Cavet
Martin Hammer
James Henderson
Dietmar Lorenz
Eric Millette
Jim Reiland
Paul Schraub
B J Semmes
Jonathan Shaw
John Swearingen
Rebecca Tasker
Catherine Wanek
WRNS Studios

Foreword

As we are all too aware, the world is facing many challenges. These include global climate change, economic inequality, homelessness and a lack of affordable housing, and a general longing for a greater sense of community.

While straw building cannot by itself solve all of these problems, remarkably, it offers solutions that have the potential to address each of them. As a highly insulating material, straw can help create housing that is comfortable, affordable, safe, and ecological for a large number of people—in California, the US, and worldwide. The thick walls are beautiful. They are also strong and fire resistant; this has been proven by earthquake testing, rebuilding efforts in Pakistan and Nepal, and the survival of several homes after devastating wildfires in the western US.

Straw is recognized as one of the few building materials that sequesters carbon. While there are other plant-based, minimally processed, rapidly renewable materials (bamboo, hemp, and, to a lesser degree, wood), none is as effective at storing more carbon than it takes to grow and process. (And, in combination with regenerative agriculture, this positive impact can be more than doubled.) Buildings can and must become part of our planet's carbon storage solution.

Straw bale construction is at a unique nexus of our many needs, and it is also poised for rapid expansion into more and bigger buildings. Along with now being in the code and thoroughly tested, this detail book is another step toward getting this esoteric but well-vetted information out there to many more people. This book is rare because it represents consensus among many professionals; it is a compilation of far-flung, hard-won knowledge. We think of this book as a new "standard," and hope it becomes a textbook for workshops, natural building programs, and green building courses taught in trade schools and universities.

We invite you to dig deep into the world of straw bale construction. The creators of this detail book illustrate numerous ways to assemble and ensure the long-term performance and durability of straw bale structures, but these are by no means the only ways. We hope you learn the lessons presented here, but also feel license to invent new ways of building using this versatile resource.

David Arkin, AIA–CASBA Director
August 2018–California, USA

Introduction

At the beginning of the 20th century, settlers in the timber-poor upper Great Plains of the United States and Canada demonstrated that necessity is the mother of invention; applying the recently invented baling machine to abundant grasslands, they produced bales, and then bale structures. Many bale houses, churches, and even a courthouse were built at this time. However, this unique building approach was bypassed and forgotten with the onset of rapid industrialization, the nationwide homogenization of building techniques, and a newly developed ease in transportation of building materials.

As the 20th century drew to a close, increasing awareness of industrialization's true costs drove emerging concerns for sustainable living. During this time of reevaluation, straw bale construction enjoyed a revival thanks to the pioneering efforts of David Bainbridge, David Eisenberg, Pliny Fisk, Judy Knox, Matts Myhrman, Bill and Athena Steen, and many others. Within the last two decades, the resurgence and evolution of straw bale building has exceeded our expectations.

The California Straw Building Association (CASBA) was formed 20 years ago on the Carrizo Plain of South Central California by a small group of dedicated and inspired building professionals. Since then, it has grown in size and influence. CASBA is devoted to furthering the practice of straw building by facilitating the exchange of current information and practical experience, promoting and conducting research and testing, and making that body of knowledge available to working professionals and the public at large.

We recognize that interest in straw building will continue to flower and hope this detail book nurtures and enables that growth. It gathers, explains, and illustrates current practices in straw bale construction. As most of California lies in moderately high seismic zones, many details in this book are especially appropriate for similar areas. However, most details will also be useful to other parts of North America—and anywhere grain crops are grown that could provide straw building materials.

Whatever the reasons for building with straw, this book will help guide decisions about both architectural and structural design, and offer proven details, methods, and insights into building with bales.

This book is not a step-by-step guide to building a straw bale structure—although it offers plenty of guidance for designing, engineering, and building one. Instead, we crafted this book to inform designers and builders about the many choices they have on the journey to creating a beautiful, healthy, energy-efficient, long-lasting structure.

The first chapter explains why people are drawn to this building system and explains the building science behind it. The second chapter covers design considerations for building with bales. Chapter 3 discusses structural issues unique to straw bale structures. The next few chapters offer practical guidance on installing utilities, stacking the straw bale walls, and finally, applying protective coats of plaster. Conveniently, the final chapter includes the often-referenced straw bale building code.

If you are already familiar with straw bale characteristics, the building design process, and plastering, or if you plan to leave the structural engineering entirely in other hands, feel free to skip directly to the sections that most interest you. Cross-references throughout the book demonstrate the interconnectedness of design and the building process.

1

Why Build with Straw Bales?

Straw bale buildings have been around since the late 1980s, and for 100 years before that, counting the original structures in Nebraska where straw bale building was born. There are now straw bale buildings in every US state—hundreds or thousands in some—and in over 50 countries worldwide. Yet despite this proliferation (along with some excellent coverage in magazines and newspapers and the outreach efforts of regional natural building groups), few people are aware that straw bales can be used to construct buildings. While working on straw bale buildings, conducting building tours, or staffing a green building show, we're often approached by curious people who say "They make buildings out of straw bales?" Yes! And each passing year sees the construction of more straw bale homes and commercial or institutional buildings. When there are so many choices for how to build, it's reasonable to wonder why people choose straw bales. The explanation includes the entire range of why people do anything, from highly emotional to completely rational.

Reasons for Building with Straw Bales

Some might check everything on this list; others just a few.

Aesthetics: Straw bale buildings are beautiful! Plastered straw bale walls offer excellent textures and a palette for natural plasters quite unlike conventional modern finishes. Light and shadow playing across a hand-plastered surface is more alive than light on the flat surfaces found in most buildings today. The deep window and door reveals that are a result of building with bales offer many advantages. They give weight and solidity to the building, while also creating opportunities for rounded corners and soft edges, greater control of how natural light enters a space, and generous window seats. Like thick adobe or stone walls, straw bale walls offer a lasting sensory impression.

Security: Many people feel more secure in a thick-walled building. Thick walls create the opportunity for prospect and refuge—you can see out while feeling secure within. Nowhere is this more evident than when sitting in a window seat nestled in a straw bale wall, where you can be on the edge between inside and out.

Design Flexibility: Straw bales can be used to create straight or curved walls, can be formed and shaped, and allow for a great variety of architectural creativity. Straw bale walls can support roof and floor loads and may also be part of a shear wall system to resist wind and earthquakes. Even when a building has a simple rectangular footprint, the walls can be sculpted with niches and rounded window and door openings that lend dynamic character to an otherwise routine shape.

Insulation: Straw bales in a wall assembly provide excellent thermal insulation; plaster covers the dense straw bale core inside and out and creates a highly effective air barrier. Depending on bale density and orientation, a well-stacked straw bale wall offers insulation values ranging from the mid R-20s to the mid R-30s—which exceeds that of most framed wall systems using conventional insulation.

Thermal Mass: An energy-efficient building also has adequate thermal mass to capture and store solar radiation or heat from another source. The thick plaster on interior straw bale walls provides excellent distributed thermal mass. The complete straw bale wall assembly buffers diurnal temperature changes and keeps a straw bale structure's internal temperature remarkably stable. Thermal mass, combined with good insulation, dramatically lowers heating and cooling costs.

Sound Privacy: Straw bale structures are particularly quiet inside. The walls effectively mute traffic, industrial, agricultural, and other nuisance noise.

Agricultural By-product: Straw bale buildings use the residue from annually harvested food crops—rice, wheat, barley, oats, rye—while usually requiring less dimensional lumber and plywood than conventional structures. Straw is the woody stems left after seed heads (the grain) are removed. Straw has other uses like mulch, animal bedding, erosion control, and additives for livestock feed, but most grain-producing regions have an abundance of straw that can be used as a building material. Every region has a common type of straw and bale size, depending on which grains are farmed locally and the baling machines used for harvesting.

Embodied Energy: Building a structure of any size, for any number of people, is an environmentally costly undertaking—among the most costly things we do during our lives. Almost half of all energy consumed in North America each year goes to constructing and operating buildings. How might we improve the world if we could reduce that amount? Think of energy the way you might think about a building budget. It requires energy to build a structure, and the building needs operating energy over its lifetime. This is a building's energy budget. Could we reduce that budget by using fewer energy-intensive materials to build with? The energy used to create, transport, and assemble the building materials—the embodied energy—will be costly if the materials are highly processed or transported from long distances, and using those materials will add to the energy budget. Even when the finished building is "net zero" (creating as much energy as it consumes) or requires little energy to heat, cool, light, and operate, a high embodied energy building's initial steps backward into energy debt delay the point at which the building becomes a net energy benefit. Locally sourced straw has one of the lowest embodied energy values of any building material. Grain crops, and thus straw bales, require minimal carbon expenditure (in the form of petroleum products like fertilizer and fuel) to plant, fertilize, and harvest. Thoughtful design and material selection can minimize embodied energy in a building and shorten the time for a net-zero structure to overcome its embodied energy and make an energy contribution.

Carbon Sequestration: Unique among building materials, straw stores 60 times more carbon than is used to grow, bale, and transport it to building sites in the same region. The carbon inputs from petroleum-based fuel and fertilizer amount to a tiny fraction of the amount of carbon stored in straw through photosynthesis. If it doesn't burn or decompose, the sequestered carbon in plant-based building materials like wood and straw prevents the formation of CO_2, the potent greenhouse gas. Sequestering straw decreases methane emissions, too, which are more damaging than CO_2 because each methane molecule has 20–25 times the short-term heat-capturing potential of a single CO_2 molecule.

What are the potential benefits? Every pound of carbon stored in growing plants prevents the formation of 3.67 pounds of CO_2. As with most woody plant materials, each straw bale is approximately 40% carbon by weight. If a two-string straw bale weighs about 65 pounds, it sequesters approximately 26 pounds of carbon, preventing the formation of 95 pounds of CO_2. The bales in a 2,000-square-foot straw bale home (approximately 220 two-string bales) sequester 5,720 pounds of carbon, preventing the formation of nearly 21,000 pounds of CO_2. Such a home sequesters almost three tons of carbon and keeps over ten tons of CO_2 from forming.

North America grows a lot of grain. If only one-tenth of the residual straw were used for building, over two million 2,000-square-foot homes could be built each year. For some perspective, there were fewer than one million new home starts in North America in 2016. If that many homes were thoughtfully built with straw each year, they would annually lock up nearly three million tons of carbon and keep 10.5 million tons of CO_2 out of the atmosphere.

Fire Resistance: Straw bale structures have survived wildfires that burned nearby conventional structures to the ground. Why? Dense straw bales covered with thick plasters resist both ignition and heat transference. A straw bale wall with one inch of exterior cement-lime plaster has earned a two-hour fire rating, and with earthen plasters, a one-hour fire rating. This equals or exceeds many conventional wood-frame walls that offer a one-hour or 20-minute fire rating. Other fire-resistant building features like Class A roofing assemblies and safe wildland fire prevention practices like clean gutters and reduced fuels around the building sites improve the odds.

Natural Non-Toxic Materials: Many modern building materials contribute to poor indoor air quality by off-gassing by-products like sulfur dioxide, ozone, acetic acid, chlorides, and formaldehyde; new chemicals are introduced each year that haven't been tested for long-term impacts. Humans have lived with natural building materials like straw, clay, and lime for millennia. These materials promote healthy indoor air quality because they don't off-gas noxious chemicals as they cure or age. When the building reaches the end of its (hopefully) long functional life, its materials can return to the Earth with no ill effect.

Community: Friends, neighbors, and family can assist with many straw bale building projects, much like the barn raisings of an era gone by. Enlisting others to help with bale wall stacking and plastering can be fun, educational, and rewarding. It may also lower building costs.

The Building Science of a Straw Bale Wall Assembly

After expressing surprise that buildings can be made with straw, people often ask "How, exactly, does that work?" They might be thinking about the childhood fable involving a wolf and three little pigs. Or they might be trying to figure out how straw fits into the modern building material framework they're familiar with. Today's building industry uses a wide variety of building methods—stud frame, insulated concrete forms, structural insulated panels, to name a few. Industry has also created a wide variety of insulation and sheathing materials that are combined in different wall systems so the buildings perform as designed. Many of these building materials are airtight and waterproof; when carefully assembled in a high-performance "green" building and combined with an onsite power source like solar panels, they often result in "net-zero" building performance.

Where do plastered straw bale walls fit into this picture? They can achieve the same "net-zero" performance, but they accomplish it differently.

To understand how a straw bale building performs, visualize the straw as tightly bundled hollow cellulose tubes that trap air in and around them, especially when the interior and exterior surfaces of the bales are covered with plaster. This composite wall *assembly* has a number of desirable properties.

Thermal Resistance

Heat transfer from a warmer entity to a cooler one happens through three mechanisms: conduction, convection, and radiation. All three have a place in understanding the performance of straw bale wall systems. The term *R-value* primarily refers to conductive heat transfer and is widely used to describe the relative performance of various insulation materials. For example, 3½" fiberglass batts are listed at R-13 and R-15. Since they are factory produced, the tested R-values are consistent from one purchase to the next. Straw bales are not produced nearly as consistently, so they generally receive conservative R-values based on test results. Appendix S in the 2018 International Residential Code (IRC) assigns R-1.55 per inch for bales laid flat and R-1.85 per inch for bales stacked on-edge (more on that later). See Figure 1-1. These values are based on the most reliable tests to date, conducted in the UK (2012), Denmark (2004), and at the Oak Ridge National Laboratory in Tennessee (1998).

Given the varied dimensions of the straw bales used in construction and other variables (such as bale density), actual insulation values of anywhere from R-27 and R-34 are possible for a straw bale wall. Visit "Resources" at the California Straw Building Association website, www.strawbuilding.org, for references to more detailed technical discussions.

Keep in mind, though, that tested R-values are barely half the thermal performance story. In stud wall construction, thermal bridges (where the studs connect the interior sheetrock to the exterior siding) can reduce a wall's effective R-value by about a third. Since straw bale wall construction has few thermal bridges, the effective R-value is not reduced. In addition, it takes heat 12 to 15 hours to pass from one side of a wall to the other—the thermal lag time—so straw bale wall assemblies usually perform better than their tested values. In most temperate climates, such as Japan's for example, winter temperatures are higher in the day than at night. This is known as the *diurnal temperature swing*. By the time heat from the interior approaches the wall's exterior, the day is warming up and heat loss is minimized. Conventional stud walls react much more rapidly to changes in temperature. In regions where the temperature drops below freezing and stays there for a week or more, straw bale walls perform more like the tested R-value; but in areas with significant diurnal temperature swings, their effective performance is higher.

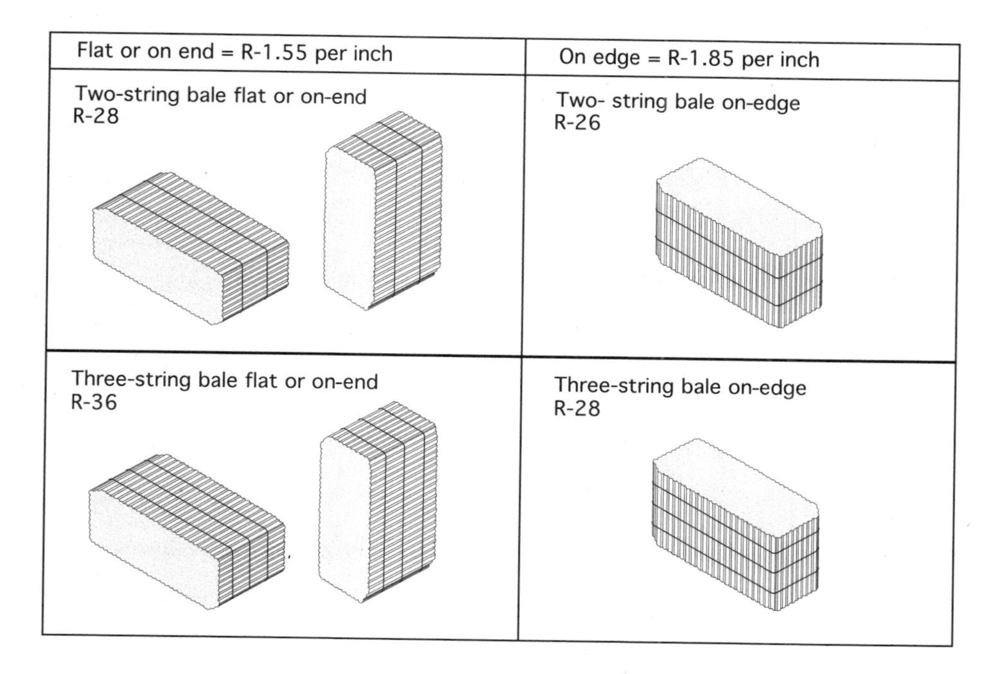

1-1. R-values of straw bales as determined using the 2018 IRC Appendix S. Note: California's energy code does not reference the IRC. The accepted R-value for straw bale walls in California is R-30.

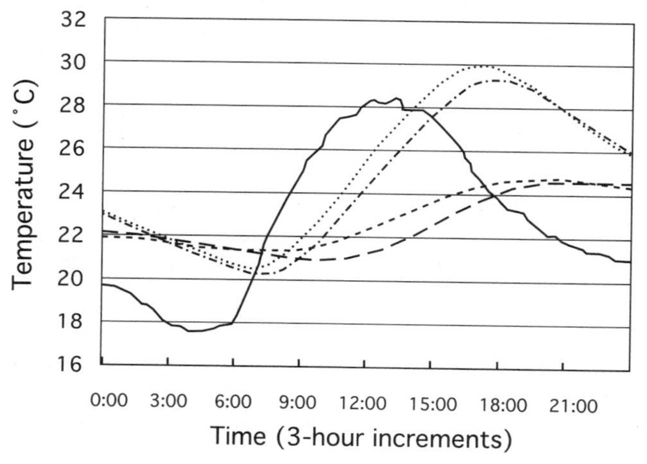

1-2. Thermal lag time of diurnal temperature swings in conventional and straw bale walls.

Straw bale walls perform similarly to keep out the heat in the summer, but the effect is less pronounced. This is because (1) the temperature differential between indoor and outdoor air is smaller, so wall R-values matter less, and (2) more of the interior heat gains are driven by radiant gains (direct sunlight) through windows and glazed doors than through the walls themselves. Two design features that help minimize summer heat gains are inset windows and wide eaves (or awnings) to shield windows from the direct summer sun.

Convective heat transfer is heat exchange from moving air. When you feel a cold draft from an open or leaky window, that's convection heat transfer. A poorly built straw bale wall has bales that aren't sufficiently dense or has gaps between or around the bales. Small convective loops within areas of loose or poorly compacted straw can significantly degrade a wall's thermal performance. When it comes to windows, keep in mind that air currents constantly wash against the wall's exterior surface. Set-in windows experience somewhat less convective loss than windows mounted flush with the exterior wall plane.

1-3. Emissivity of various materials. Data courtesy of Mikron Instrument Co., Inc.

Surface Material	Emissivity
Aluminum	0.028
Building Bricks	0.450
Cast Iron (smooth plate)	0.800
Copper	0.260
Gypsum	0.930
Human Skin	0.985
Paper	0.940
Plaster	0.910
Plastic	0.950
Soft Wood	0.820
Steel (flat, unpolished)	0.95-0.98
Hardwood (oak)	0.890

Heat transfer through *radiation* is always occurring, everywhere. You are constantly radiating heat to the things around you and the house itself—or they are radiating heat to you. Imagine standing between a wood stove and a single-pane window on a cold night. Since the stove is warmer than you, it gives heat to you, and you are warmer than the window, so you give heat to it. An object's heat content—its temperature—and its emissivity—primarily governs its ability to give off heat. The higher the emissivity, the better the material radiates heat. Most materials found inside the home, with the exception of metals, have similar emissivity. In particular, the plasters on straw bale walls and sheetrock walls have nearly identical emissivities.

Thermal Mass

The straw bale wall's capacity to accept or store heat also affects radiative heat transfer. A simplified way of thinking about this complex subject is contained in the term *thermal mass*.

All else being equal, a home with too little thermal mass will experience much greater temperature fluctuations than one with the appropriate amount of mass. Think of thermal mass as a storage locker for heat that it gains from solar radiation, HVAC operation, appliances, and even the people in a home. To be effective, thermal mass needs to be inside the thermal envelope (insulation), or, in certain cases, part of the thermal envelope. Cooling that mass at night with open windows and skylights allows for a cooler feeling the following day because people are then radiating more heat to the mass than it is giving to them. Mass on the outside of the insulation is of little value in the winter and may actually increase cooling loads in the summer.

For typical stud frame homes, ½" sheetrock is the only significant thermal mass the walls provide. The insulation has effectively zero thermal mass, and the widely spaced studs add only marginal mass. Interior straw bale wall plaster, however, is at least ¾" thick, and its increased thickness and density provide one and a half times as much thermal mass as standard sheetrock. The plaster can be as thick as 2", providing four times the thermal mass of sheetrock. This large surface area of considerable thermal mass allows for a broad heat exchange with interior air, so it moderates interior temperature variations. Thus, it performs like a wall with a much higher R-value (see *Design of Straw Bale Buildings*, Bruce King, Green Building Press, 2006, page 189) and substantially reduces heating or cooling requirements and costs.

Additionally, effective thermal mass in straw bale wall assemblies, just like in log homes, goes deeper than the surface finish. On a pound-for-pound basis, the bales themselves have a heat capacity similar to hardwoods (.48 Btu/lbF) and about double that of concrete or gypsum. However, straw bales weigh much less per volume. They are roughly 1⁄16 the density of concrete. Consequently, they have about ⅛ the thermal mass capacity per inch of thickness. Since straw bale walls are much thicker than stud walls and are uninterrupted instead of spaced, even without plaster, they have 13 times greater heat capacity per lineal foot than a 24" o.c. stud wall with fiberglass insulation.

Realize that walls are only one—and perhaps not the most important—component of an efficient energy design. In a typical house, the major sources of heat loss, from largest to smallest, are fresh air changes, air-leaking windows and doors, poorly insulated ceilings or roofs, then walls, and finally raised or on-grade floors. An energy-efficient design must control air leaks and provide appropriate insulation for the ceiling, floor, and all other spaces or voids. Both the structural and mechanical (HVAC) system designs should be designed to maximize insulation and minimize thermal bridging and gaps. Windows and doors, even those with good U values (U-value is the reciprocal of R-value; it measures the rate of heat transfer) offer very little insulation, and they often have high infiltration.

Moisture Capacity

We know that it is good advice to keep damaging moisture out of walls, but what constitutes "damaging moisture" and how to deal with it is greatly dependent on whether stud walls or straw bale walls are being discussed. Following is a brief overview; for more information, consult the resources listed at www.strawbuilding.org, particularly those by John Straube and Kyle Holzhueter.

"For moisture-related problems to occur, at least four conditions must be satisfied:

1. A moisture source must be available,
2. There must be a route or means for this moisture to travel,
3. There must be some driving force to cause moisture movement, and
4. The material and/or assembly must be susceptible to moisture damage."

(From *Design of Straw Bale Buildings*, B. King, et al., 2006.)

Straw bale walls in a well-built home will absorb moisture under certain circumstances, from interior as well as exterior sources. They also have a comparatively huge capacity for storing moisture before damage occurs, especially if moisture can leave through diffusion in sufficient time. In general, straw bale wall construction should prevent moisture intrusion, but not stop moisture movement. Examples of appropriate prevention features include locally appropriate roof overhangs and relatively tall foundation walls to prevent rain and splash from excessively wetting the walls. Allowing moisture movement means avoiding relatively impermeable membranes (e.g., polyethylene, aluminum-faced paper) and finishes (e.g., Portland cement, veneer tile, or stone). More on this, below.

Bale walls can hold moisture either as vapor, water, or ice. Water and ice (after it melts) will almost certainly cause problems unless the water escapes as vapor through the surface finishes. Straw bale walls can have a higher vapor moisture content than most conventional walls in buildings with the same relative humidity before saturation (the point at which the vapor becomes liquid). At 80% relative humidity (RH), straw can have approximately 60% higher moisture content than fiberboard, and 25% more than plywood.

However, liquid water left in place over time will cause mold, mildew, and straw decomposition, just like it does with wood. And, straw will decompose faster than wood due to its lower density and greater surface area. Leaks from plumbing, poorly executed detailing around doors, windows, and other penetrations often cause problems when moisture enters the walls through these poorly detailed openings and edges faster than it can leave.

What Is Vapor Permeability?

"If a material has a perm rating of 1.0, 1 grain of water vapor will pass through 1 square foot of the material, provided the vapor pressure difference between the cold side and the warm side of the material is equal to 1 inch of mercury (1 inch Hg).

"As temperature and relative humidity (RH) go up, vapor pressure gets higher. The greater the vapor pressure differential across or through the material, the greater the tendency for water vapor to migrate from the high pressure side to the low pressure side." (University of Alaska, Cooperative Extension Service, 2017)

Vapor Permeability

When talking about straw bale wall finishes, builders often use the term *breathability.* They are referring to how well the wall permits water vapor movement. To be considered vapor permeable, all of the wall materials, especially the finishes, must allow for any moisture that enters the wall to escape. When interior temperature and vapor pressure force water vapor in a straw bale wall to move toward the exterior, it needs to escape. If it hits an impermeable surface (e.g., plastic sheeting or certain paints) at or below dew point, it will condense back into water and accumulate, creating conditions for the straw to decompose.

If using paint of any kind on a plastered straw bale wall, make certain that it is highly vapor permeable even after several coats are applied. Paint invites more paint (as new owners decide to change colors), and additional layers of latex paint may make the wall surface increasingly impermeable. Consider clay paint or milk paint for interior surfaces, or mineral paint for either interior or exterior walls.

Vapor Transfer

Straw bale walls plastered with clay, lime, cement-lime, soil-cement, or gypsum plasters remain vapor permeable; water in vapor form can move relatively easily through the wall, passing through the straw and both plaster surfaces. Because of the way water vapor diffuses—driven by differences in vapor pressure and temperature—it may be constantly moving in and out of a wall assembly. Water vapor

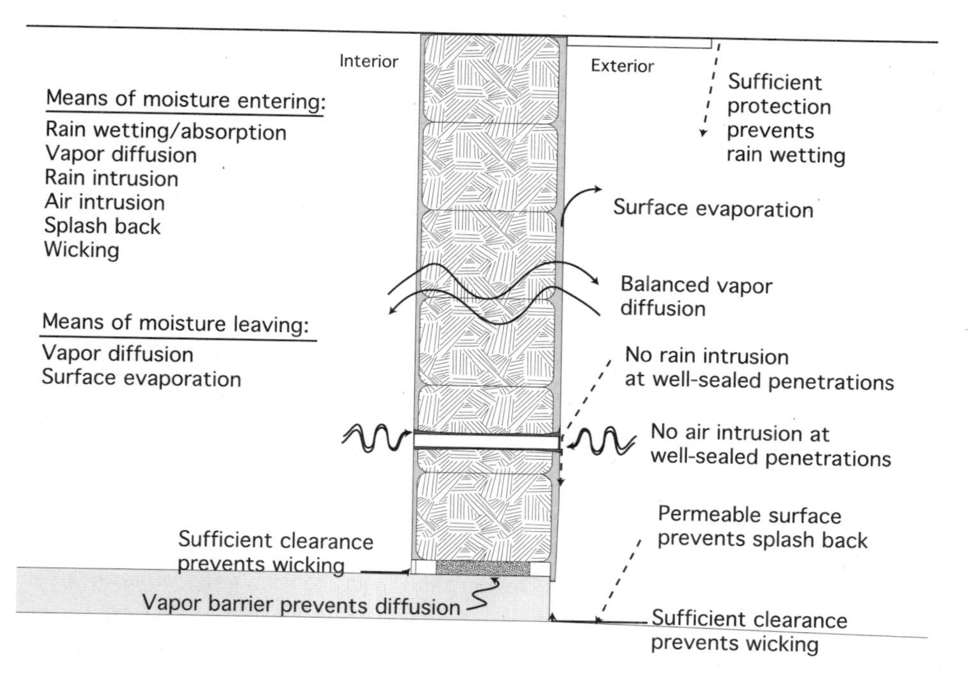

1-4. Moisture and straw bale walls.

moves, or diffuses, through vapor-permeable materials from areas of higher vapor pressure to lower, and from warm areas to cold. When the interior of a structure has a higher vapor pressure and temperature than outside, moisture will tend to move from inside to outside. This describes typical winter conditions in temperate climates. In summer, the moisture path may be reversed, depending on outside vapor pressure (humidity) and interior temperature. When the interior of a structure has a lower vapor pressure and temperature than outside, moisture will tend to move from outside to inside.

Different plasters have different permeances. The relatively thin (1"–2") layers of plaster on each side of a straw bale wall must be vapor permeable. Clay plasters are among the most permeable, rated at 11 US perms for a 2" thickness. A lime plaster has 9 US perms for a 2" thickness. Adding cement to the lime at a 1:2 ratio (1 part cement to 2 parts lime) maintains 9 perms at 1.5" thickness, but increasing the amount of cement steadily reduces permeability. A straight cement plaster at 1.5" thickness is rated at 1 perm, so it is prohibited by code on straw bale walls (IRC Appendix S requires a minimum of 3 US perms).

In dry regions and temperate climates, it's common to use a lime exterior plaster for durability and a clay interior plaster for easier application and maintenance—they have similar permeance. In cold climates, it's best for the exterior plaster to be more permeable than the interior plaster because when water vapor driven by moist, warm interior conditions passes through a more permeable interior plaster, it may slow and condense where it meets a cold, less permeable exterior plaster.

Although straw and wood are both cellulose materials, straw is more susceptible to damage from free moisture—moisture not bound in the material—than wood is because the countless straws in the wall have much more surface area than the wood framing. This same feature also allows the straw to absorb more moisture, and, so long as that moisture stays below 20% (the level at which microbes become active and can begin to digest the straw), the walls will do their job.

Exterior plasters protect the straw bale walls from bulk moisture intrusion from wind-driven rain, but only to a point. Lime and lime-cement plasters are porous and can absorb water. How much depends on thickness, aggregate size, and application. Lime plasters can be relied on to protect the underlying straw bales as long as the absorbed moisture can regularly evaporate to the exterior on dry, warm days. Lime plasters subjected to repeated wetting with little opportunity to dry have failed; the plaster becomes saturated, and moisture wicks into the straw bales themselves. A few minutes of wind-driven rain is generally not a problem so long as the walls can dry between storms. Days, or even many hours of heavy wind-driven rain can overwhelm the plaster. If a given location is known to receive wind-driven rain without opportunities for drying, avoid relying on plaster alone to protect the walls. Design the structure with generous roof overhangs, install gutters, wrap the straw bale walls with porches, or consider using a rainscreen—siding over the plastered straw bale wall with an air gap between. See *Straw Bale Rainscreen Detail* in Chapter 2.

Exterior clay plasters behave differently when wetted, but they are also vulnerable to excessive exposure to water. Clay is hydrophilic; it attracts moisture. When wet, it swells, effectively sealing a wall from further penetration. However, it can also become soft, and if clay plaster receives frequent rain, it will eventually erode, exposing the straw bales to the elements. The same design features used to protect lime plasters are effective with clay plasters too: large roof overhangs, porches, and rainscreens.

The vapor transfer topic has been explored both through scientific testing and a lot of hands-on experience. But research continues, and the last word has not been written. Better understanding of how straw bale wall systems manage moisture will result in better buildings.

Keep liquid water out of the walls. Assume water vapor will be driven into the walls, and make sure that nothing stops its diffusion and evaporation from the wall surfaces. Understand the local climate and design for it. Understand internal moisture loads; a family of four will produce about two gallons of moisture per day, but the number can be much higher. According to the Canada Mortgage and Housing Corporation, an average home must manage 400 to 2,000 gallons of water in the air during a typical heating season. Use mechanical methods to remove moisture. Codes often require venting of range hoods, clothes dryers, and bathrooms to the outside.

Types of Straw

Sometimes people wonder if any kind of straw can be used. Five grain crops—wheat, rice, barley, oats, and rye—produce what we know as straw (not to be confused with hay, which should *not* be used for building). Straw bale building codes don't differentiate between straw bales made from different grain crops, so long as the bales meet minimum requirements for density and moisture content. But when builders have choices, they may prefer one over another.

In the United States, rice straw is available mostly in Northern California, the Mississippi River valley, and the Gulf Coast. It contains silica, which makes it more rot resistant when exposed to moisture. This is one reason rice farmers once burned their fields to clear straw after harvest—it can take years to decompose. The high silica content also makes rice straw unpalatable to ants and termites. Because of how rice straw lays on the ground when it is cut, bales tend to be dense and dusty, with intertwined stems. Compared to other types of bales, rice bales hold their shape and compression better when notched, and the bales separate into more distinct flakes when the strings are removed, making resizing and retying easier. Rice stems don't splinter as much as wheat straw, making them less likely to embed tiny fragments in exposed skin. On the downside, silica more quickly dulls cutting tools, and the silica in rice straw dust is irritating to breathe. Anyone cutting bales should wear a dust mask, particularly while working with rice straw. Some people have an allergic reaction to rice straw dust, and it seems to have a cumulative effect. Finally, rice straw bales may be heavier, requiring two people to lift and handle them. A three-string rice straw bale can weigh over 80 pounds, and a two-string bale weighs up to 65 pounds.

Wheat straw bales are more widely available throughout continental North America. They are less dusty, although wearing a dust mask is still prudent, especially when cutting bales. Wheat straw bales don't dull tools as fast, and they tend to be lighter, but they may still require two people to lift and place safely. Wheat straw will deteriorate more quickly if it comes in prolonged contact with moisture. As with any straw bale, it is imperative to keep them dry! Wheat bales have sharper edges because the straw tends to lay front-to-back instead of tangled like rice straw, so it can more easily scratch bare skin. Wear long sleeves when handling wheat bales.

Barley, oat, and rye straw are less widely available, but they are often used in straw bale structures. Their qualities are similar to wheat straw, and as long as the bales are

Organic Bales or Conventionally Grown Bales?

Finding organic straw bales can be a challenge for those who want a completely natural, chemical-free house. Organic farmers usually till straw back into the soil to recycle nutrients, control weeds, and lessen the need for other soil amendments. Still, conventionally grown rice, wheat, barley, and other grains and their straw are relatively free of pesticides. For example, rice farmers use commercial nitrogen fertilizers, usually in the form of hydrous ammonia applied to the soil prior to planting. When soil samples indicate deficiencies, farmers apply certain soil amendments prior to planting. They use various weed-control herbicides, usually within the first 30 days of the rice plant's growth. At this time, a rice plant is less than a foot tall, and by full maturity, this part of the plant makes up a small portion of the total mass because the plant develops new tillers until it reaches the seed-production stage. Rice reaches maturity in 135 to 165 days, so these herbicides, applied in very small quantities (usually a few ounces up to a pound of active ingredient per acre), have long been dissipated and broken down to very low levels. The pictures you see of crop dusters trailing a fog from a spray boom are actually the aforementioned amount of herbicide diluted in 10 to 15 gallons of water, applied per acre. Pesticides are seldom used on rice crops, and if used, they are applied very early in the plant's growth. Fungicide might be the only chemical applied, if conditions warrant, at mid-growth. The same holds true for wheat production, which usually requires even less herbicide application. In recent years, however, the striped rust fungus has forced many farmers to double the amount of fungicides, applied right up to wheat's mid-maturity stage.

sufficiently dense and dry, they are entirely suited to straw bale construction. For a more detailed discussion about sourcing, ordering, handling, shaping, and cutting bales, see Chapter 5, Stacking Straw Bale Walls.

Historical Note

After the practice of burning rice straw was banned to improve air quality, California rice farmers began flooding fields to hasten the decomposition process, worrying fisherman who thought too much water might be directed away from fisheries. In 1993, the fishermen teamed with the newly formed California Rice Commission to initiate the first state code guidelines for straw bale construction, which were adopted in 1996.

Bale Size

North American farmers gather the vast majority of straw in big bales because in that form they are more economical to produce, handle, and ship. These bales are usually 3'×4'×8', or less commonly 4'×4'×8', and they weigh from 1,000 to 1,500 pounds. Their large size generally makes them unsuitable for straw bale construction.

The bales used in straw bale structures are tied with polypropylene twine to make two-string and three-string bales. Though baling equipment can vary, and the bale dimensions are not precise, three-string bales tend to be about 15"×23"×46", and two-string bales tend to be about 14"×18"×36". The length dimension varies the most. Some farmers bale their own straw; others hire contract baling operators who also bale hay from alfalfa and other forage crops. Shifting crop and land values can impact the amount of grain grown in any particular area; as land values rise or more profitable crops displace less profitable ones, straw availability can fluctuate.

Straw Harvest, Baling, and Availability

Most small bales made from wheat straw are suitable for straw bale construction because they are dry and dense. Wheat, barley, oats, and rye are harvested in summer from completely mature crops, so the straw baled from these fields tends to be very dry. In fact, in extremely dry-summer regions in the West, many farmers bale the wheat straw in the morning or evening when dew increases the moisture content enough to make the straw stems more pliable and able to resist shattering as they are baled. Wheat straw typically comes from stripper-harvested wheat fields baled from the swathed rows after the stripper header harvests the grain and leaves the seedless plants standing. Because the harvest occurs during the summer building season, wheat straw bales are readily available directly from the field. Established markets maintain a high demand for clean wheat straw—it doesn't sit around long after harvest.

Rice plants grow in standing water. Rice is harvested from a plant that is mature, though still green. Weather plays a major role in producing rice straw bales for building purposes. The harvest takes place in the fall after the water has been drained and the fields are allowed to dry for three or four weeks. The ground is still damp just below the surface; the fall days are shorter, and morning and evening dew is heavier. This shortens the available time window when rice straw can be baled because conditions must be as dry as possible. The harvested straw must lay in the sun for at least four days for plant stems to lose their green moisture. After the initial dry-down period, the straw is raked into windrows to be baled. Before baling, windrowed straw is tested with a moisture meter to determine that it's dry enough to bale—it must be at 10% before baling commences.

The longer-stemmed straw preferred by straw bale builders requires the longest drying period. Flail-chopped straw or straw partially shredded by rotary combine-harvesters dries faster because moisture exits the stem via a shorter path. This straw is most often used for erosion control, and more recently, in fiberboard products. Builders prefer longer-stemmed rice straw bales made by older-style cylinder and straw-walker combine-harvesters because they damage straw residue the least—but these have become increasingly scarce. Although rice crops are grown in Gulf states like Mississippi, Arkansas, and Texas, higher humidity makes baling dry straw even more challenging.

The fall rice harvest also comes late in the building season, when winter weather threatens. The major markets for rice straw—erosion and livestock feed and bedding—take delivery before winter is over, so you may need to take delivery directly from the field or arrange for dry winter storage. Since rice straw harvesting varies, work with a supplier who knows the difference between bales used for erosion control and bales used for building.

Environmentally Sustainable Design

Besides using a super-insulating, low-embodied-energy, carbon-sequestering agricultural by-product in the walls, consider other steps to ensure your building consumes less energy and has a smaller carbon footprint.

Regions graced with a Mediterranean climate or mild winters make it practical to live in smaller buildings wrapped with porches and patios for outdoor living. Smaller buildings consume fewer building materials, are faster and less expensive to build, and they cost less to heat and cool. If the structure is built in a forested area, consider locally harvested trees for posts, beams, and other dimensional lumber. Also consider using site-excavated or locally available clay-rich soils for plasters. All else being equal, using local materials is less energy and carbon intensive than using materials from far away.

The concrete in a typical foundation is a large percentage of a building's carbon footprint. Look for ways to trim the amount of concrete to what is actually needed, and inquire at the local mixing plant how much of the cement in the mix can be replaced with fly ash, a waste product from coal-burning power plants.

Note that these choices can affect the building schedule, structural system, and building maintenance. For example, high-fly-ash concrete is stronger, but it sets more slowly, and exterior clay plasters may need additional weather protection.

Those drawn to building with straw bales often incorporate other aspects of sustainable design:

- Build in a walkable neighborhood to reduce car use and make aging in place easier.
- Renovate an existing building, rather than building from scratch.
- Design with a potential addition in mind.
- Incorporate passive solar features.
- Install energy-efficient lighting and appliances.
- Use recycled-content ceiling insulation, like treated cellulose or cotton.

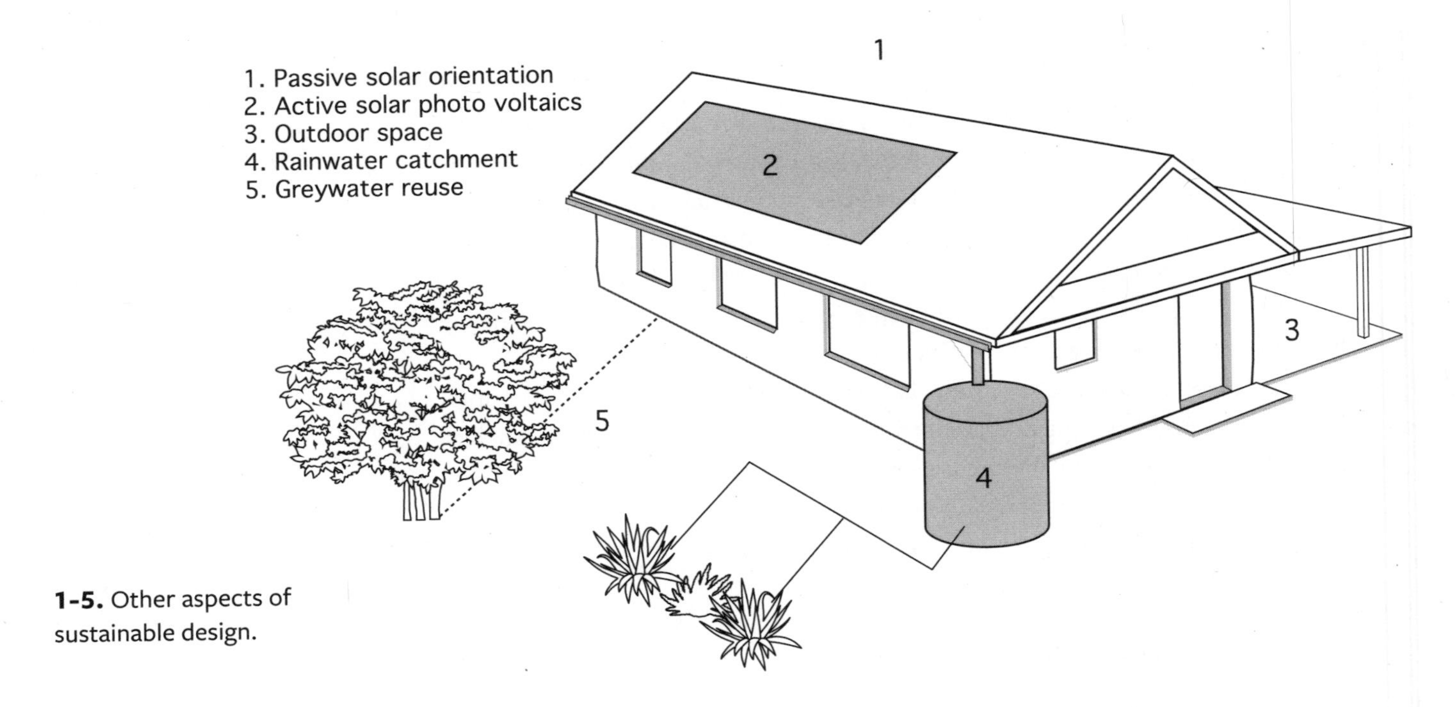

1-5. Other aspects of sustainable design.

- Use salvaged materials.
- Use solar electric power and heat pumps.
- Harvest rainwater from the roof for landscape and garden use.
- Plumb for greywater use.

Many of the items in this list can be incorporated into the design, and some can be added later. An environmentally sustainable design will consider these options in the building's short- and long-term environmental cost.

Conclusion

We've covered why people build with straw and touched on the building science of how a straw bale wall assembly insulates, stores heat, and handles moisture. Subsequent chapters explain design and construction as it pertains to building with bales: general design considerations, structural design issues, electrical and plumbing, stacking straw bale walls, and plaster options, including preparation and application. The last chapter contains the International Residential Code's (IRC) Appendix S: Strawbale Construction, along with its informative commentary. Finally, the appendices address fire safety during and after construction, and straw bale work parties.

If you are already familiar with the subject in any of these chapters, feel free to skip directly to the chapter(s) that interests you most. Cross-references throughout the book illustrate the interconnectedness of the building process and demonstrate how design choices can impact the finished structure.

2

Designing with Straw Bales

Aesthetic Traditions and Issues

Matts Myhrman, one of the straw building revival's early leaders, famously quipped: "You can have anything you want in a straw bale house, except for skinny walls." A wide range of successful designs have shown there is no single appropriate aesthetic for the thick walls possible (or, inevitable, depending on how you think of it) in bale buildings. Any design style that effectively uses thick-wall construction can use bale walls. A distinct "Southwestern style" is popular in the West and Southwest—where many adobe, Mexican hacienda, and territorial-style structures have been built with bales. Craftsman, frontier, prairie, and Pacific Northwest styles have also been built with bales. Architects have successfully integrated bales into modern and organic-style buildings as well.

There are limits. Avoid building styles with minimal or zero roof overhangs—parapet roofs provide too little protection for the bale walls. Otherwise, choose any style suited to thick-wall construction and appropriate to the site. Aesthetic considerations influence material, finish, and detail choices, too.

A straw bale wall's thickness, mass, imprecise dimensions, and plaster finish present a new design language to those familiar only with wood-frame construction. The bales and plaster are more plastic, resulting in three-dimensional surfaces that are not typically flat. Straw and plaster much more easily shape rounded corners, curved walls, and flared window and door reveals. Decorative recesses, called *niches*, can be carved into the walls, and stone or timber shelves can project from them. Plaster finish choices affect how the bale walls interact visually with intersecting partitions of other materials. Given the beautiful informal surfaces bales naturally create, it's tempting to make interior partitions with bales as well, but that sacrifices interior space. Interior partitions of straw-clay, plaster on lath, or intentionally uneven plaster on drywall can achieve a character similar to bale walls.

2-1. Examples from a wide range of styles.

Photography (*left to right, top to bottom*): Rebecca Tasker, Edward Caldwell, Jim Reiland, Jim Reiland, Paul Schraub, Jim Reiland

Moisture Considerations

Although covered in the previous chapter, this bears repeating: straw, like wood, begins to break down when sufficient water is present for a long-enough period of time. Water can enter a straw bale wall in two forms. As liquid, it can leak through roofs, windows, and cracks in plaster; it can be absorbed from rain and snow-wetted plasters; and it can wick up from the foundation. As vapor, it can condense on cold or impervious surfaces inside the bale wall assembly. Since water damage to straw bales, wood posts and beams, and door and window frames can be costly to repair, it is important that you design and detail to minimize opportunities for water to enter a building.

Give bale walls *a good hat, good boots, and a coat that breathes.* Protect bale wall assemblies with appropriate roof overhangs, vapor-permeable plaster skins, carefully detailed openings, a suitable elevation above grade, and include a way for water to drain away at the building's base. See Figure 2-3. Adhering to these principles and details keeps the straw's moisture content well below damaging levels.

Straw Bale Characteristics

Bale Sizes and Orientation

We categorize buildings by the primary materials in their walls: a stone cottage, a brick house, a log cabin, a straw bale house. Straw bales are a relatively inexpensive part of a building, yet they define its character and greatly influence the overall design. Knowing bales' unique properties will help with both design and construction.

Bales for building must be dry and dense. Appendix S stipulates a minimum bale density and moisture content because less-dense bales, or bales with a higher moisture content, could compromise a building's performance goals. See IRC Sections AS103.5: Density, and AS103.4: Moisture Content (given in Chapter 7) for more information on evaluating bales. When working with a straw bale supplier, be sure to specify the minimum density and maximum moisture content, or you might receive a load of "fluffy" or wet bales.

Because straw bale dimensions vary, confirm bale dimensions with your supplier to avoid surprises. For example, in California, straw bales with two strings

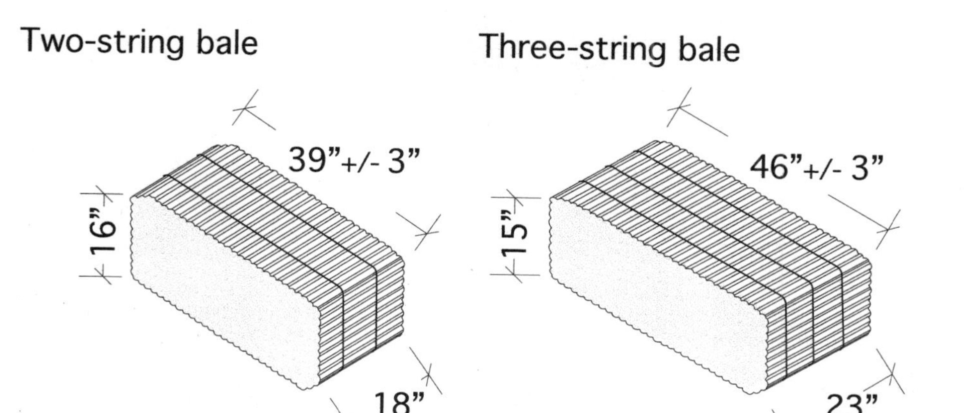

2-2. Approximate dimensions of straw bales in California.

2-3. Good Hat, Good Coat, and Boots

Straw bale construction creates well-insulated, fire-resistant buildings from a sustainable, affordable, and readily available agricultural by-product. Straw bale walls must be kept dry. Appropriate design and proper detailing prevents moisture intrusion and allows water vapor to escape. Provide a wide hat (roof overhang), breathable coat (plaster), and tall boots (foundation).

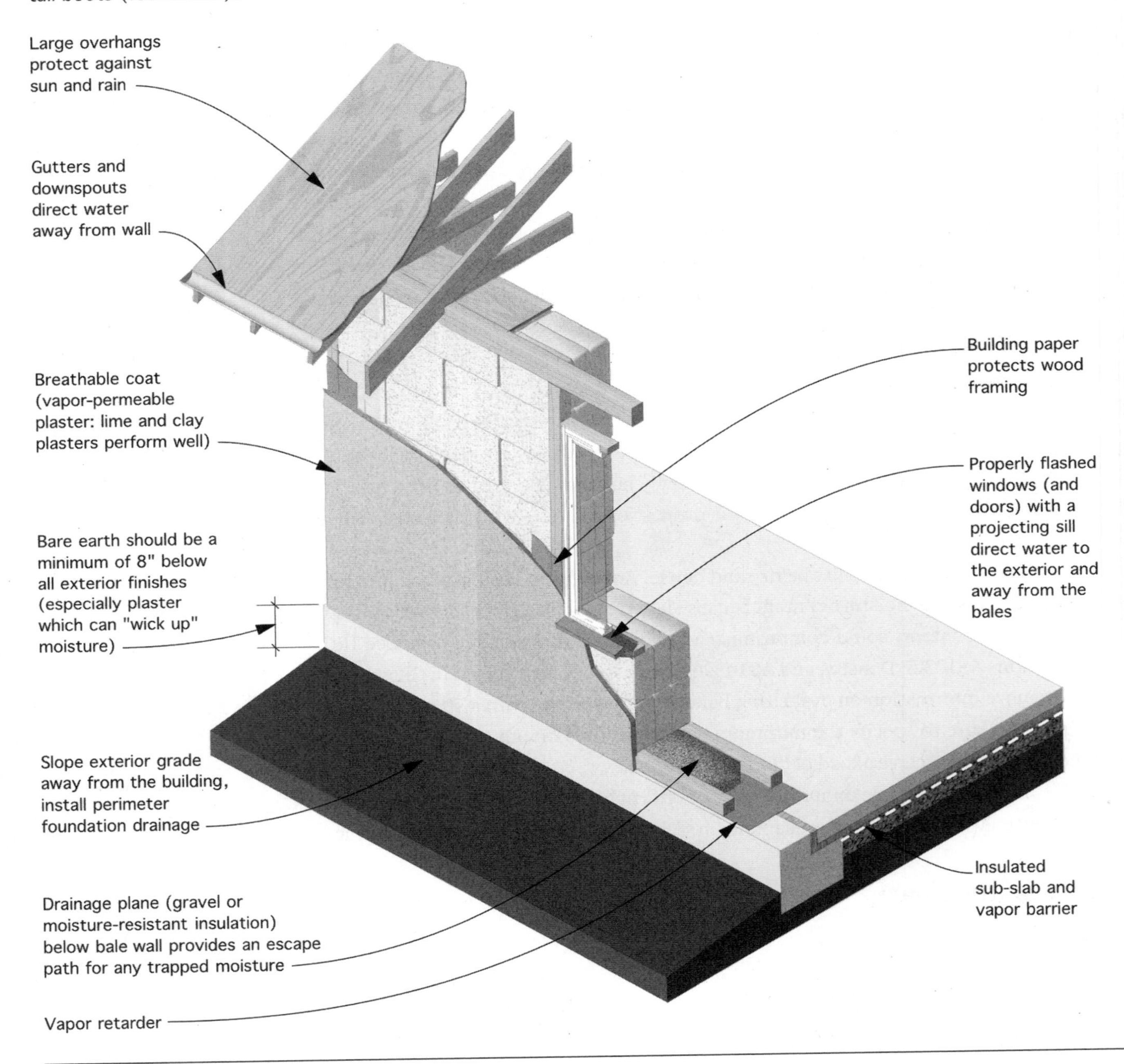

NOTES:

- Additional care should be taken when developing a plaster mix with high cement content. While sometimes necessary structurally, cement mixes are far less vapor permeable and can lock moisture inside the bale wall assembly.

- Detail the floor package and bottom of bale walls to prevent moisture intrusion from below (e.g.,install a vapor barrier below bale walls in slab-on-grade (S.O.G.) installations; ensure proper crawlspace ventilation to prevent moisture concerns in wood-framed floor conditions).

tend to be 18" wide × 16" tall, and an average 39" (± 3") long. Three-string bales are about 23" wide × 14½" or 15" tall, and an average 46" (± 3") long. See Figure 2-2.

Bale Orientation: Laid Flat or On-Edge?

Bale orientation in the wall affects the insulation level, wall thickness, and your ability to shape the bales. If R-value weighs heavily in the decision to use bales laid flat or on-edge, note that differences within the R-26 to R-36 range are usually overshadowed by siting (solar orientation), windows, doors, floors, and ceiling/roof assemblies (*Design of Straw Bale Buildings*, B. King, Green Building Press, 2006, page 193). See Chapter 1 for discussion of the straw bale wall assembly thermal properties and other factors influencing energy efficiency.

Most builders lay bales flat, with the strings on the top and bottom. Bales laid flat are more stable during stacking and before the wall is plastered. Because the strings are recessed 4 to 6 inches from the exposed surfaces, bales laid flat can be notched for inset posts and sleepers for mounting cabinets, kerfed for bracing poles, and carved into for niches and other sculptural detailing. Bale corners at doors and windows can be rounded, but flared window reveals generally involve more shaping than can be done in those 4 to 6 inches. See Figure 3-11: Load-Bearing Wall, and Figure 3-13: Post-and-Beam Wall.

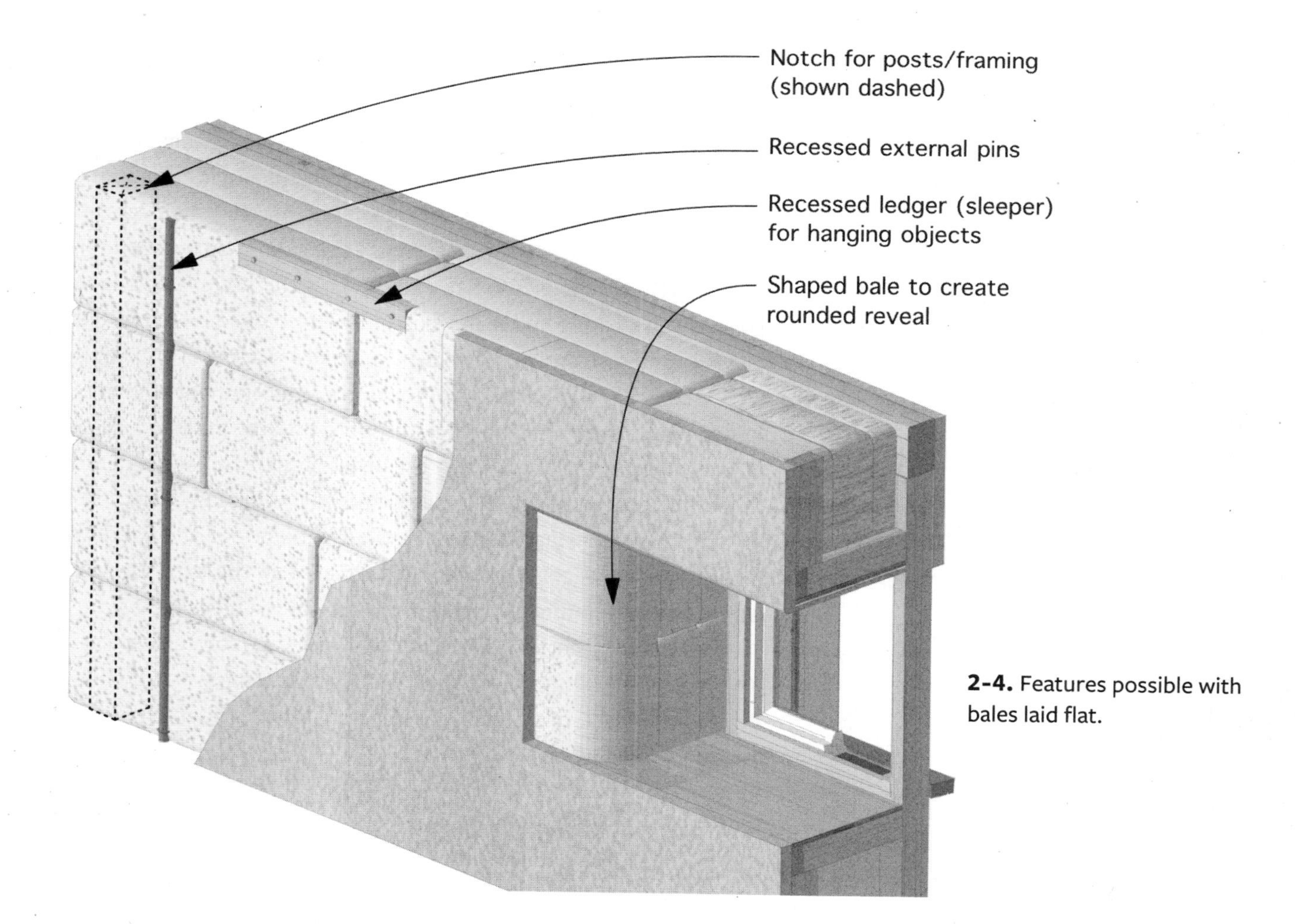

2-4. Features possible with bales laid flat.

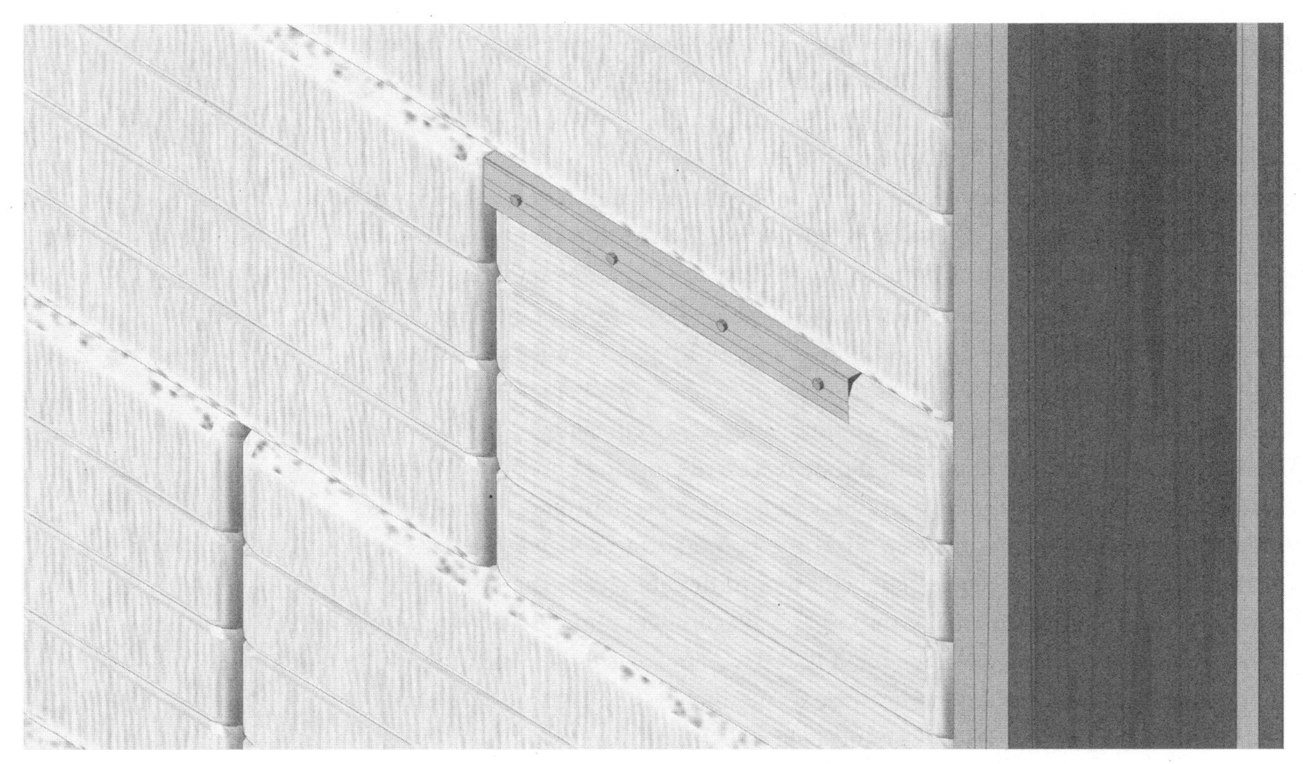

2-5. Sleeper notched into bales on-edge.

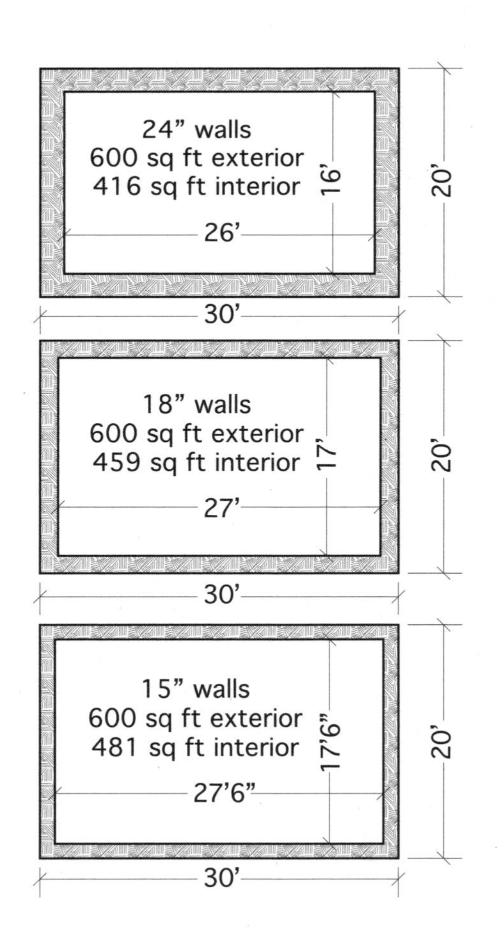

Bales on-edge are less stable during stacking because of their narrower footprint. However, fewer courses are needed to achieve the same wall height, so this partly compensates for the instability. Because on-edge bales aren't as stable to stand alone, they aren't often used for load-bearing structures. And, because their strings are on the exposed surfaces, bales on-edge cannot be notched for posts, so they are most often used in I-joist structural systems (see Figure 3-16: I-Joist Wall). Because there's no need to notch the bales—only resize some of them—the bales stack quickly. Bales can still be notched horizontally for sleepers that support cabinets or furring strips for siding. In addition to offering greater insulation per inch of thickness, they take up less square footage, and the top of the wall can be more easily trimmed to fit snugly under the box beam or top plate, or under a raked gable. Door and window reveals in on-edge walls tend to be squared to the exterior wall surface unless framing creates flared or rounded reveals.

A variation places the bales "on-end" between carefully spaced framework; this minimizes the need for notching. As with bales on-edge, a stack of on-end bales is not stable enough for load-bearing systems.

2-6. Bale wall thickness and interior space.

Two-String Or Three-String Bales?

The choice of two- or three-string bales—and how to orient them—depends on your design goals. If the ability to sculpt the walls and create deep window and door reveals inside or outside is important, thicker walls and bales laid flat may be preferable. If maximizing insulative value and/or the amount of interior space is paramount, bales on-edge may win out. Thinner walls permit more interior space for a given exterior footprint.

Bale weight may also be a design factor. Smaller bales weigh less and are easier for one person to handle, but they can't be stacked as high as larger bales.

The Bale Module In Design

The contemporary wood-frame construction system benefits from decades of research and development. This efficient system, organized around a 4' module, is conveniently divisible by 24", 16", 12", etc. Most sheet goods like plywood and drywall come in 4'×8' sheets, and 2× studs are commonly available in 8', 10', and 12' lengths. Other building material dimensions conveniently synchronize with wood construction dimensions.

Conditioned as we are to the certainty of the modular wood-frame construction system, working with straw bales can take some getting used to. There isn't an "industry standard" for straw bale dimensions. While bale heights and widths may be locally standard, lengths vary by as much as 12". Always confirm bale dimensions with the supplier early in the design process.

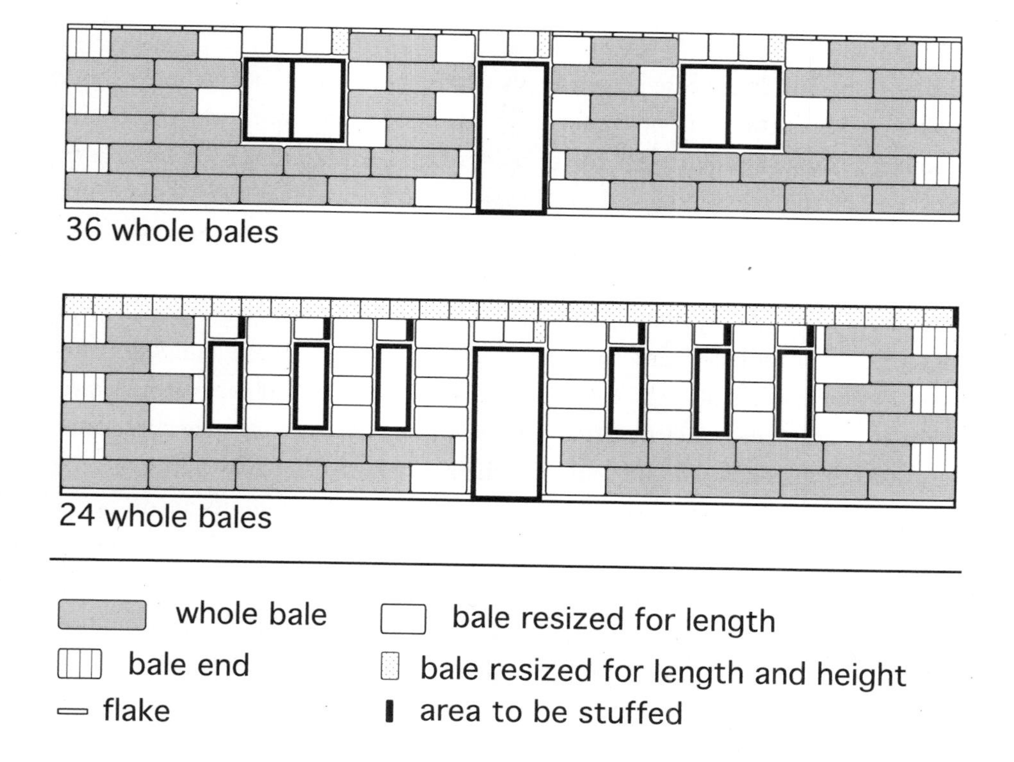

2-7. Utilizing the bale module.

It is possible that bale dimensions may conveniently fit into a wood-frame modular system. For example, a 2× sill plate with six courses of 16" tall bales equals a conventional 96" stud wall height. But a 4× plate—used quite often in seismic areas—makes the same wall 2" taller, no longer the conventional "module" size. And neither example takes into account bale compression—the fact that a stacked wall of bales may "settle" or be mechanically compressed (this is discussed in Chapter 5, Stacking Straw Bales).

Take advantage of the bale module by using full-size bales whenever possible. If you calculate windowsill, header, and top plate heights based on your bale module, you can minimize the need to customize bales. This results in a building that stacks faster and has fewer seams between bales, likely improving the walls' thermal performance.

Where other design considerations override the benefits of keeping to the bale modules—the right tools and techniques shape bales to fit the available space. See Chapter 5, Stacking Straw Bale Walls—Resizing Bales. Extremely energy-efficient homes have been built with parts of the wall made of partial bales, e.g., a top course of custom-modified bales or compressed "loose flakes" stuffed below a box beam, top plate, or raked gable wall. But in general, the more bale customizing required, the longer it takes to build, and the more likely that the insulation value will suffer.

A drawing of each wall's bale layout showing where bales will need to be customized alerts designers to areas where bales may not be practical, like in narrow spaces between close-set windows. See Figure 2-7.

Design Details

Where to begin? Straw bale structures have much in common with most other buildings: foundations, floors, walls, roofs, windows and doors, electricity and plumbing. Those familiar with designing conventional buildings will be able to imagine different ways to handle the design details of a new building system, but why reinvent the wheel? For 30 years, designers and builders have been working out some of these details; here we offer ways to handle each feature unique to straw bale buildings depending on bale orientation, the framing system (if any) that supports roof loads, and the design's aesthetic goals. Many building plans don't show these details, which leaves it up to the builders. If they have experience with straw bale structures, they may have a system to address each issue not described in the plans. But if building with bales is new, they may unnecessarily puzzle over this building system's unique challenges. Working it out on paper costs a lot less than trial-and-error in the field. The next few pages will cover some of these challenges.

Beginning at the wall bottom. Well, not the very bottom. For a discussion of footing and foundation systems, see Chapter 3, Structural Design Considerations—Foundations.

2-8. Wall base.

The straw bale wall base must perform a variety of critical functions: support the bales above, anchor the wall to the foundation, support the bottom of the plaster skins, serve as a location to attach lath, direct moisture away from the wall, and possibly play a role in a braced wall panel assembly or in out-of-plane loading.

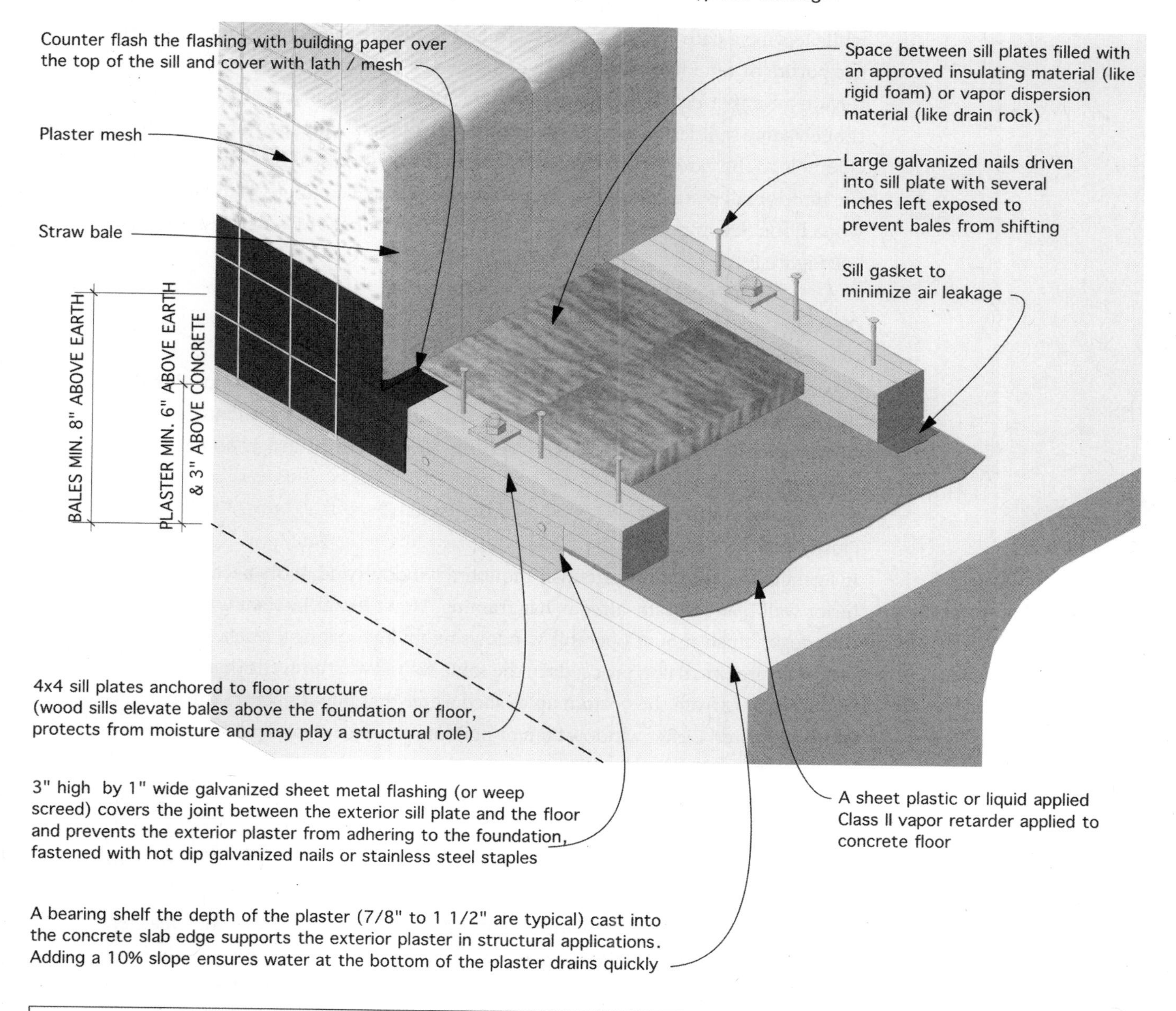

NOTES:

- 4 x sills are typical where they play a structural role; 2 x and 3 x sills can be used in other applications.

- Use pressure treated wood when the sills are placed over concrete or near grade. Pressure treated wood is not necessary over a raised floor.

- Some builders lap a liquid film or waterproofing membrane up the inside surface of the interior sill and under the exterior sill to direct liquid water away from the interior.

- Kerfing the exterior sills with a 1/4" saw cut at 2' centers channels water away from the wall.

Wall Base

Most straw bale walls rest upon a double sill plate that supports the exterior and interior bale edges and elevates the bales above the interior floor to protect them from water damage caused by a broken water line. The space between the sills must be filled with materials that bear the bale weight and insulate the cavity while leaving a path for any water that might enter the wall to drain out. You can use perlite or rigid foam alone or in combination with drain rock. Most builders rely on an imperfect seal between the exterior sill and foundation for drainage, though some builders kerf the underside of the outside sill plate to create runoff channels to the exterior. Air leakage concerns argue for a sealing gasket between the interior sill plate and the floor. Anchor the sills to a concrete foundation or a wood floor. See Appendix S: AS105.3 and AS105.3 for base of wall requirements. See Figure 2-8.

Outside

Windows and Doors

You can install windows and doors flush with the straw bale wall's exterior surface, as in most conventional construction, or recessed some depth into the wall. While most designers choose flush or recessed for the entire design, some designs use both.

Flush Mount

It's generally easier to install flush-mounted windows and doors because the exterior wall plane usually already has framing. However, at kitchen counters and over desks, flush-mount operable windows might be too long a reach. Flashing to prevent water intrusion proceeds in the same way as with conventional construction, shingling from the bottom up to shed water that might migrate through the exterior plaster. Either window trim or building paper, lath, and plaster cover the window's mounting fin. See Figures 2-9 and 2-11.

Recessed

Viewed from the outside, buildings with recessed mounts put the thick wall on display as shadows highlight the windows and doors. Recessed mounting better protects the heads and the windows and doors themselves from the weather, which reduces wear. Studies suggest that recessed windows help maintain interior temperatures better because warm or cold outside air doesn't wash across them as much as flush mounted. It's also possible that this feature offers better fire resistance for much the same reason—super-heated air from a wildfire may flow across the wall and bypass the window or door. But there are trade-offs. Recessed windows create exterior sills that require careful flashing to prevent rainwater from entering the bale wall below. On the other hand, operable windows set closer to the interior wall surface are easier to reach, especially across furniture or a kitchen counter. See Figures 2-10 and 2-12.

2-9. Flush window with no sill.

Setting the face of the window flush with the exterior wall eliminates the challenges of waterproofing a recessed sill and simplifies window mounting where framing supports roof loads and/or windows.

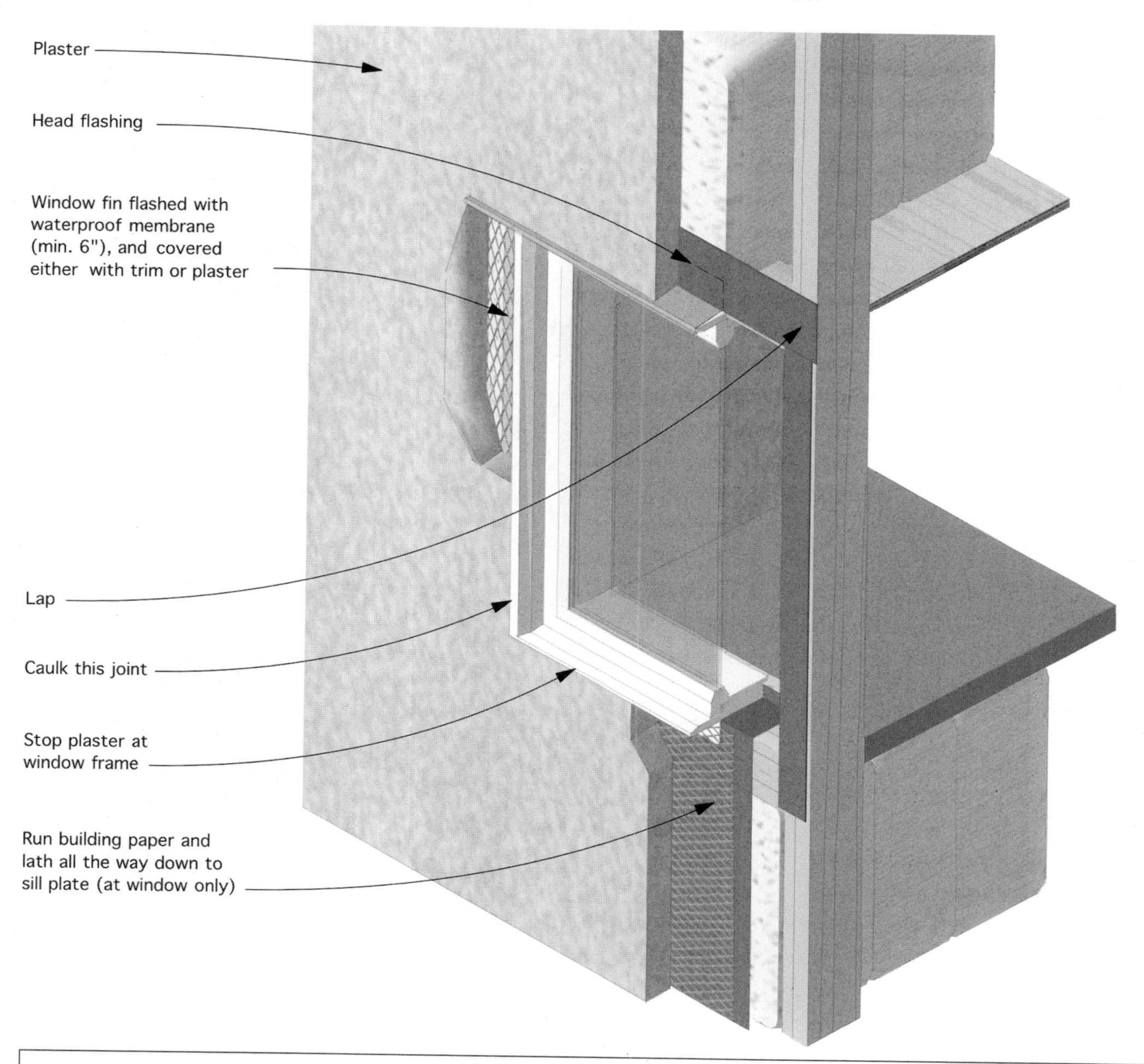

NOTES:

- Flush windows eliminate the shadows created by recessed windows as viewed from the outside.

- Useful along walls that will see very little wind-driven rain. Water that sheets off the window will drain down the plaster below the window where it can soak through the plaster and into the bales unless a vapor-permeable building paper extends to the ground. At the very least,it may cause the plaster there to weather differently, resulting in erosion and streaking. See Figure 2-11: Flush Window with Sill.

- Flush windows expose the vulnerable top of the window to weather. Where this is a concern consider adding a small awning over the window. See Figure 2-33: Flashing for Roof Joining a Straw Bale Wall.

2-10. Recessed window–integrated sill.

Windows set behind the wall's exterior surface shield the vulnerable window head from rain and create shadows that emphasize the bale wall depth. Recessed windows demand careful detailing to prevent leaks into the bales below. The recessed sill is especially vulnerable as water gathering on the sill may seep into the bale walls below. A common solution is to flash the window head and use a self-adhesive waterproof membrane all around the windows, with the top lapping the sides and the sides lapping the sill.

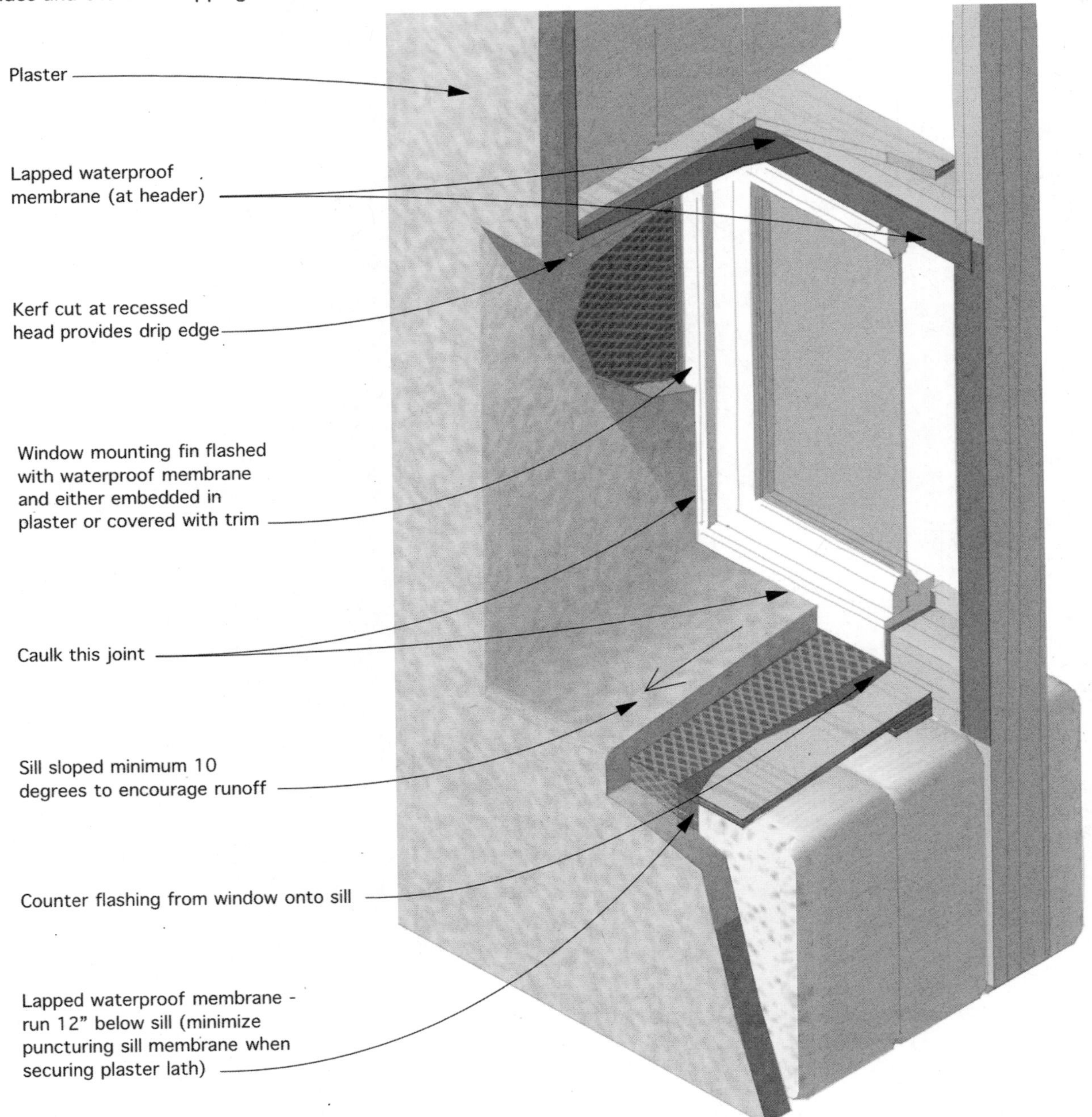

NOTES:
- Recessed windows trade off interior space that could be used for window seats, or window shelves.
- Plasters are not waterproof. If recessed windows are facing heavy weather, use a window sill.
- Best used under a porch or ample roof overhang, or in a very dry climate.

2-11. Flush window with sill.

Sills below a flush window direct water that gathers on the window surface away from the wall.

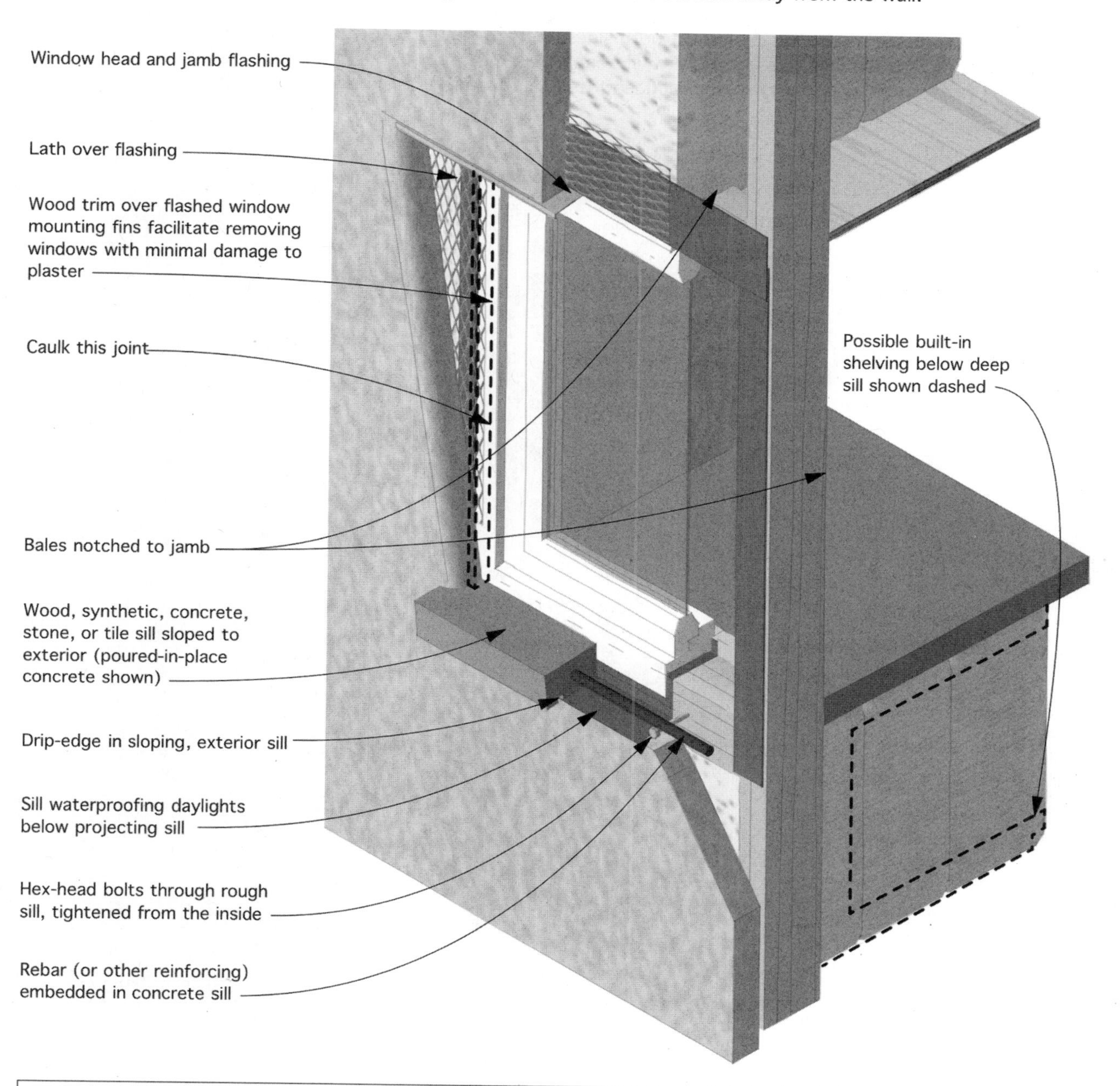

NOTES:

- Forms needed for poured-in-place concrete sills.

- Wood sills require greater maintenance. By installing casing bead around the sill, sides and bottom, a wood sill can be removed and replaced without damaging the plaster.

- Where bale wall depth is used for recessed shelving / niches (e.g., below deep window sills, as shown above), you sacrifice the insulative and acoustic benefits of the straw. Using a higher R-value insulation such as fiberglass batts or rigid foam will help minimize thermal loss at these skinnier walls.

2-12. Recessed window sill.

Plaster is not waterproof, and even a waterproof coating over the plaster will not help when the plaster cracks. Therefore, window sills should be detailed like small roofs. Before applying the sill material, place a waterproof membrane that extends beneath the window and turns up on the interior and at both sides of the recess. Extend this sill waterproofing out past the wall plaster surface to "daylight" on the exterior of the wall, beneath a projecting finish sill. The finished sill should project several inches beyond the wall surface to prevent the concentrated runoff from eroding and staining the plaster.

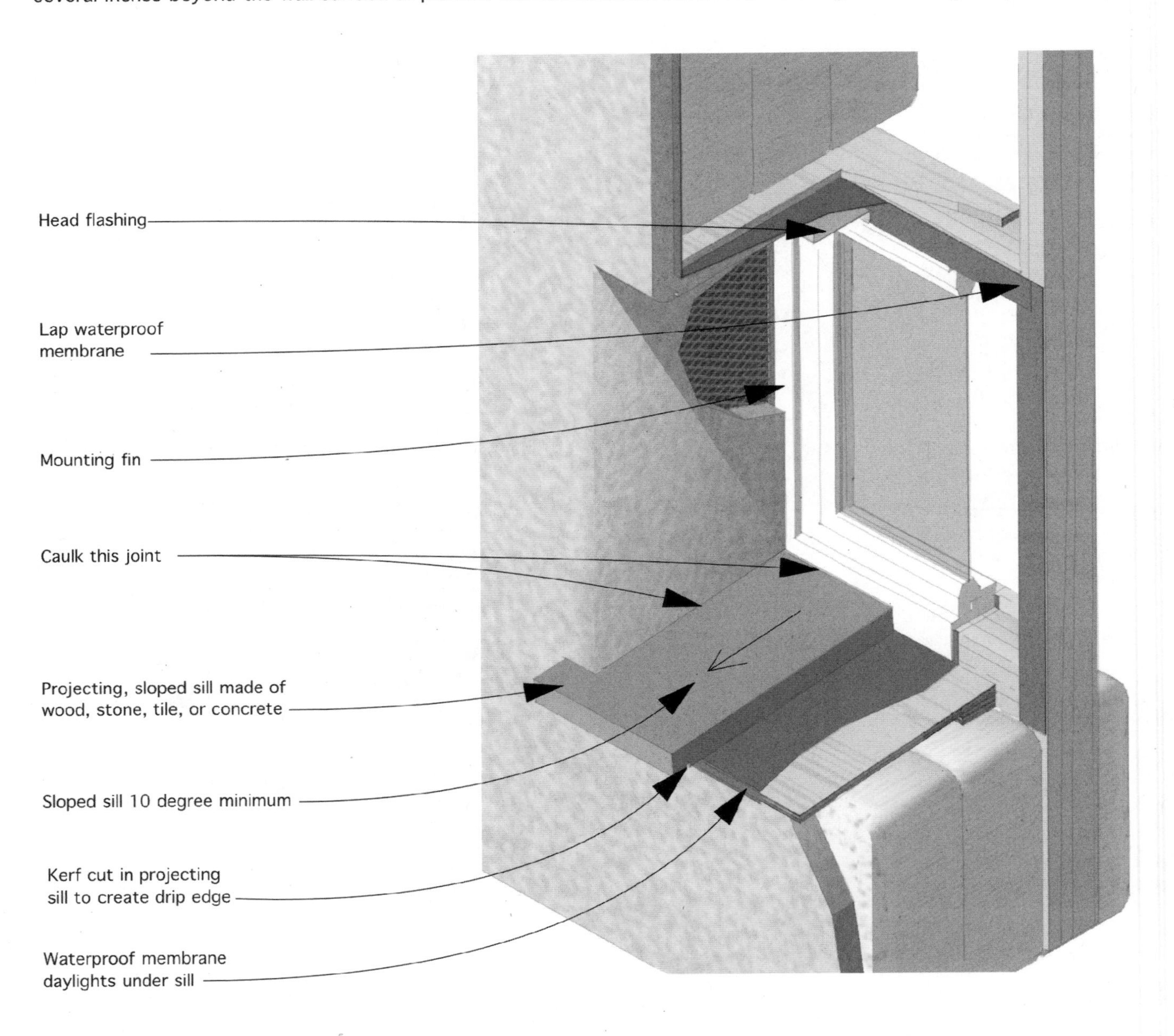

NOTES:
- Avoid puncturing the waterproof membrane while installing the sill.

- Slope tile or concrete sills away from both the window and the reveal walls to direct water away from these joints.

Sills or No Sills? Glass doesn't absorb water, so wind-driven rain that strikes the window or drains onto it from the wall above will collect at the window's bottom where it can either run off the plaster surface...or soak into the plaster. When water soaks through the plaster, properly installed pan flashing—either metal or membrane—directs it away from the bales beneath the window. However, unless the flashing extends from the bottom of the window to the base of the wall, there's a chance that liquid water could seep into bales beneath the window.

Window sills direct water away. Make sills of an impervious material like metal, synthetic lumber, painted or sealed wood, sealed concrete, stone, or glazed tile. An integral part of the window flashing, install sills before plastering. Sills should slope at least 10 degrees and protrude a few inches beyond the exterior surface of the plastered wall. A ¼" × ¼" drip edge along the underside of the outer edge of the sill prevents water from adhering and returning back to the wall. Sills may not be necessary when windows don't receive wind-driven rain or are located under sufficiently sheltering roof overhangs.

When in doubt, install window sills. They don't add much to the building's cost and are much cheaper than repairing water damage to plasters or straw bales beneath the window, or installing sills afterwards.

Inside

Windows and Doors

Reveals and soffits can be rounded, squared, or flared. They can be uniform or irregular. These shapes can be formed by the bales, by hand-sculpted straw-clay, lath and plaster, or by framing materials. For reasons discussed below, many

2-13. Shape of window reveal and effects on light.

2-14a. Window reveal (interior) shaped with straw-clay.

The window reveal (at the jamb and head) and stool (sill) can all be shaped with straw bales, then filled in with a straw-clay mix. The dried straw-clay surface requires no additional lath unless mesh is needed for plaster reinforcement. This technique can also be used to shape doorway reveals.

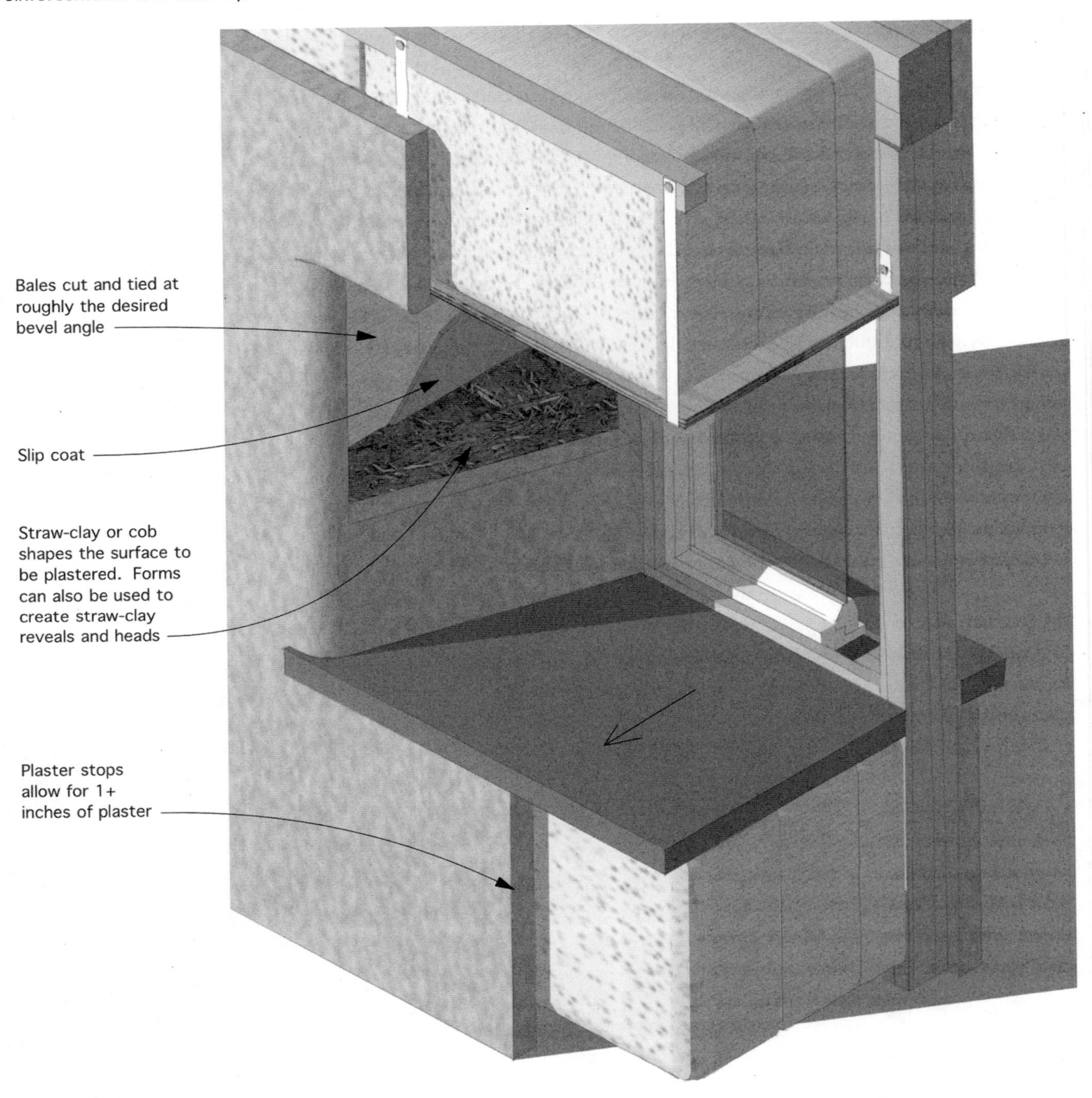

NOTES:
- If the bales are not carefully shaped, this technique becomes labor intensive.

- Good choice for projects where a more informal appearance and using natural materials is preferred.

2-14b. Window reveal (interior) shaped with lath.

This method quickly produces reveals, although consistency depends on the care taken to install the lath and secure it to the bales; the plaster skins will follow the plane of the lath. Loose straw or straw-clay fills in behind the lath both to improve insulation and to provide sufficient resistance so the reveal doesn't flex when the plaster is troweled on. This technique can also be used to shape doorway reveals.

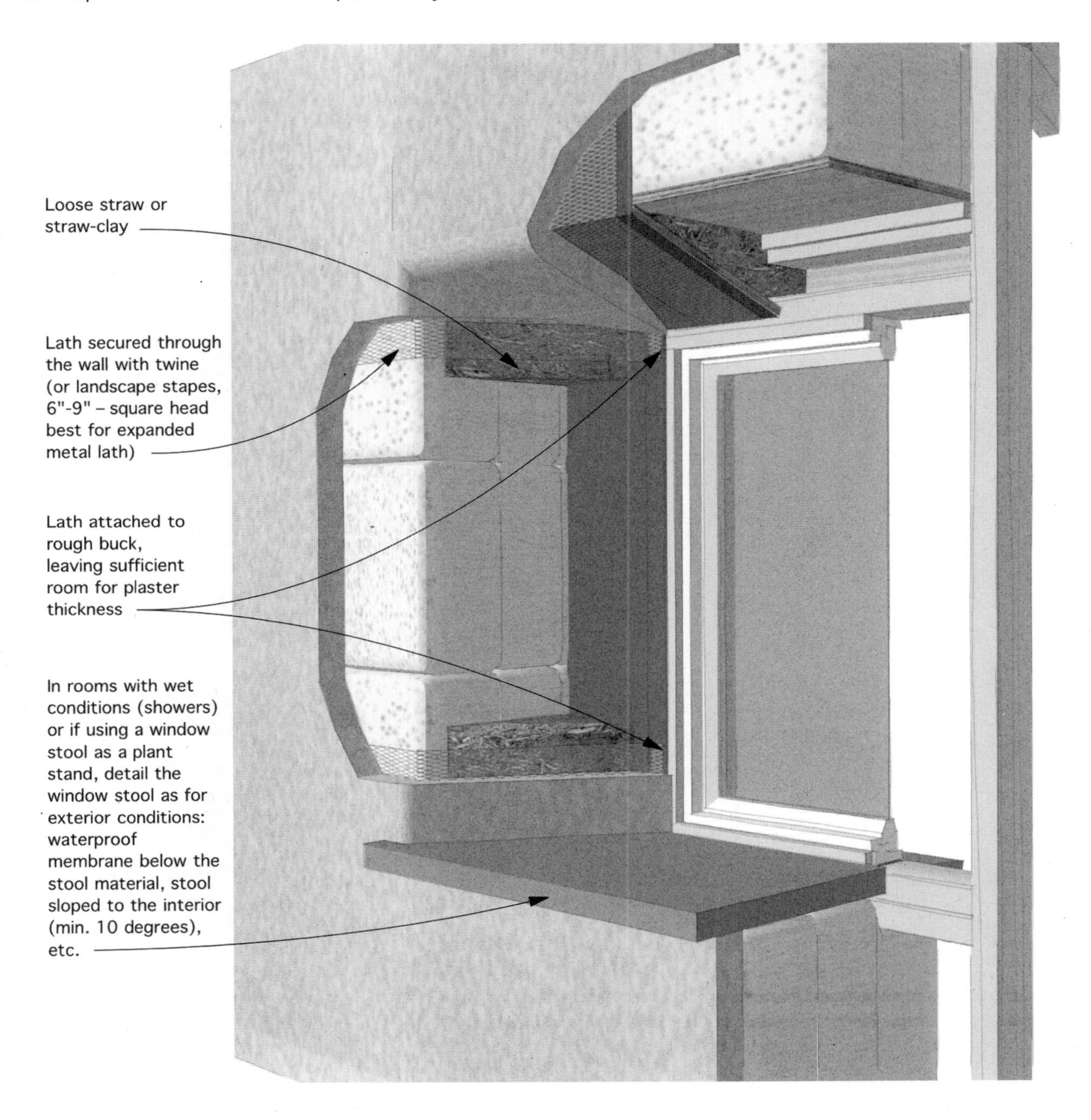

NOTES:

- Know where the plaster will stop at the window frame.

- Over-stuffing the reveal cavity bows the lath inward toward the window.

2-15. Rounded reveal with 2x studs.

When a larger-radius curve is desired at corners and openings, 2x studs covered with building paper and lath can be used to create a broad range of curvilinear reveals.

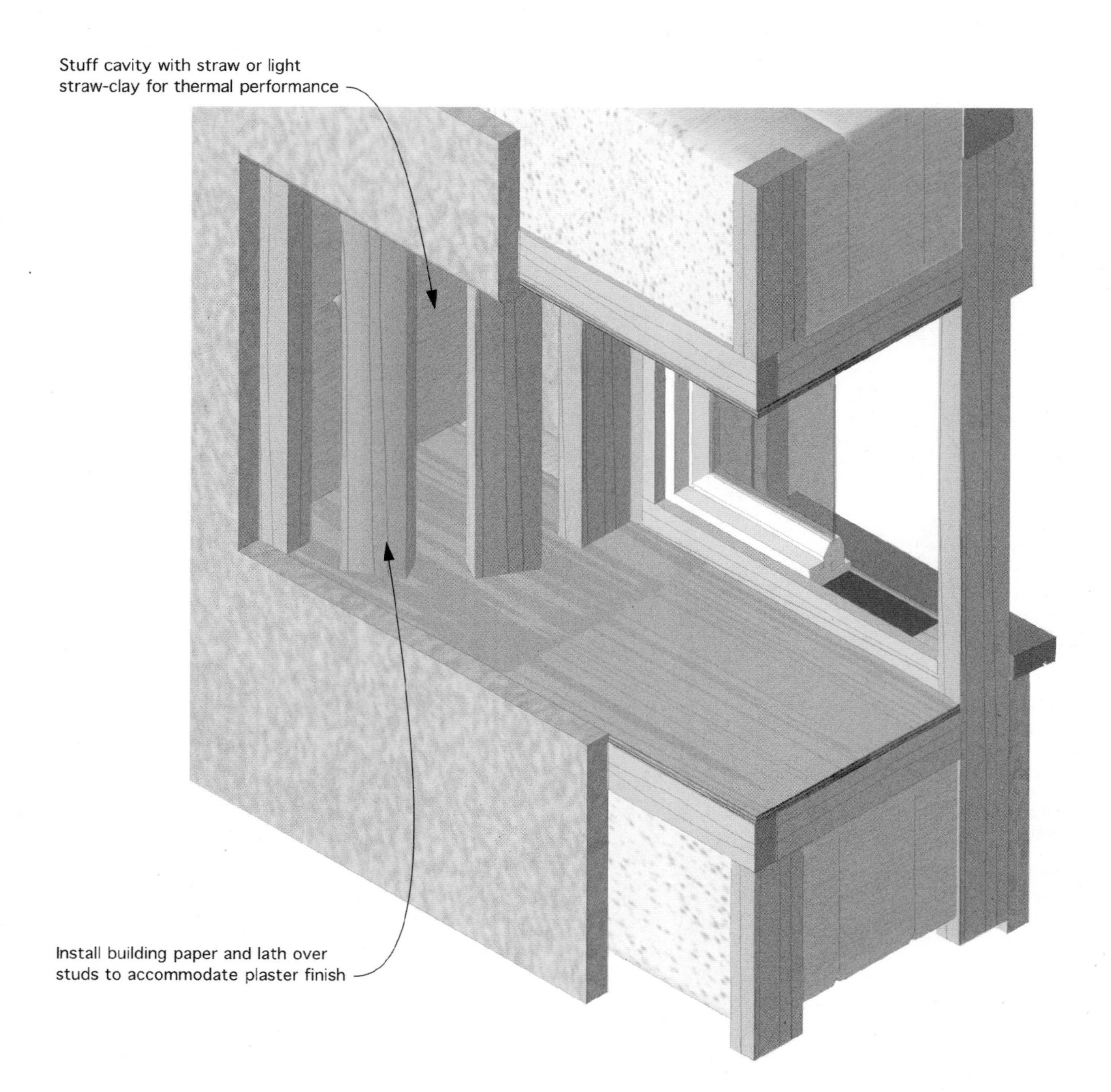

NOTES:
- Efficiently creates uniform radius curves.

- Stuff cavities with loose straw; use straw-clay only within a few inches of the wall surface.

building designers employ a variety of reveal treatments. Efficient methods have evolved for each option and may be included in the plans.

Rounded

Here, the reveal is perpendicular to the window or door and curves until parallel to the interior wall plane. The light on the rounded portion of the reveal surface reads as midway between the bright outdoor light and the darker interior wall surface. Because it transitions between the two, there's less contrast, and hence less potential for glare. This is probably the most common look in bales-laid-flat buildings, and in most cases it's the easiest to create. Since the bale strings are generally about 5" from the surface, these curves can have up to a 4" radius. Simply cut off the outer corner of the bale. If lath will also cover the wall, pulling it tightly across the corner helps to shape it. Straw-clay can also be used to form the radius, which can be template perfect or informally irregular.

Somewhat more difficult are large-radius curves. Bale strings can be moved inward a few inches, but beyond that the bale loses compression. Use framing materials to shape curves approaching an 8" radius. Affix 2×4s along the radius to stool and header framing. For doorways, the 2×4s span from sill to the door header. With larger 2× spacing, span the gaps with thin plywood bent to the radius. Cover with building paper and lath suitable for the plaster. Stuff loose straw behind the forms. Plasters build out the curve.

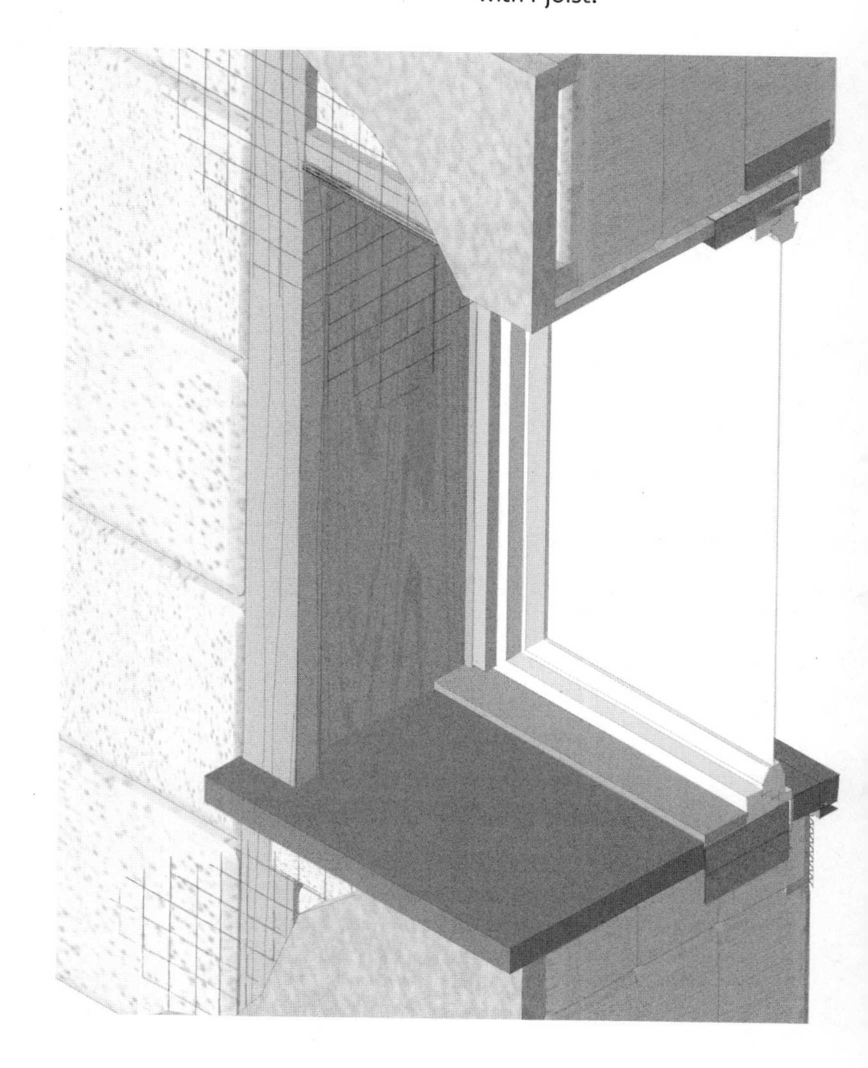

2-16. Square window reveal with I-joist.

Squared

Some people prefer the aesthetic of a window or door reveal that is perpendicular to the exterior wall plane. The bright window and reveal offers greater visual contrast with the interior wall surface. A squared reveal can have a very small radius or a crisp edge for a "modern" design aesthetic. If a window seat is meant to be used "chaise lounge" style, squared window reveals on the lower portion are more comfortable to lean against. When plans call for a squared doorway reveal, be sure to locate the door frame far enough from the reveal so the latch doesn't gouge the plaster as the door swings.

Bales laid-flat or on-edge easily make squared window and door reveals. With

2-17. Window reveal (interior) shaped with plywood.

A method to quickly produce consistent reveals throughout a building involves framing the window stool, reveals, and header in 2x and plywood before bale stacking begins. As bales are stacked adjacent to the window reveals, stuff loose straw in the triangular cavity between the plywood and the bale. The joint where the plywood ends and bales begin may be papered, then lathed for plaster. This technique can also be used to shape doorway reveals.

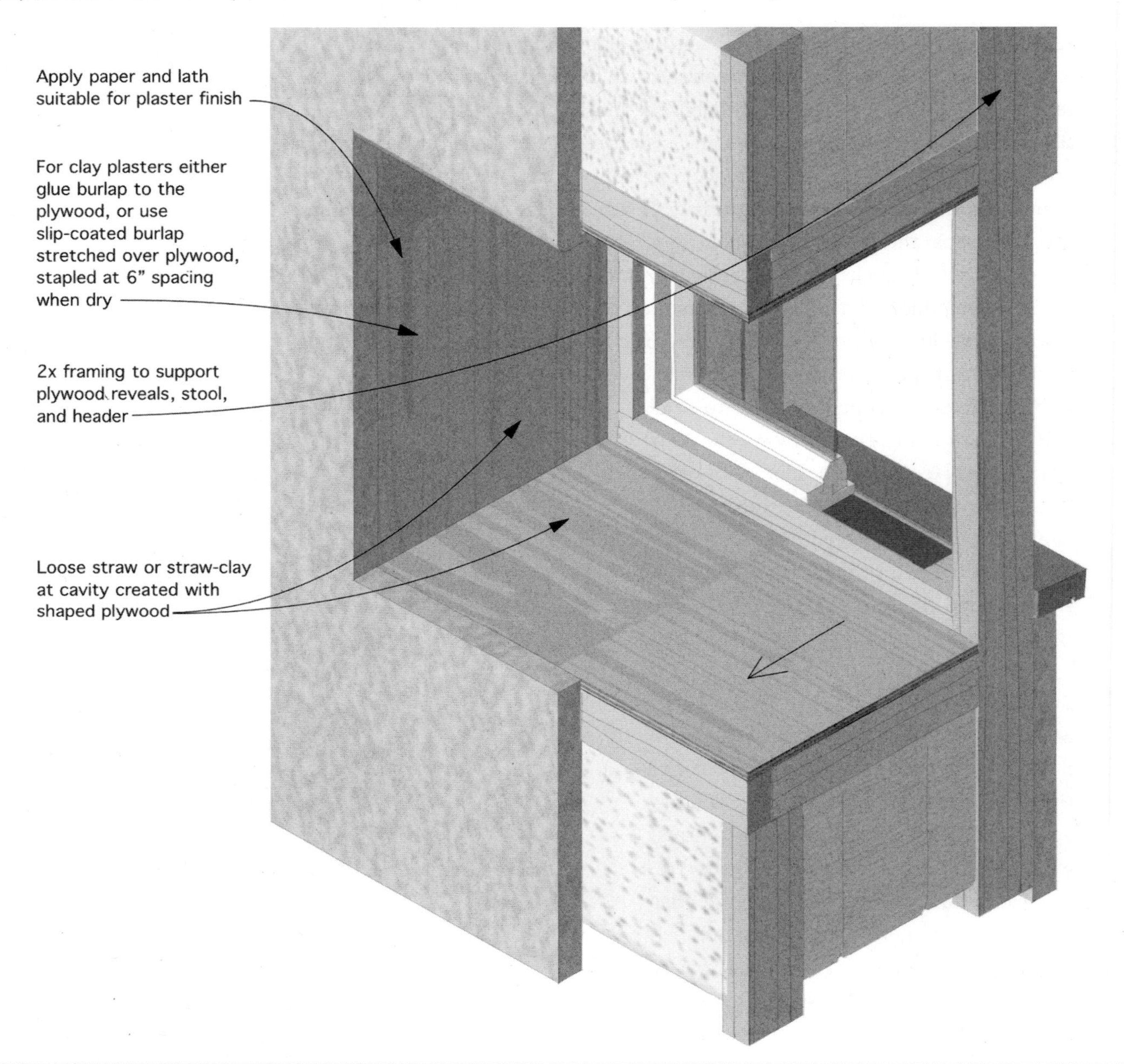

NOTES:
- Know where the plaster will stop at the window frame.
- Be careful not to overstuff the cavity – it's easy to bow the plywood inward toward the window.
- Window stool materials like wood, concrete, and tile wear better than plaster.
- Optional trim on window.

laid-flat walls, either install a 2× or 4× stud to shape the inside reveal edge, or don't cut the bale edge and build the desired profile with lath and plaster. In on-edge wall systems, the I-joist or laminated veneer lumber (LVL) "posts" are often part of the window and door framing and provide this reveal shape. To reduce plastering costs and provide material contrast at windows and doors, some designs finish the reveal with wood panels attached directly to the I-joist or LVL posts.

Flared

Flared reveals between 15 and 30 degrees create a large transitional surface midway between the outside light and the interior wall. These are generally created with 2× framing and plywood fastened directly to framing that supports the window. In laid-flat wall systems, a bale placed alongside the flared plywood box leaves a triangular-shaped void that must be stuffed with loose straw. With on-edge systems, create a flared reveal by attaching plywood between the window frame and interior edge of the LVL or I-joist post. Install the plywood in sections and stuff straw behind as the sections attach. Then prepare for the desired interior plaster. Lath can also be stretched from the window frame to the corner, and straw stuffed behind, building up to the top of the reveal. Pack the straw as densely as possible without bowing the lath or plywood outward. Don't worry too much about sacrificing insulation value here—this narrow area transitions between an R-27+ wall and the window itself, which has a much lower R-value.

Alternatively, shape the bales as much as possible (rectangular bales resist becoming triangular) and build out the surface with straw-clay. Although more labor intensive, straw-clay offers more sculptural possibilities.

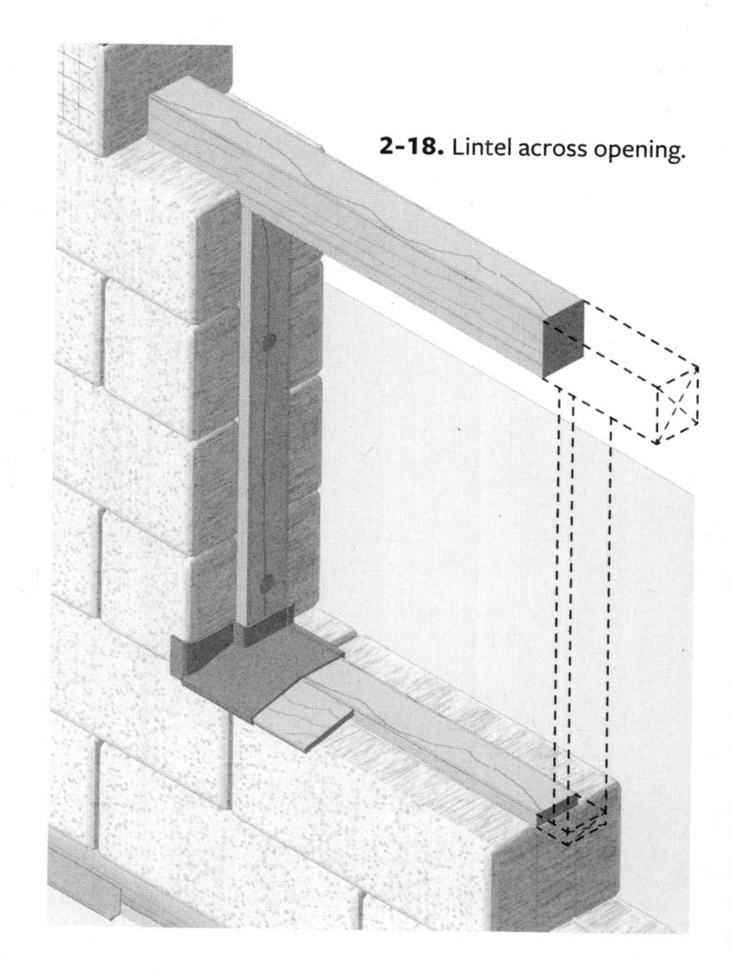

2-18. Lintel across opening.

Window and door soffits

The window and door interior ceiling—the soffit—can also be rounded, squared, or flared. However, bales over openings need support. In load-bearing wall systems, a wood window buck fabricated from 2× lumber and plywood floats in the wall. Narrow window bucks are often adequate to support the loads above them. With larger windows, plan for a lintel across the opening. Lintels have been fabricated from I-joists, 2× and plywood, 4× lumber, and even angle iron. Alternatively, the bales above openings can be tied or strapped to the box beam or top plate.

2-19. Supporting bales above openings with straps.

Where a flatter window head is desired, support the bales on a lintel with a shallow plywood shelf and metal straps secured to ceiling framing.

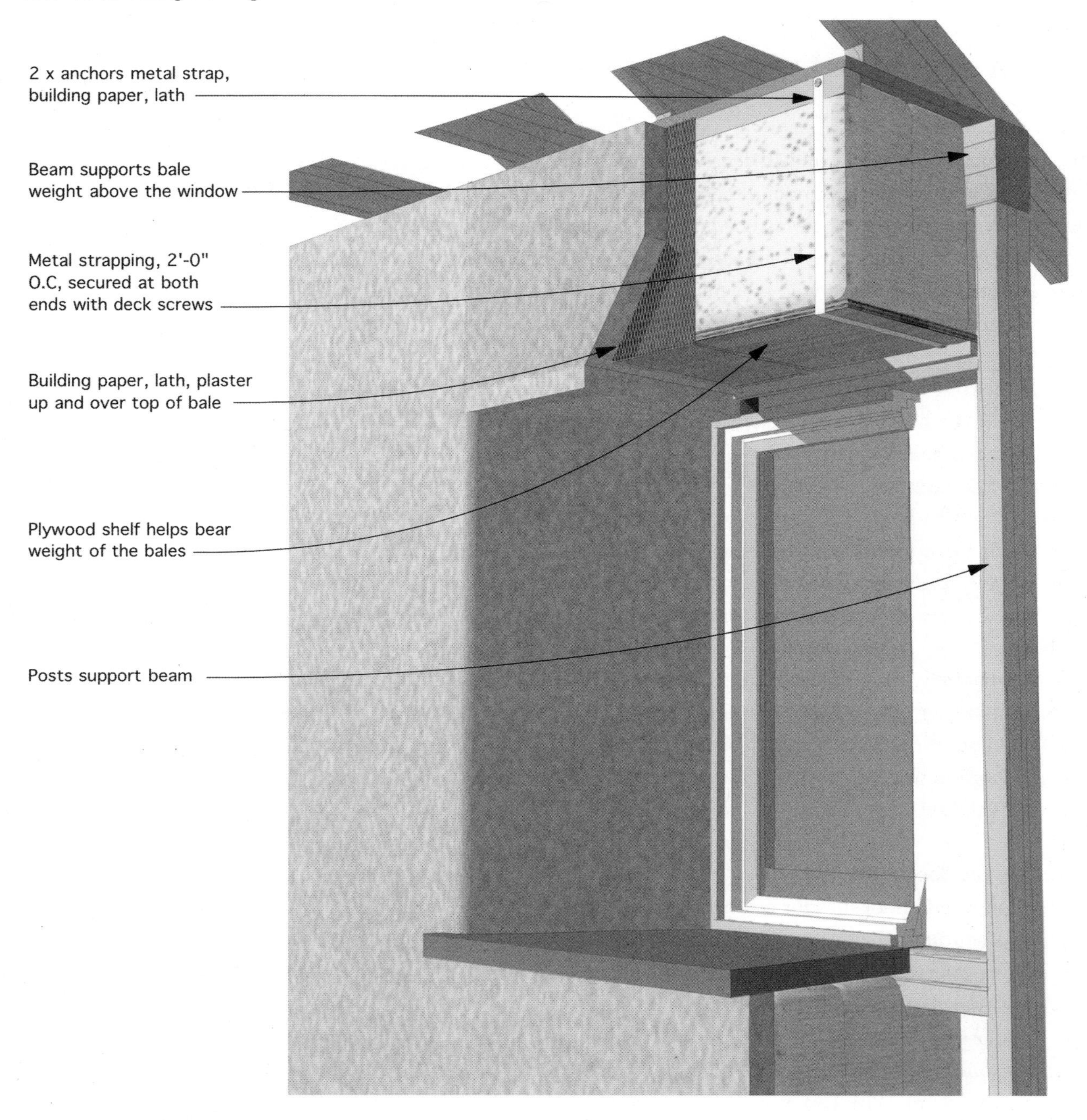

NOTES:

- Window lintels must handle the load and span; be sure the lintel is sized appropriately.

- This system works well where the reveals are also formed by plywood; the header shelf is supported by the reveal plywood on both sides.

- This may not be suitable where there are more than one or two bale courses above the window.

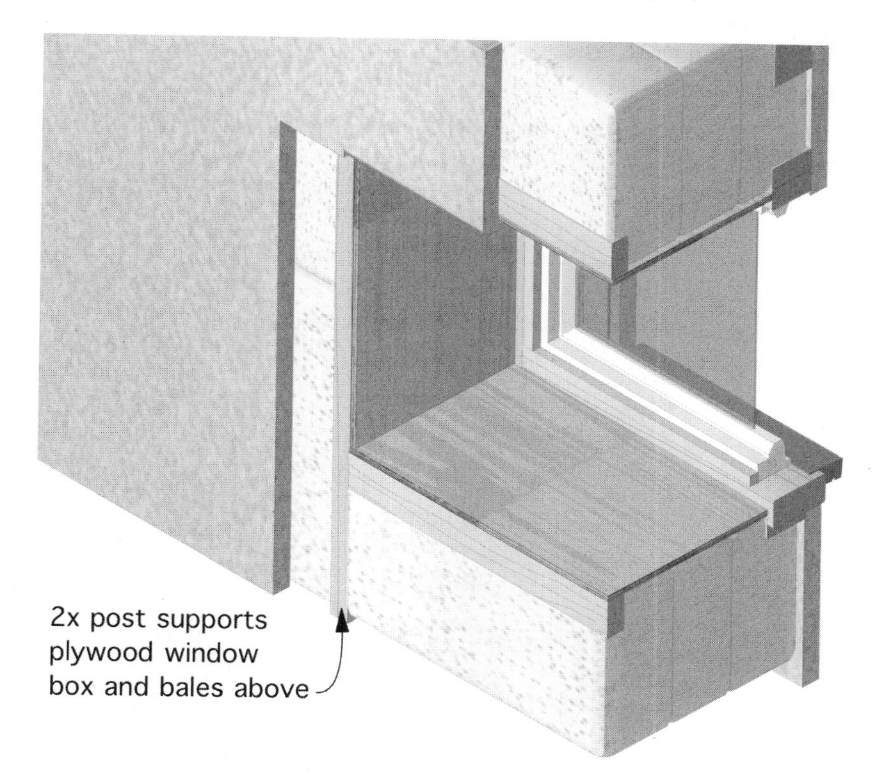

2-20. Plywood shelf supporting bales at header.

In post-and-beam wall systems, windows are often attached to structural posts that carry roof and/or floor loads. Window headers can carry much of the bale load above the window; depending on the window span, some designers attach a plywood shelf to the header, which is then supported by 2× lumber from above or below. Install the 2× framing during the bale stack to facilitate fitting the bales. Metal strapping attached to the window header and stretched under the bale and up the interior bale surface to the box beam or ceiling-roof framing often helps support this load.

Window Seats

Window seats allow you to take advantage of the extra wall thickness. Comfortable window seats that support intended functions like enjoying an outside view or a conversation circle will see more use than seats that are too high, too narrow, or built without forethought.

Top of the Wall

Box Beams and Roof Beams

In load-bearing construction, rafters or trusses rest on an insulated box beam placed over the bales. Fabricate box beams onsite from 2× and plywood. Build them in sections on the ground, lift into place, and fasten together. A wood beam or a box beam caps the wall in a post-and-beam system. In both systems, design the wall height with the bale module in mind—the top course of bales should fit snugly beneath the beam. If there's a narrow gap, fill it with straw flakes or stuff it with loose straw.

Lath or Mesh?

These two terms are often confused by designers and builders. Lath gives the plaster a means of attachment or bonding to the wall. Many materials can function as a lath on a straw bale wall: expanded metal lath, woven or welded wire mesh, fiberglass mesh, burlap, or reed mat. Because a straw bale's rough surface is more than adequate to hold most plasters, lath is often used only over the wood framing members of a straw bale wall. Mesh reinforces the plaster to prevent cracks or to reinforce the plaster in a structural role. It's almost always embedded in the plaster's inner half. Some of the same materials used as lath also function as reinforcing mesh: 17-gauge woven-wire stucco netting, 2×2 welded wire mesh, polypropylene mesh, etc. Lath holds plaster; mesh makes it stronger.

2-20a. Window seats.

Window seats in a straw bale wall make good use of the extra wall depth. Design and build them depending on how the seat will be used. Comfortable seats with cushions are usually no more than 18" high and 18" deep. The backrest of the window seat is typically sloped 10-15 degrees from vertical, with a shallow 2-5 degree seat slope. Window seats can combine features from both drawings below.

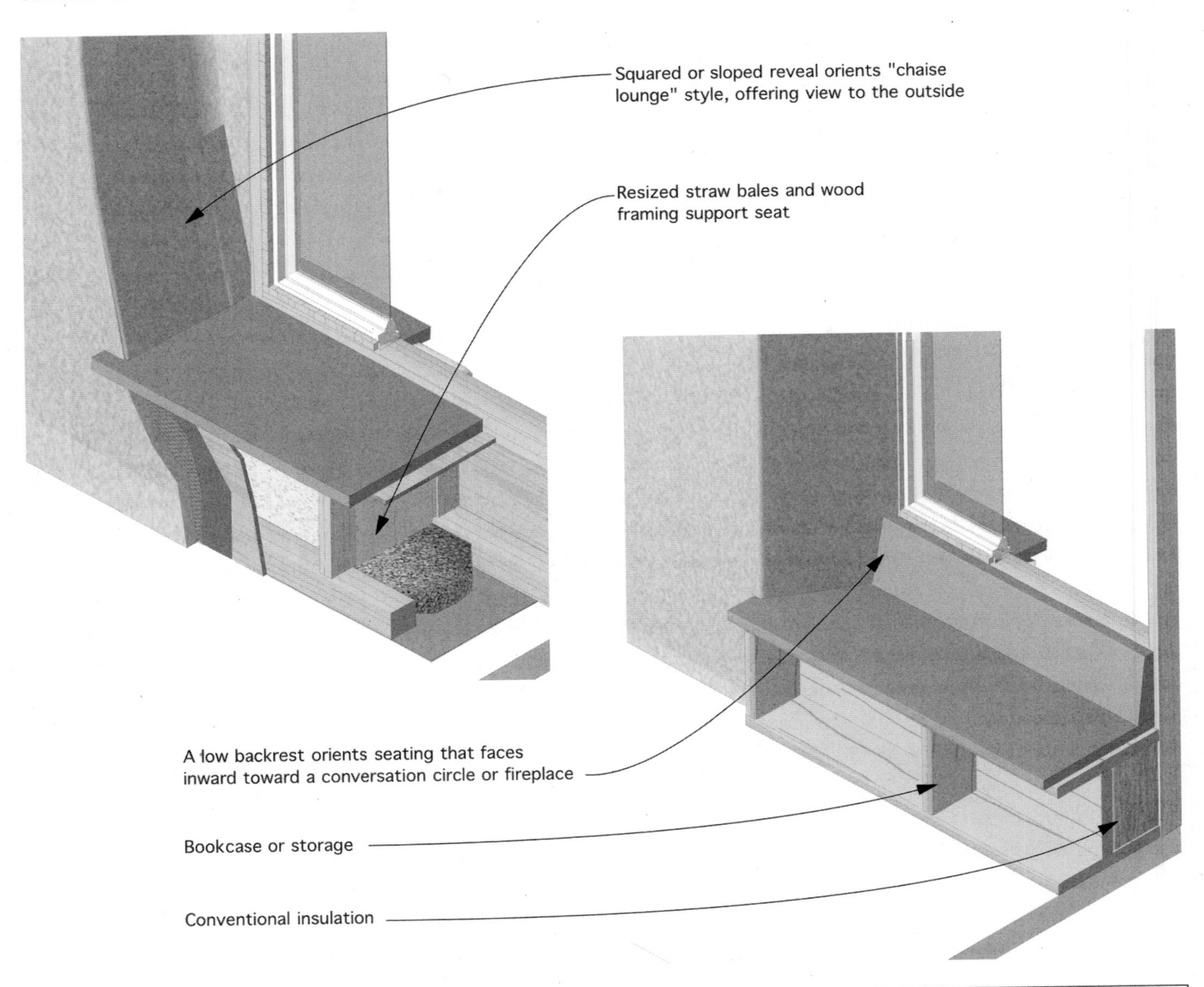

NOTES:
- If a bale is too high for comfortable sitting, use framing to create the desired seat height and fill the space below with bale flakes or resized straw bales.

- Window seats in 2-string laid-flat walls must be "bumped out" for comfortable sitting depth (shown).

- Make seat surface from wood, oiled cob, poured-in-place concrete, or tile. Plasters can work, but must be extra thick and quite hard to resist wear.

- Pillows and seat cushions make sitting more comfortable.

2-21. Box beam.

Also called "roof bearing assemblies" (RBAs), box beams transfer roof loads to the bale walls, and may perform other important roles: attachment surface for interior and exterior lath and mesh, cabinets, etc. Using the foundation as a template, build RBAs on the ground before stacking the bales.

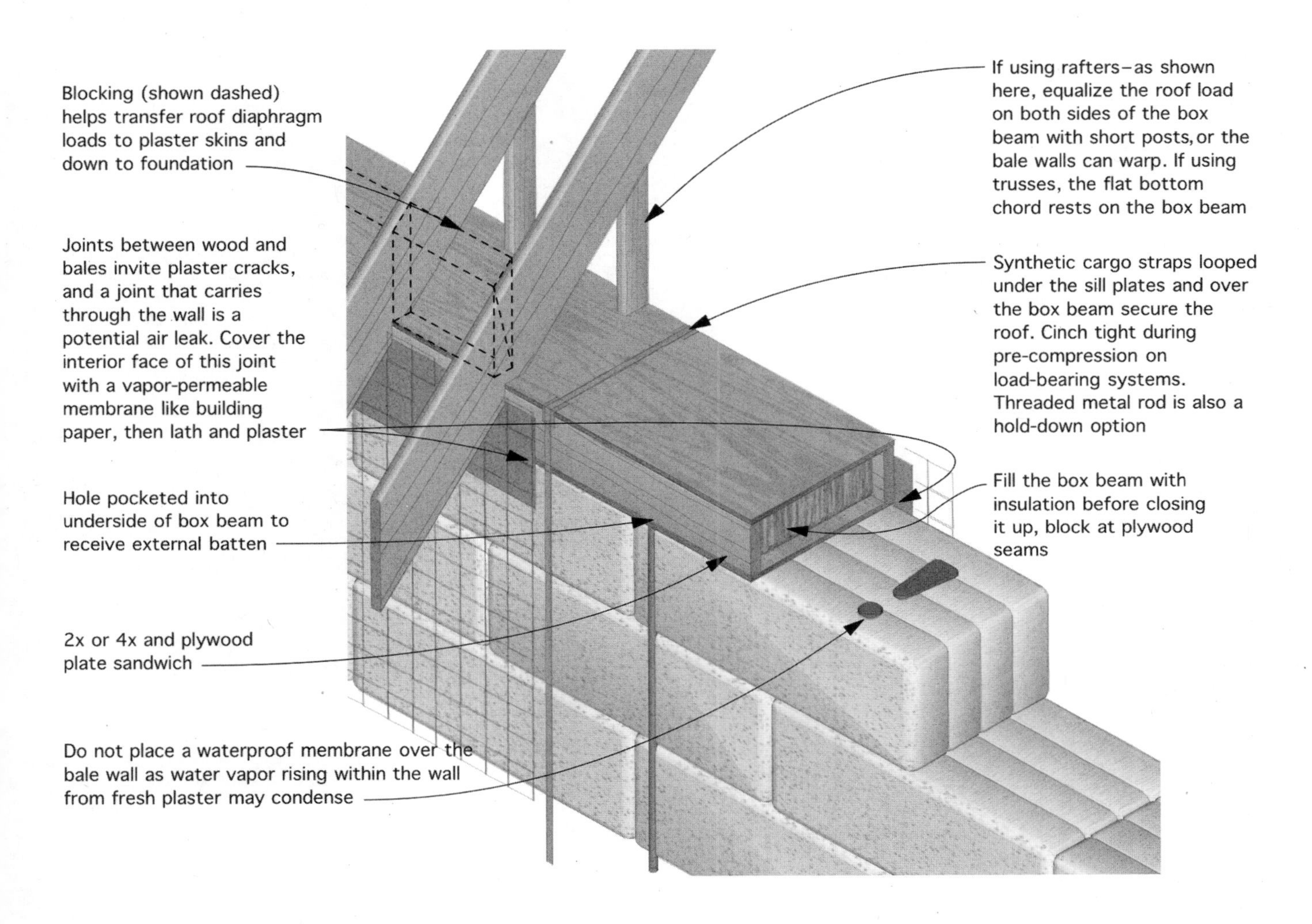

NOTES:

- Although they are more wood intensive than a single 4 x top plate, box beams are often used in post-and-beam structures because they can simplify ceiling-wall connections, securing bales over windows, and installing a fire-resistive barrier.

- Box beams use smaller dimension lumber than most 4 x top plates.

- Plywood box beams function as a fire-stop between the attic and wall assembly; some builders add a layer of sheetrock or plaster to increase ignition resistance.

- Blocking between rafters can double as a plaster stop / finish trim at the top of walls.

2-22. Roof beam notched into bales.

Post-and-beam structures often use a single 4x top plate to efficiently and economically transfer roof loads.

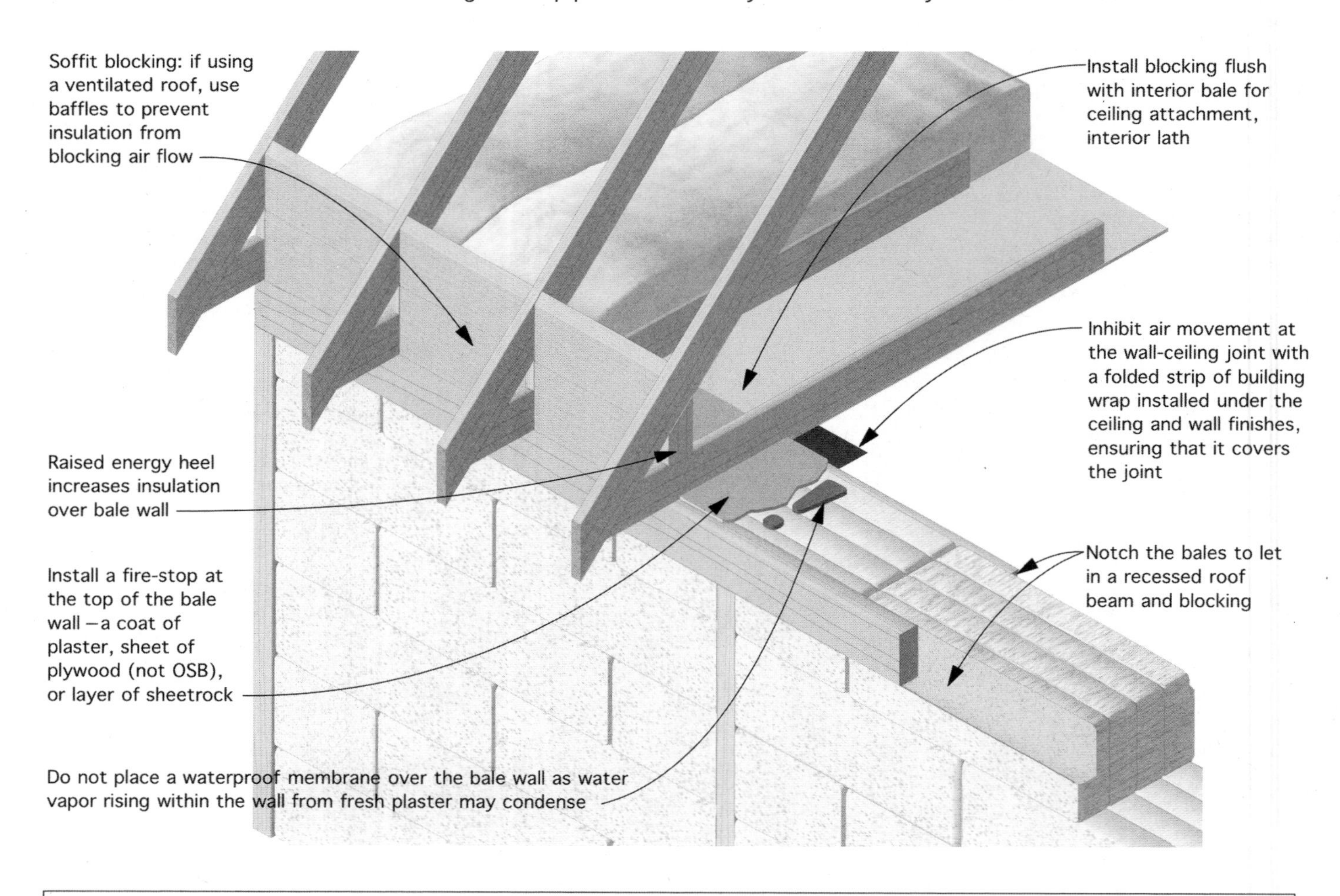

NOTES:

- When repeating the same cut in many bales (like a roof beam notch), a plywood jig can help guide the saw to create uniform notches.

- Do <u>not</u> notch the lower chord of the truss.

- This system works well with a variety of structural designs.

Box beams provide a secure surface for lath and mesh attachment on both sides of the wall. The plywood offers some fire resistance between the attic and the bale wall and vice versa, which is easily supplemented by a layer of gypsum board.

Many bales laid-flat structures employ a single 4× roof beam to carry roof loads. Plywood supported by blocking seals the wall from the ceiling insulation space above. The top bales need to be notched on one side. Avoid leaving the top of the bale wall open to the ceiling insulation space.

Gable Walls

Straw bales can be placed in the upper portion of a raked gable wall. On-edge wall systems with fairly low-angle rakes are the simplest because cutting the bales to follow the roof angle doesn't cut all the strings. This isn't possible with bales laid flat. Stack full and partial bales, then stuff flakes and loose straw into the remaining triangular cavities. Thermal performance may suffer. To avoid this, many designers continue a level top plate around the entire building, then use a conventional wall system above the bales. For example, an 8"-wide double-stud wall with cellulose fill carries the straw bale wall's insulation value from the top plate up to the roof. This transition must be carefully detailed to protect the lower walls from moisture.

Rainscreens

In regions where heavy wind-driven rain can be constant for days, the wetted walls may not have an opportunity to dry out. If water overwhelms the plaster's ability to absorb and release it, moisture can wick into the straw bales. If designing a structure in an environment with these conditions, consider wrap-around porches to protect the walls. Where that's not possible, use a rainscreen. Incorporate supports for furring strips into the exterior bale surface (see Figure 2-31: Attaching/Hanging Objects from Straw Bale Walls—For Heavy Loads). The exterior bale surface is usually sealed with at least two coats of plaster, then furring strips are attached to the walls. Vent screens at the bottom and top prevent insect and rodent intrusion and allow air to circulate and escape. Attach siding over the furring strips. If the walls play a structural role, make certain the rainscreen assembly is detailed in the engineered drawings.

Adding Exterior Insulation

In extremely cold climates, a straw bale wall assembly benefits from additional insulation on the exterior surface. Water vapor migrating from the warm, moist interior toward the cold, dry exterior could condense as it nears the exterior wall surface. Although there are several approaches to this, one is to stack the bale walls to the inside of a framing system. Blown-in cellulose against the bale surface fills and air-seals the framed cavity. A vented drain plane and siding follows. This approach moves the "dew point"—the location within the wall where water vapor could condense—outside of the bales.

2-23. Wood walls above bale walls.

Bale walls can absorb and release moisture that penetrates the exterior plaster, but avoid adding additional water that often runs down the plaster or siding on gable wall ends. Direct this moisture to "daylight" onto the plaster surface over the bale walls.

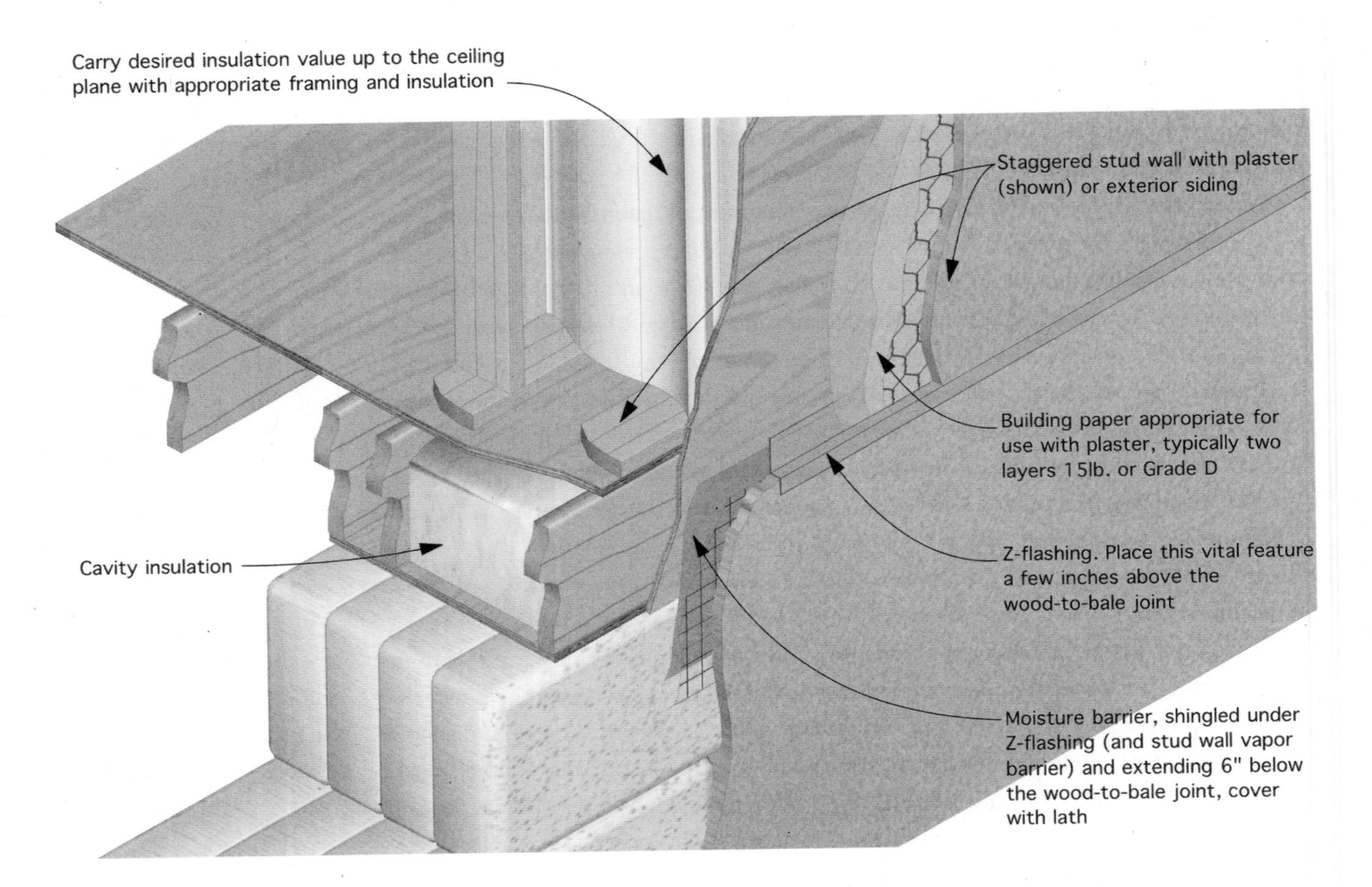

NOTES:

- Many builders use conventional siding materials on a second-floor gable wall to lower building cost.

- This drawing is meant to show moisture and insulation detailing between a bale wall below and wood wall above. The absence of posts does not suggest load-bearing wall systems can support second floors.

2-24. Rainscreen over straw bale wall detail.

Rainscreens ensure that wind-driven rain won't overwhelm a plaster and soak into a straw bale wall.

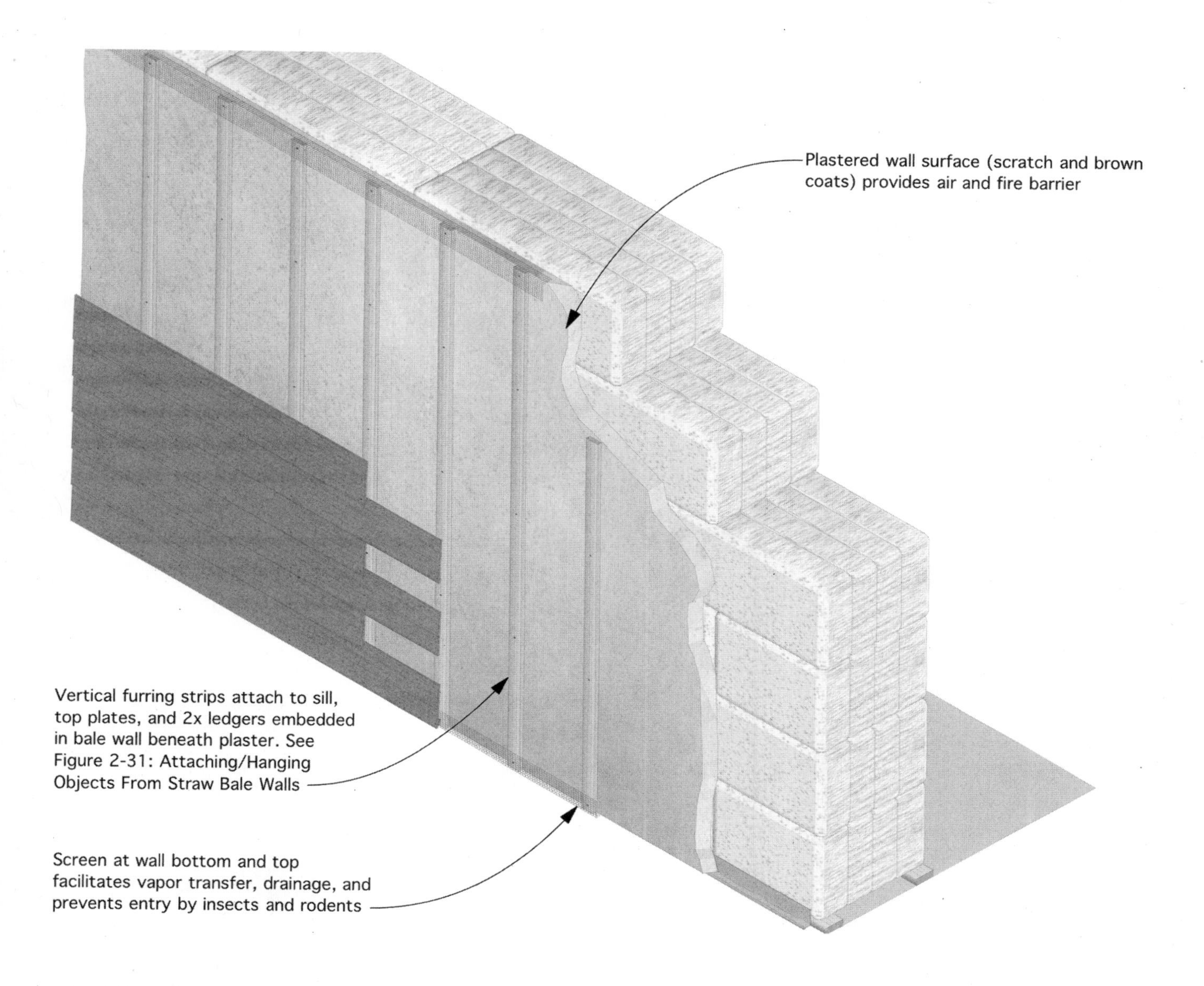

NOTES:

- Design wall system to support rainscreen weight.
- If part of a structural wall, pay attention to load paths.
- For furring strips to lie flat requires very carefull attention to achieving a flat, plumb wall surface during bale stacking and plastering.
- Some builders "wet set" furring strips after the brown coat to ensure a flat siding application.
- Rainscreens must be carefully integrated into window sill and flashing detailing.

2-25. Exterior insulation over straw bale wall detail.

Additional insulation can be added to the straw bale wall assembly when required to increase thermal performance or move the dew point out of the straw bale wall core.

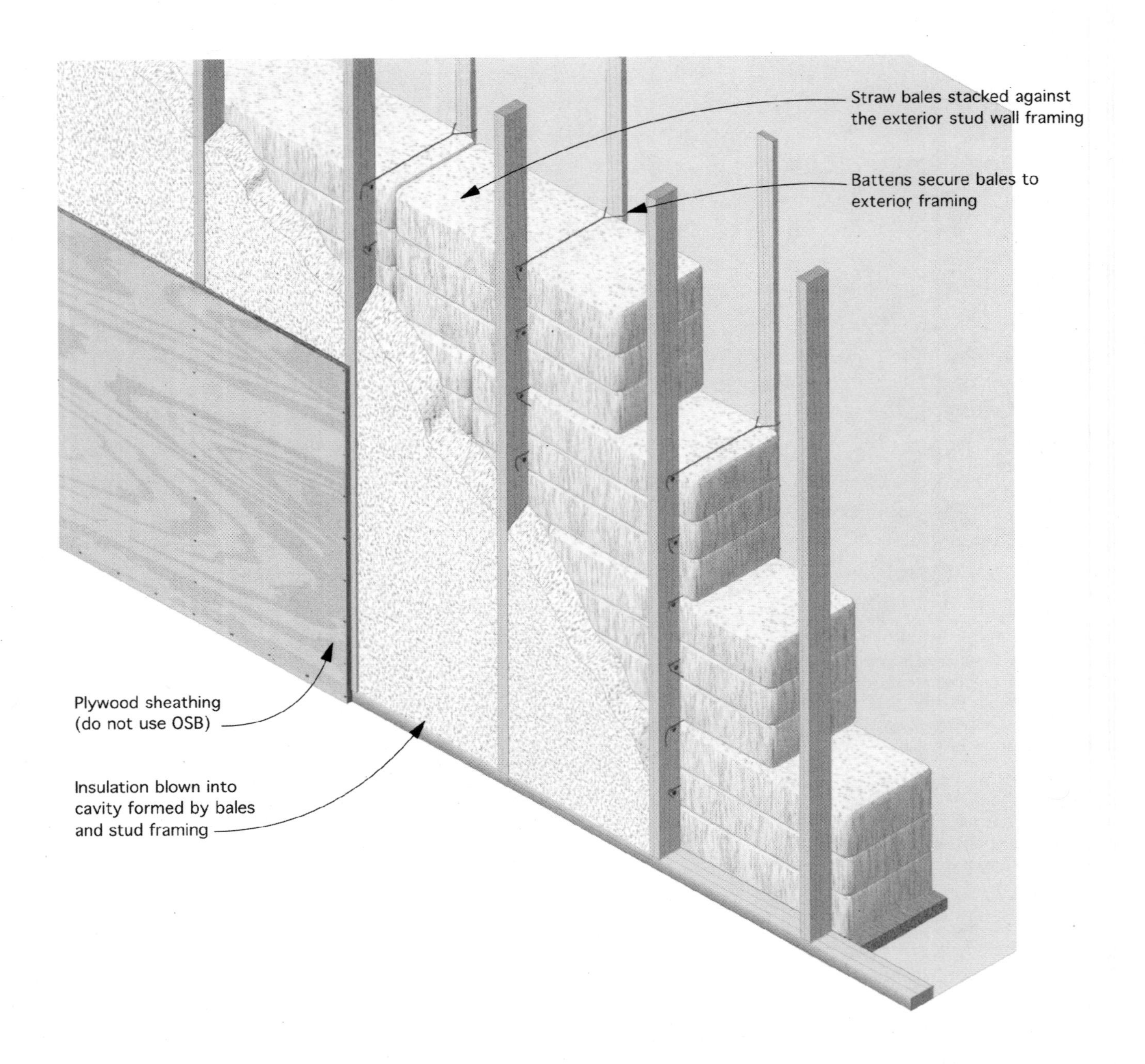

NOTES:
- Design wall system to support the weight of the additional insulation and sheathing.

- If part of a structural wall, pay attention to load paths.

Floors Above

Lofts

You can support second floors in a straw bale building in a couple of ways. Make the attic space within the roof habitable, which requires a steeper roof pitch to create more headroom. Interior partition walls, floor joists attached to the top plate or box beam, and free-standing posts have been used to support attic floors. Making single-story walls a few bales taller adds headroom.

If the second story has walls of its own, many designers specify stud walls with conventional insulation on the second floor. Particularly in seismic areas, putting much heavier plastered straw bales on the second floor can increase structural design and construction costs.

Platform-framed Second Floors

Several methods have been used to support second-floor straw bale walls. In an I-joist wall system, fasten floor ledgers directly to I-joist flanges, keeping the attachment entirely inside the wall's thermal envelope. See Figures 2-27 and 3-16. Second-floor joists have been attached to 4× top plates, with the first- and second-floor bales "sandwiching" the top plate/rim joist. This area needs to be carefully insulated to maintain thermal performance. Second floors have also been supported by a separate post-and-beam system set to the interior of the straw bale walls. Be aware that taller walls have more exposure to rain, so detail accordingly.

2-26. I-joist wall with floor attached.

2-27. Platform framed floor.

Wood and straw seasonally shrink and swell at different rates. This means that wood floors that extend to the exterior need a plaster control joint to prevent cracking. This joint requires careful flashing, especially in two-story conditions where exposed walls get little rain shelter from a roof overhang.

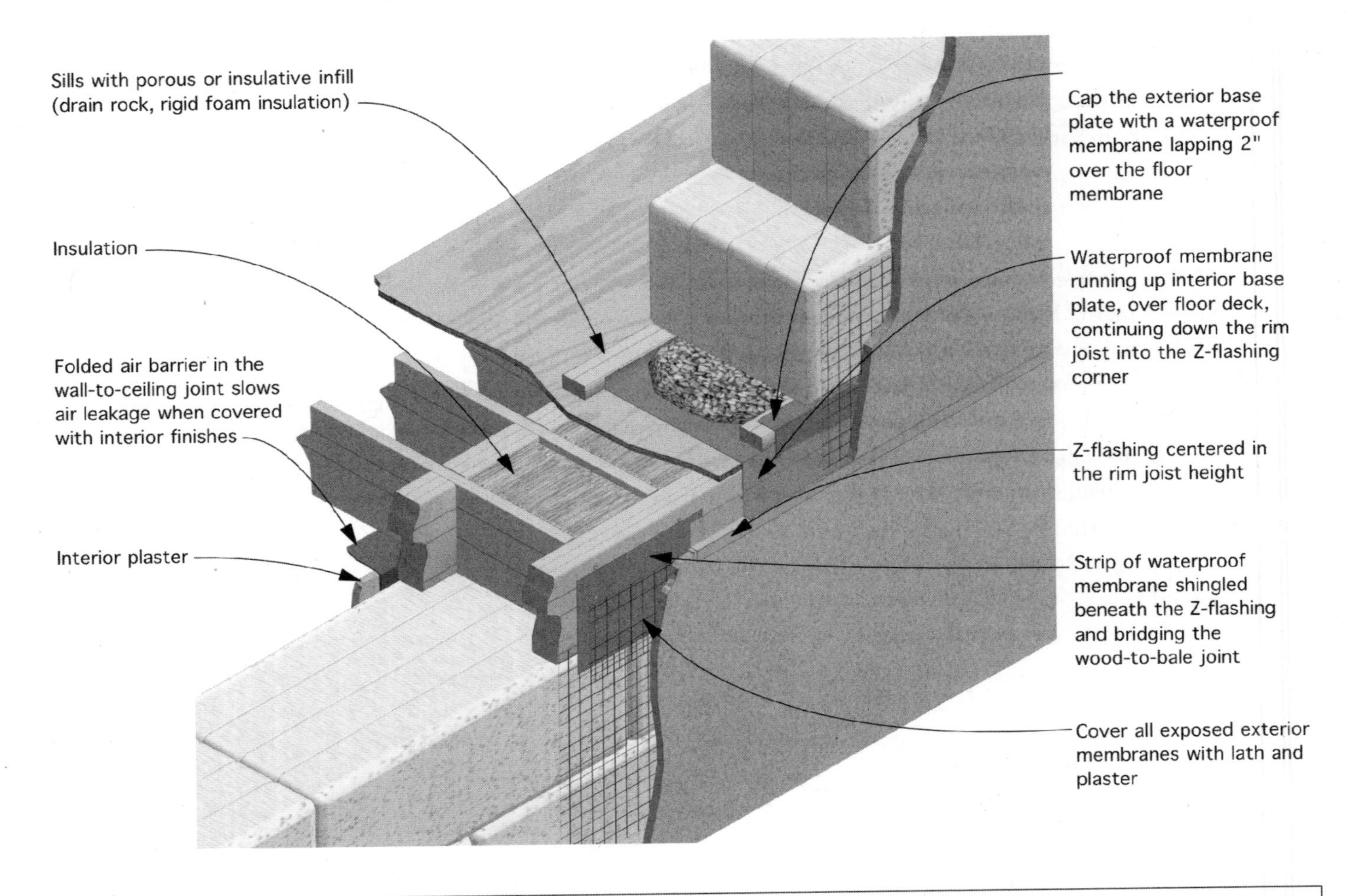

NOTES:

- Detail is intended to show the flashing required with a control joint. It does not suggest that two-story buildings bear solely on bale walls.

- Z-flashing is optional if the wall isn't exposed to excessive weather, but a control joint should be considered when plastering 2 or more stories.

Transition from Straw to Other Wall Systems

Many straw bale buildings aren't made entirely with bale walls. Some designers employ straw bales in areas where their properties provide the greatest benefit—e.g., walls with the fewest openings, or those facing a noisy road, the hot afternoon sun, or the cold north wind. Designers often switch from bales to wood framing on the south side of a passive solar design because there it may be impractical to place bales in the small spaces between large areas of windows and doors. Wood framing is also easier where there are baths and kitchens because you can avoid routing plumbing in bale walls or detailing for wet interior environments. Carefully detail the joint between two different wall systems, whether or not plaster covers both building materials. See Figures 2-28 and 2-29.

Partition Wall Connection to Straw

Poorly anchored partition walls will develop plaster cracks and may be a source of air leaks.

In bales laid-flat wall systems, place two studs at the wall joint—a 2×6 stud let into the bale wall and sitting on the bale wall's interior sill plate, and a 2×4 stud set on the partition wall sill plate. Center the 2×6 stud so material exposed on either side of the 2×4 can be used to attach lath. Some builders substitute ½" or ¾" plywood for a 2×6. If the interior face of the straw bale wall plays a structural role, consult with an engineer about interrupting the structural skin's continuity. Folded building paper stapled into this corner will slow air movement and separate the wood and plaster, which prevents movement cracks from telegraphing through the plaster. Some builders drive 12" long ½" or ¾" wooden dowels through both studs into the bale wall to anchor it, usually three or four per wall. When the partition wall has a door, some builders will further secure this connection by driving all-thread rod completely through both studs and the bale and anchor the exterior end with a plywood plate washer.

In on-edge bales systems, a stud let vertically into the wall cuts the bale strings. To avoid this, designers land partition walls on I-joist or LVL posts and add blocking for lath or mesh attachment. If the partition must land on bales, let in short lengths of 2× or plywood between the strings at intervals along the joint. Although not full length, this still allows for a folded building paper air barrier and a nailer surface for lath staples. Again, check with an engineer if this joint interrupts a structural skin. See Figure 2-30.

In post-and-beam systems, framing precedes placing bales in the wall. With bales laid flat, posts and beams on the exterior sill facilitate placing bales from the inside. Use screws to secure partition studs adjacent to straw bale walls so the studs can be temporarily removed before the stack and reattached afterward. In on-edge systems, partition walls that land on an I-joist or LVL post shouldn't impede progress; stack bales from either side of the wall.

2-28. Straw bale wall and conventional wall system transitions (exterior conditions).

There are places in a straw bale structure where it makes sense to use a conventional wall system. For example, where plumbing and vents are more easily installed in kitchens and bathrooms, where windows placed for solar gain negate adjacent super-insulated walls, or at transitions to below-grade wall systems. Careful detailing where the two systems meet can create seamless aesthetics and performance between them.

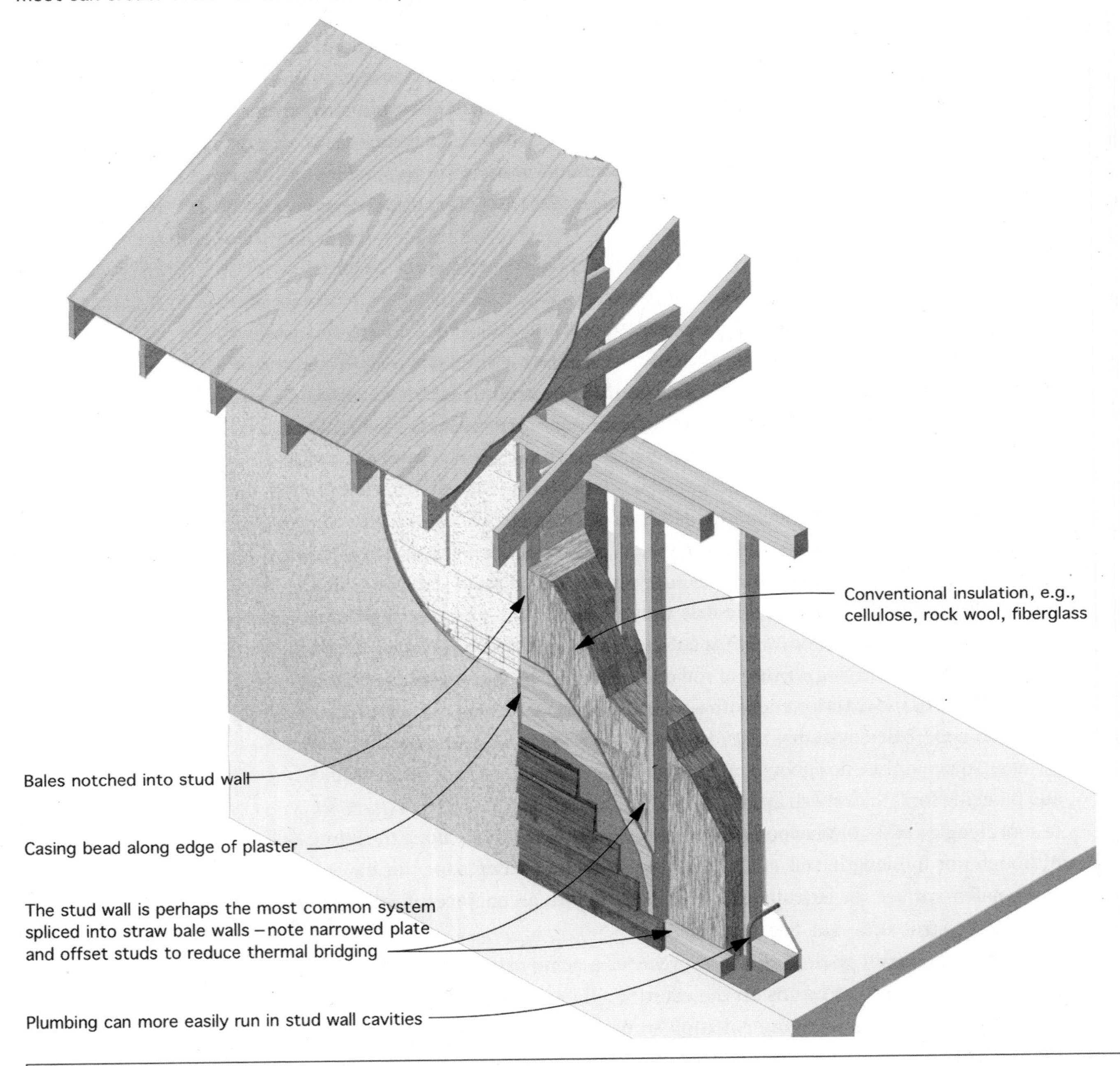

NOTES:
- Doorways can be a convenient bridge between two wall systems as well.

- Lath depends on the surface finish. If the exterior is plaster, paper and mesh should lap onto the straw bales. Where siding is used on the conventional wall's exterior, paper the joint between the two wall systems, and install a plaster stop where the two surfaces meet.

2-29. Straw bale wall and conventional wall system transitions (interior conditions).

As described in Figure 2-28, transitioning from straw bale walls to a conventional wall system has some advantages. From the inside, it allows for deeper kitchen counters and/or a shorter reach to operable windows in the wall above counter tops. Careful detailing ensures seamless aesthetics and performance between the two wall systems.

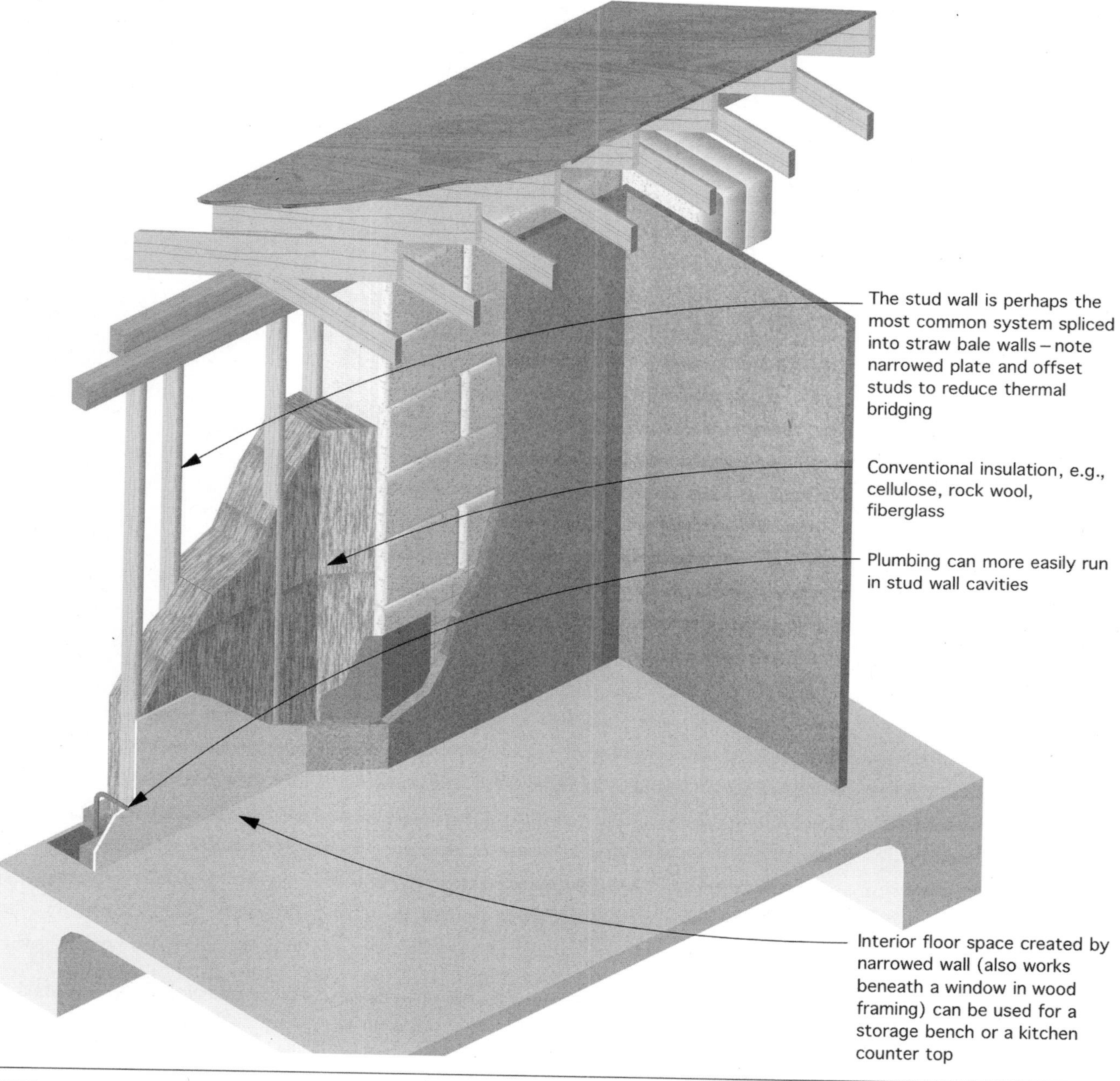

NOTES:

- Doorways can be a convenient bridge between two wall systems as well.

- Lath depends on the surface finish. If the exterior is plaster, paper and mesh should lap onto the straw bales. Where siding is used on the conventional wall's exterior, paper the joint between the two wall systems, and install a plaster stop where the two surfaces meet.

2-30. Partition wall connection to straw.

Detail the partition wall connection to facilitate lath and mesh attachment and reduce cracking.

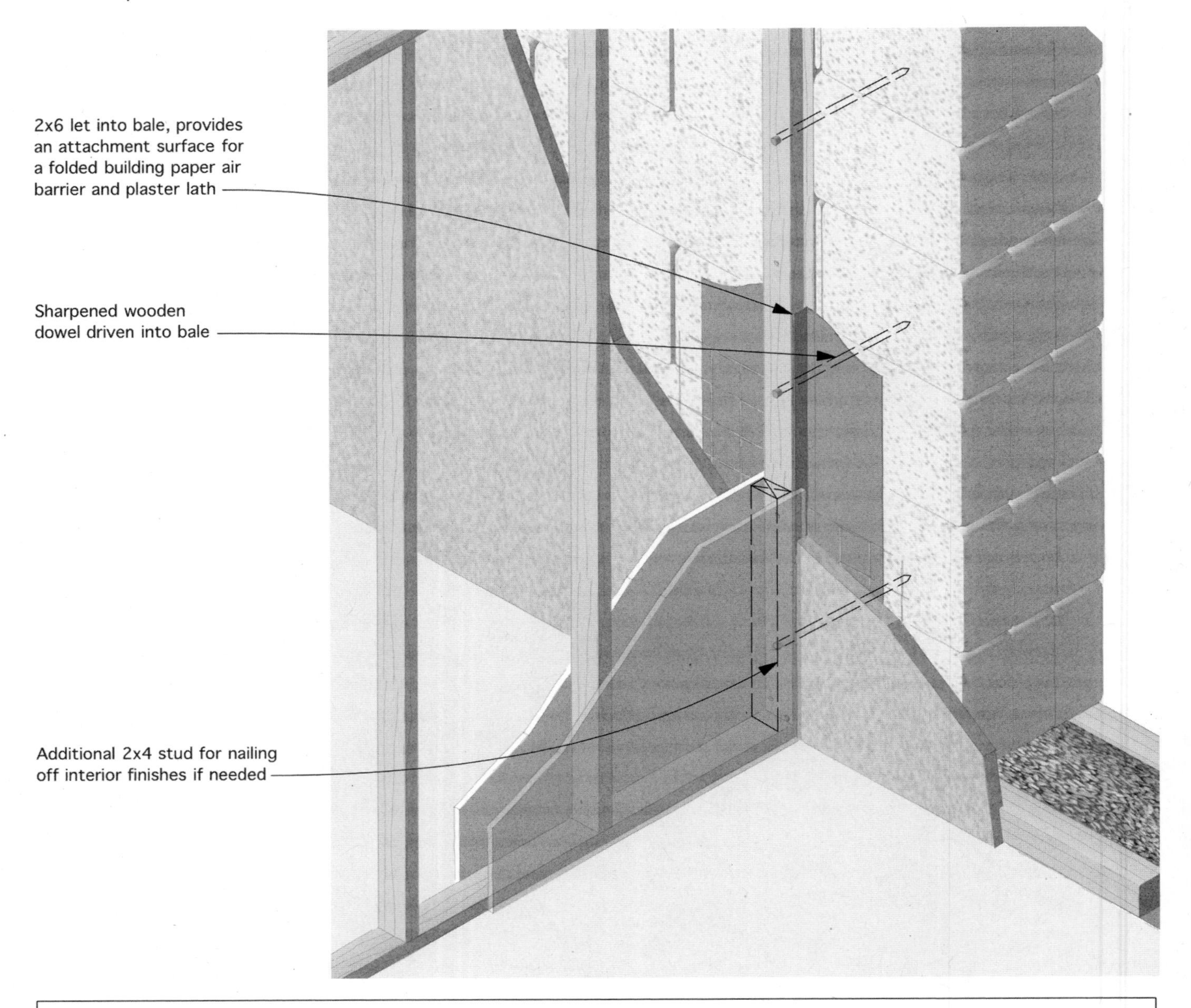

NOTES:
- If a wall plays a structural role, consult with an engineer about how to attach the partition wall without interrupting the structural panel.

- 2 x 6 let into a bales laid-flat wall should be secured to the sill and top plates.

- Some builders install a plywood strip several inches wider than the partition wall stud for paper and lath attachment. Thin plywood can be used in bales on-edge walls where partition walls don't align with an I-joist post.

- Instead of sharpened wooden dowels, all-thread rod can be used to anchor the partition wall, secured to the exterior wall with a plywood washer.

2-31. Attaching/hanging objects from straw bale walls.

Straw bale walls may need to support objects large and small: art work, small cabinets, stair stringers, even entire floor packages. How to hang these depends on the object's weight and depth: does the object just pull down on the hanger, or does it also pull out from the wall?

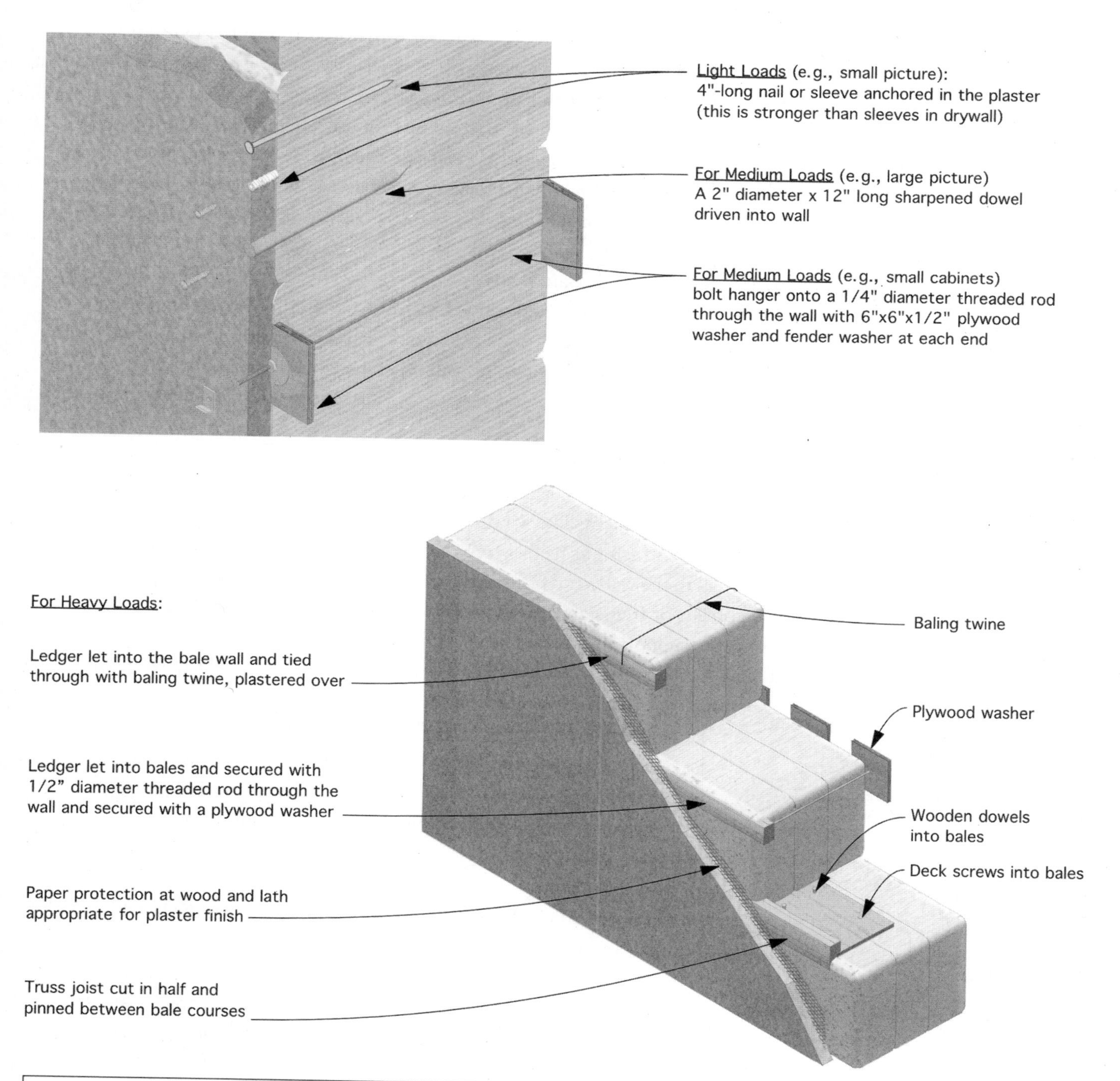

NOTES:

- Where the interior plaster is also a mesh-reinforced structural skin, the ledger(s) should be recessed into the bale so as to not interrupt the structural surface.

- Make sure to note the ledger height before plastering over so you can find them later.

Hanging Heavy Objects on Bale Walls

Install cabinet blocking in the straw bale walls of kitchens, baths, and utility rooms. Note the blocking heights and locations on the plans to ensure this isn't overlooked. A 2× set flush horizontally into the bale wall, then tied through with twine is often adequate to support cabinets, especially if the walls have mesh on them. Some builders secure the 2× with all-thread rod run to the exterior wall surface with a plywood plate washer, while others use an I-joist section cut lengthwise and inserted between bale courses. It's much easier to let blocking into a straw bale wall along the top or bottom edge of the bales. Carefully document the support locations because they may be difficult to find after the finish plaster conceals them! See Chapter 5, Stacking Straw Bale Walls, for a discussion on cabinet blocking, partitions, niches.

Hanging Lighter Objects on Bale Walls

Hang most small- to medium-sized framed pictures (or similar items) from a nail driven into the thick plaster. For heavier items, drill a hole in the plaster and drive a wooden dowel into the straw bale wall. Then hang the item from a screw centered on the dowel.

Roofs Joining Straw Bale Walls

Use a shed roof attached to the bale wall to create outdoor living spaces or protect vulnerable building elements like windows and doors. Shed roofs separate from the main roof can attach in several ways. Where the wall is tall enough, fasten the shed rafters to the wall's top plate. Deeper roofs may need additional support from knee braces or free-standing posts. Or attach a shed roof to a ledger connected to posts supporting the main roof. Carefully detail these connections to protect from moisture intrusion. See Figure 2-33.

2-32. Design Process Diagram.

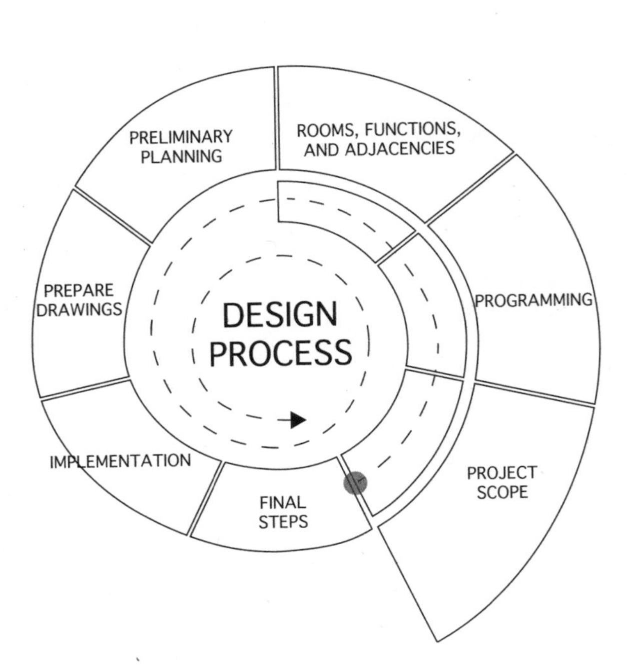

The Design Process and Implementation

Following a design process usually results in a better building. The process outlined here is one example of a building's design phases—follow a process you are comfortable with.

Project Scope

Energy-efficient design starts with the site. Locate the building to optimize solar exposure, and consider climate factors (prevailing winds, temperatures), views, privacy, and zoning. It can be challenging to balance competing site characteristics (e.g., the best solar orientation may conflict with zoning setbacks or the best view). A frequent siting mistake is the misplacement of the front door. Remember that the door seen from where one approaches the building is the one visitors will treat as the front door. You will need to prioritize all the issues you can identify, and strike workable compromises.

2-33. Flashing for roof joining straw bale wall.

Shed roofs offer greater protection to exposed gable walls, extend outdoor living areas, and shelter windows and doorways, but they need to be properly flashed to avoid moisture leaks.

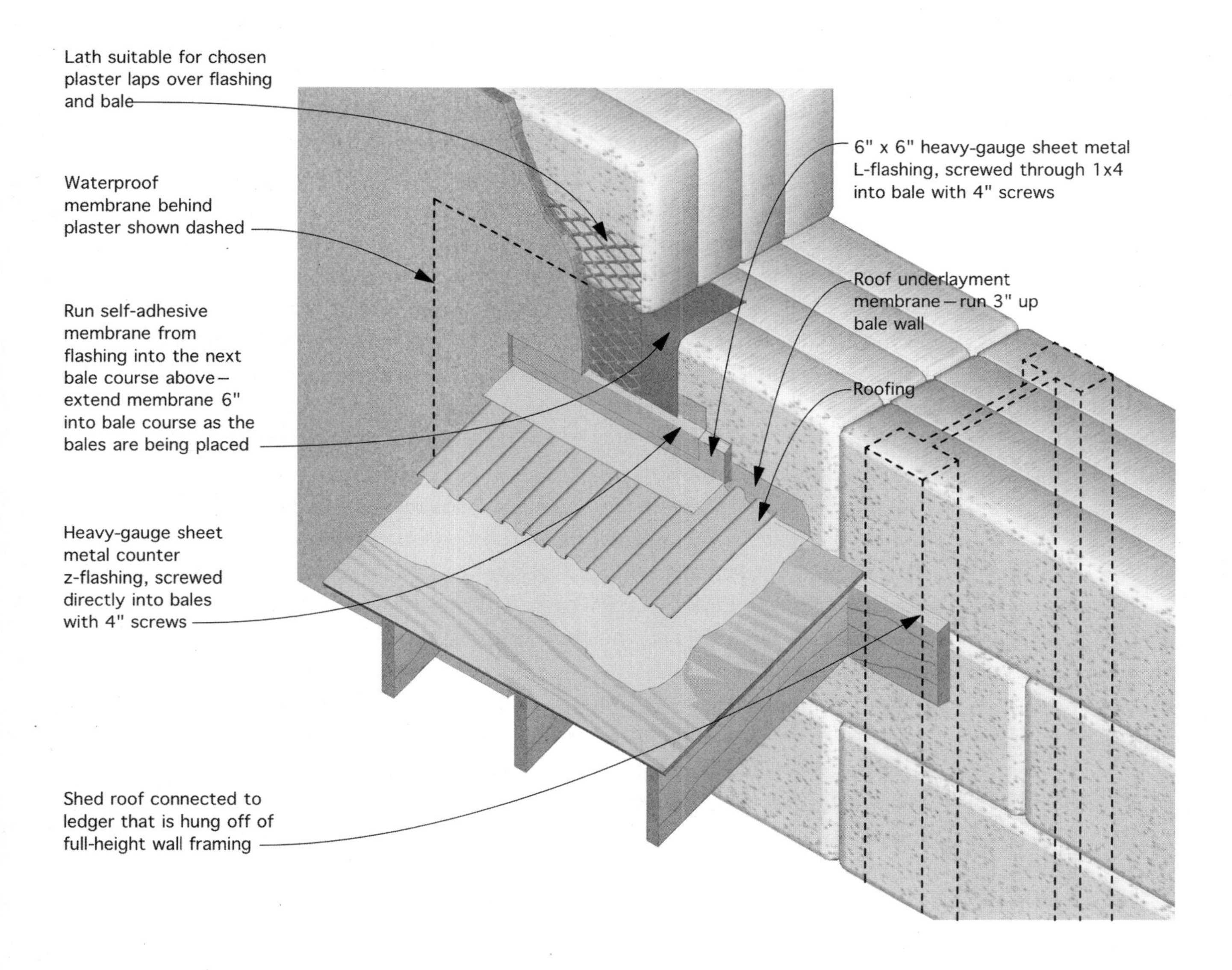

NOTES:
- This is not an appropriate feature on a structural wall that uses the exterior mesh and plaster as a continuous structural element.

- A roof connection would likely attach to posts in the exterior wall surface, and a longer, heavier shed roof would require additional post support at the roof's lower end.

- Preferred if fasteners for trim and flashing secure to wood framing let into the bale wall. See "Attaching/Hanging Objects from Straw Bale Walls."

Thermal mass and insulation work together to create a comfortable indoor space. The distributed mass qualities of an interior plaster were discussed in Chapter 1; additional thermal mass can be added using a finished floor with thermal mass (concrete, gypcrete, tile, earth) and interior thermal storage walls.

In many code jurisdictions, if the building is going to connect to any utility (including a propane tank), the plans submitted for a permit must demonstrate that the structure meets minimum energy-efficiency standards. This analysis is referred to as the *Energy Calculations*, or, in California, the *Title 24 Calculations.* The latest Title 24 rules recognize straw bale wall assemblies. An energy consultant will be familiar with local energy-efficiency standards but may not be familiar with straw bale buildings—both this book and others listed at the CASBA website, www.strawbuilding.org, will help with the calculations. Energy consultants can offer guidance on how to improve other aspects of a building's performance, too.

Begin thinking about size of the project, a time schedule, and a budget. All of these will surely evolve as the process unfolds, but find a starting place. What is desired, and what is realistic? Some projects may take several years to get through design, engineering, plan check, and construction. Others start and finish in less than a year. Consider the level of involvement expected for all parties, from an owner-builder and volunteers during bale raising and plastering, to architects, engineers, and contractors.

Recognize that the project scope may change as the design takes shape. For example, an owner-builder might decide to take on more, or less, of the project work depending on size and complexity. Or, instead of initially building a house, garage, and guest cottage, the design might be downscaled to a guest cottage first (if it meets primary dwelling requirements), followed in later years by a larger house or an addition.

Learn about building costs in the area. There can be a big difference between buildings that meet minimum code requirements and custom energy-efficient structures. Rural areas often have lower building costs than urban areas, too. What you learn about cost may influence your next steps.

To illustrate a design process, we'll use two examples:

- Example 1: Joe and Ann want to build a small two-bedroom, passive solar, energy-efficient house in a rural area, using mostly locally available materials.
- Example 2: Larry and Sandy want to build a quiet, energy-efficient four-bedroom house for their growing family on an urban lot.

Programming

Create a site plan to show the building location relative to its surrounding landscape. Visit the building site to take pictures and make notes about views, solar orientation, first impressions on soil and geological conditions, climatic conditions (e.g., wind, rain, snow), and neighborhood activities. Investigate zoning, fire engine access, and other local requirements. Learn if there are other completed straw bale buildings in the area. Visit them if possible, and talk with the owners or builders.

Develop the project's "character." Owners might think about other spaces they've been in that left favorable impressions (e.g., the sunny and inviting wrap-around porch of a friend's house, the breakfast area of a parent's cottage, the mysterious nooks and crannies in an auntie's house). Look at pictures in books and magazines to identify features that resonate (or, if you are the designer, ask your clients to do this).

EXAMPLE 1

Joe and Ann's rural building site sits on a wooded south-facing ridge with relatively high wildfire danger. The site receives precipitation between November and May, much of it wind-driven from the southwest, with occasional heavy snows. The structure will be occupied by two adults who will also have home offices there. The slope lends itself to a walk-out basement "suite" suitable as a guest room, or an in-law apartment for an aging parent or a caretaker. Zoning setbacks limit the building envelope.

Joe and Ann like elements of both Craftsman and Southwestern-style structures. They want to save money by being their own general contractor. Having attended several straw bale building workshops, they intend to handle the building phases they have experience with: bale raising and plastering. They renovated their last home and feel comfortable handling ceiling insulation and interior finishes, and they have contacted local subcontractors to get a handle on local building costs. Space isn't an issue, so they plan to use three-string bales laid flat in a post-and-beam wall system.

EXAMPLE 2

Larry and Sandy have found a 50'×100' vacant lot in an urban neighborhood they enjoy (having rented a house there for a few years). It is close to several bus lines that connect to a light rail system, and it is within walking and biking distance of stores, restaurants, schools, and other amenities. They have two children, but want to "age in place" and live in the house when they are no longer able to drive. Both are working professionals and don't envision doing any construction work. They have discussed building costs with a local builder who specializes in energy-efficient homes. A busy street a few blocks away creates traffic noise, and they are counting on the bale walls and triple-paned windows to minimize noise inside the house. To reduce the square footage taken up by the bale walls, they'll be using a bales-on-edge wall system. They visualize the house as a demonstration of ecological design and a retreat from their busy lives, so they also want to minimize maintenance chores.

Rooms, Functions, and Adjacencies

Develop a list of spaces and functions the building should contain and fulfill. Which of these should be next to or near one another? Which should be remote? Consider sun, shade, views, and climatic conditions as you define adjacencies and spatial orientation. For example, late sleepers may not want early morning sun pouring into the bedroom; laundry appliances can be conveniently located near a mudroom entrance; and pantries should be near the kitchen. Be as specific as possible about features, approximate room size, and other requirements. Summarize what you learn. To continue with the example projects:

EXAMPLE 1

Mature oak trees closely border the building site; Joe and Ann want to preserve the trees to screen wind-driven rain and shade the building during hot days. Ann loves to cook and entertain guests and fondly remembers a childhood kitchen with a walk-in pantry. Because parties always end up in the kitchen, they want it to be the focal point of a "great room" area that includes dining and living space. They both like to sleep in on weekends. They want higher ceilings in "public" spaces and prefer lower ceilings for quiet areas. They have sited the kitchen near a garden area and would like a doorway between them. They also enjoy reading and bird-watching. Although they have some building experience, they want to keep the design as simple as possible to reduce challenges and costs.

EXAMPLE 2

Larry and Sandy's vacant site faces north onto an east-west street; this would allow for the more private south face to have lots of glass for passive solar heating. Storms also come from the south, so wind-driven rain has to be designed for. Setbacks limit where they can build on the property, and there are height and bulk limits as well, but they want to demonstrate with this house that the simplest way to make a house ecological is to keep the overall size modest. They plan to convert the childrens' bedrooms into a rental/caretaker unit when the kids no longer live at home. They prefer to socialize outside the house, but recognize the kitchen as the house's activity core, so they want it to include space for big, comfortable chairs. Important aspects of the design include private spaces for each of them—a workspace for Larry and a study for Sandy.

Preliminary Planning

For the design process to produce a strong solution, start with an idea or theme that can bring the project together. Start sketching! Sketches help explore and define conceptual guidelines and are a wonderful way to communicate complex ideas between different members of the design team—the architect, engineer, owner, and builders. Develop a preliminary plan either with a computer program or with pencil on tracing paper and evaluate it with members of your design team.

Continuing with the example projects:

EXAMPLE 1

Joe and Ann's building evolved into a two-story house with a separate garage and workshop that form a courtyard. An 800-square-foot walk-out basement contains utilities, a small office, a guest room and bath, and a television/library/exercise room. An open staircase connects it to the main floor. The 1,200-square-foot (1,000-square-foot interior) straw bale main floor contains an entry, laundry, master bath, and bedroom closet along the north wall; a small bedroom on the southwest corner of the house; and a large, open south- and east-facing room that serves as a kitchen/pantry, dining, and living room. There are generous window seats in the dining and living room space and in the bedroom. To save space, they used pocket doors leading to a bathroom that serves the main floor and master bedroom. Windows on the east, south, and west sides frame mountain views and admit light; an 8' deep porch runs the length of the north (entrance) wall to function as a shaded outdoor space on hot summer days, and the bedroom opens onto a small patio.

EXAMPLE 2

Larry and Sandy decided that while they want to live in town, they want as much connection to the outdoors as possible. They made the house long and narrow—just one-room wide in most places—making it easy to achieve good natural light, ventilation, and privacy in every room. They took advantage of the mild climate and put the covered staircase outside the house to save interior space, increase privacy, and make later division simpler. Larry and Sandy's bed is on casters so it can be rolled out onto their private terrace to sleep outdoors whenever the weather permits. The terrace outside the kitchen is detailed as an outdoor room, with dappled shade, reading lamps, electrical outlets, and blankets stored in weathertight bins. Spaces in the house overlap and condense functions wherever practical. Sandy's office has a closet and a futon couch, so it doubles as a guest room. Larry's workspace also has a couch, and it doubles as the media room. Smaller spaces function as alcoves opening off of larger rooms. For example, the childrens' bed lofts over the bathroom and closet open onto their shared play space.

Preparing Drawings

Once you have arrived at a preliminary solution, start the process of obtaining formal drawings for the building permit application and contract documents. These drawings must be made to scale (e.g., ¼" = 1'), and the building department officials may have specific requirements for drawings and details. The set of drawings typically includes a site plan, floor plan, elevations, sections, roof plan, material specifications, and construction details. There may also be several pages showing structural details supplied by an engineer. While not required by the permitting agency, it's good to include *bale elevations* or *structural elevations* which coordinate the bale courses and structural elements like wall chases, windows and door openings, and porch ledgers. These elevations help everyone to see the whole picture.

Once you have drawings, you can begin gathering more accurate information about project costs. Many builders have a ballpark cost per square foot, but accurate estimates require drawings. If you need a structural engineer for the project, they'll need drawings, too. If they have worked with straw bale wall systems before, they may have a system they are comfortable with, but they may also be interested in using a different approach more suited for the project.

Details for straw bale construction can differ significantly from conventional buildings, and the plans need to show these differences. Note that design decisions impact costs, and final drawings form the foundation for developing an accurate budget. The trade-offs of many design choices have been discussed in this chapter.

- Preferences in floor treatment affect the heights and detailing at the sill plates. For example, pouring a 1" earth floor over a subfloor affects door jamb height differently than laying ¼" tile over ½" backer board.
- Decide whether wood trim will frame the doors and windows or if plaster will abut them directly, and show this on the plans.
- Baseboards cover the joint between the wall and floor and protect the lower wall surface from kicks and vacuum cleaner bumps. They are simple to install along flat walls, but the rounded corners common in straw bale wall doorways are challenging.
- Include features that allow for building in phases or later adaptations. For example, place a structural header for future door openings, or ledgers for future shed roofs. Plaster and bales can later be cut out for doorways if the structural system permits that modification. Porches can be converted to rooms if footings have been designed to support the wall load.
- Consider interior and exterior plaster thickness as you design. Since plan dimensions are usually measured to bale and stud surfaces, it's easy to overlook the roughly 1½" that earth or lime plasters occupy. When planning for tight clearances, this can make a big difference. Pay close attention to this when designing exterior entryways, enclosed stairs, and spaces that must fit standard-sized fixtures and appliances like prefabricated showers and bathtubs.
- The wall character is most apparent at window and door openings; give these details special attention. It might seem straightforward to have reveals perpendicular to the wall surface, but this can make smaller windows feel cramped, and it restricts both the spread of daylight into the room and the portion of the room from which the window view can be enjoyed. Be aware that partitions close to the window make wide reveals impossible.
- Similarly, thick walls make the window head an important design consideration. Straight window heads are simple, but they can feel visually heavy, especially for larger windows. Flaring or arching the openings relieves this weightiness and highlights the opening as a feature.
- Straw bale walls may not be appropriate in some locations. It's not difficult to detail bathrooms and kitchens along bale walls, but these locations require special attention. Using stud walls in the kitchen simplifies cabinet and utility installation and saves floor space, but note that windows look different set in a flat wood-frame wall than they do recessed into the more organic surface of a bale wall.
- If you want an uninterrupted straw bale wall, plumbing for bathrooms and kitchens can come through a cabinet bay or along the interior sill plate to avoid passing through the bales.

- Casement windows with crank handles facilitate operating windows across a kitchen counter and straw bale wall.
- Showers on a bale wall need water- and vapor-proofing. See Appendix S: AS105.6.2 Vapor retarders. If there is a window, slope the stool to shed water. Typically, you would secure tile backer board to a 2× frame proud of a plastered bale surface, and then treat it as in any conventional shower or bath area. See Figure 2-24: Rainscreen over Straw Bale Wall Detail.

Implementation

A set of plans must be submitted for permit approval. This can be a straightforward process if your state or local jurisdiction has already adopted the IRC's Appendix S or if other straw bale buildings have previously been approved in your county or municipality. If your project is the first, this book should be useful in helping building department officials become familiar and comfortable with straw bale construction.

Familiarize yourself with the permit and building process. To gain insight, talk to other straw bale builders who have recently completed construction. Attend a straw bale workshop or conference to gather information from builders, designers, and homeowners. Visit the building department to ask about their process—and realize, as with any organization, that you may get different information from different people in the same department. While building with straw bales is no longer experimental, it may not be familiar to the people you need approvals from. If the initial reaction is resistance, be aware of the following:

1. Building codes do not prohibit alternative methods of construction. Under the section titled "Alternative Materials, Design and Methods of Construction," a design or construction method not specifically mentioned in the code can be allowed if the permit applicant demonstrates it is at least equivalent to what the code prescribes. This *can* be a simple procedure, but sometimes it can be onerous, costly, and time consuming, depending on the proposed alternative and parties involved.
2. Straw bale construction is included as an appendix in the 2015 and 2018 International Residential Code (IRC). The IRC is only a model code, but it is the basis for the residential building code in virtually every jurisdiction in the United States. However, many jurisdictions have not yet adopted this appendix, often referred to simply as Appendix S. Even if your jurisdiction has not adopted IRC Appendix S, it can be proposed on a project basis through the Alternative Materials, Design and Methods of Construction section. IRC Appendix S, along with its highly informative commentary, is included in Chapter 7, Straw Bale Construction and Building Codes, and is available for download at www.strawbuilding.org.

The construction documents also allow for a more accurate cost estimate—an important concern however the project will be financed. Once approved for permit, the building team—architects, general contractors, subcontractors, building inspectors, and owners—will use these documents. Set a schedule that outlines when suppliers and contractors will be onsite for various construction phases. The time line will probably change over the course of construction, but it at least provides a starting point. A project supervisor—a role often filled by the general contractor or played by an involved owner or owner-builder—must be available onsite to coordinate work activity; ensure adherence to schedule, drawings, and specifications; and to address unforeseen problems as they arise.

Final Steps

Once construction is complete, passing the final inspection results in an occupancy permit. The owners move in and write down any issues that remain in order to really complete the project—this is called a *punch list* in the building industry. The list might include such details as an unfinished wall, or an unconnected hose-bib, or a door that doesn't latch, or a window that won't close. After the tremendous effort of getting a house built, don't put off minor details. Resolve them quickly.

After occupancy, please register the building on the International Straw Bale Building Registry (see Resources at www.strawbuilding.org). Opening the home periodically for green building tours may inspire others. The reasons that draw us to build with straw bales will appeal to others, and our experience adds to the growing body of information that moves straw bale construction forward.

Property Taxes, Permit Fees, and Square Footage

At this time, most jurisdictions base permit fees and property taxes on external square footage. A few California counties recognize that nobody actually lives in the extra thickness of super-insulated walls, and they reward the energy savings of thick-wall construction by assessing square footage as if interior space were enclosed by conventional 2×6 walls. This alternative would, for example, reduce the assessment of a 1,000-square-foot exterior to 874 square feet of stud wall equivalent. Propose that your building code officials adopt this more equitable approach to assessing your structure. See Figure 2-34.

Estimating Cost

How much does it cost to build a straw bale house? This is perhaps the most frequent question posed to builders and architects. And it's nearly impossible to answer accurately without a lot more information. Different methods are used in different regions, but most divide a project into three parts: soft costs, site and infrastructure costs, and construction costs. *Soft costs* are those you might pay for an architect or designer, engineer, and permits. They could include surveys and soil testing. *Site and infrastructure costs* include those related to developing the land you're building on—roads, wells and water lines, septic systems, power and communication connections, etc. These, too, can vary widely—power lines to a building site from a nearby power pole cost less than pulling power a mile or more. City water hookup will likely cost less than drilling a new well. *Construction costs*—what we'll talk about next—are those associated with actual building construction.

Ask around your area to learn what the average cost per square foot of a "green" custom home is. Straw bale homes are comparable to "green" custom homes both in cost and in that they are designed to exceed energy performance codes.

A local homebuilders association is a good resource for this information, as are local builders who specialize in green building construction. Of course, the figure you hear may be higher or lower than the cost of the home you want because there are so many variables. To take one example: a home located on flat suitable ground may have a quarter of the excavation and foundation costs of the same home built into a steep hillside. And local average cost figures may include both custom and production homes, which may range from large and complicated to small and simple.

Without a set of engineered drawings showing foundation, framing, window, door, and many other details, it's very difficult to estimate the cost of any structure. At one end of the scale, building a custom green home can cost as much as 20% more than a conventional home of equivalent size and finish—especially if you hire someone to build it. With straw bale homes, some of the cost increase may come from extra materials—more concrete in the foundations and slabs that support the wider walls, a somewhat larger roof, the higher labor costs of interior and exterior plasters, and the detailed finish work needed at door and window openings.

At the other end of the scale, a straw bale house can cost much less... if:

- it is built in an area where permits and building inspections aren't required.
- there are no earthquakes, high winds, or snow loads to design for.

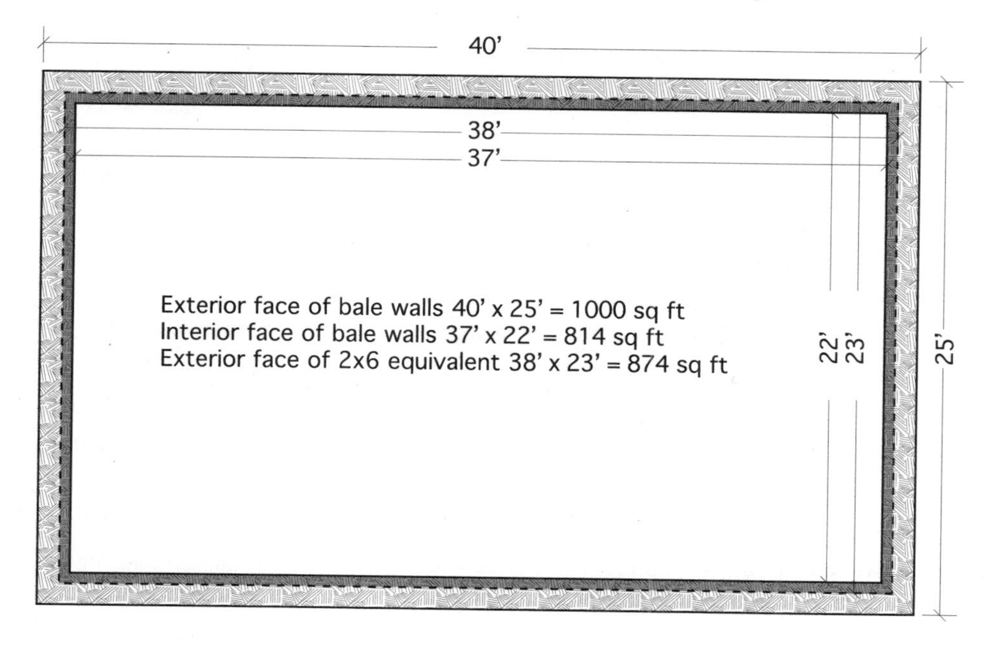

2-34. Calculating square footage based on a 2×6 wall equivalent.

- there's no public utility connection or local fire department service.
- the structure has a very small and simple kitchen and bath.
- the owners salvage most of the building materials, appliances, doors, and windows.
- the owners do all or most of the work themselves or with friends and neighbors.

Then there's the middle ground, where owners hire much of the work out and do some of it themselves. Bale raising and plastering work parties take advantage of the user-friendly nature of these construction phases; they can also draw friends and neighbors into your project, but they need to be carefully managed. Badly run work parties can cause more grief than good, resulting in the need to restack walls and reapply plasters. Disastrous work parties can also needlessly sour the plans of potential straw bale building enthusiasts. See Appendix: Managing Successful and Effective Work Parties.

To arrive at an accurate cost estimate, complete a set of plans to the level of detail required for a building permit. This will often require structural design details. Make materials, appliance, and fixture selections—or at least determine if these will be low, average, or high-end items. If you are applying for a construction loan, the application will require a detailed cost breakdown—the bank uses these to gauge draw requests. It's a very useful exercise even if the project is self-financed. A contractor with experience in straw building will be able to contribute figures on those construction phases unique to the straw bales—stacking, plaster preparation, and plastering. Most contractors have a rough idea of a square-foot cost based on the region, level of finish, and wall height, but it takes considerable time to arrive at an accurate estimate for an entire construction project.

Design Features That Drive Up the Cost of Any Building—Conventional Construction or Straw Bale

Large Homes: Paring the house down to what you will actually inhabit is the single most important way to keep costs down. Although the price per square foot usually drops as buildings increase in size, large houses still cost more. If you are building in a mild, temperate climate and you like being outdoors, consider building a not-so-big house and surrounding it with covered porches, patios, and garden courtyards. In 1950, the average American home provided 258 square feet per member of the family. As we end the second decade of the 21st century, that average is 1,072 square feet. How much space do you really need?

Complexity: Complex designs cost more to build than simple designs. In general, the longer it takes to draw the design, the more costly it will be. Rectangular buildings with shed, gable, or hipped roofs are simpler to build than those with unique footprints and roofs to match. Steep or complicated roofs, cathedral ceilings, high-end kitchens and baths, and curving walls add cost.

Uneven Ground: It takes longer for work crews to carry materials and equipment, climb ladders, and use tools when their footing is uncertain. Level the ground at least four feet out from the exterior wall immediately after the foundation drainage and downspout drains have been inspected. A crew working on the wall needs to focus on the vertical surface—not the ground. Uneven ground, open trenches, and construction debris alongside the wall inhibits progress.

Building sites without convenient access to delivery trucks, electricity, or water drive up construction costs.

Basements: It costs more to create space underground. There are rational reasons for below-grade spaces, but note that production builders avoid them because basements complicate foundation drainage, and the walls must be properly insulated.

Second Floors: They make efficient use of the foundation and roof expense, but the effort to carry materials upstairs and to work safely overhead on ladders and scaffolding roughly doubles labor costs. Think about creating a habitable attic or loft as an alternative.

Clerestory Windows: These create great daylighting and might be part of a passive solar design (though they may be a net loss since rising heat escaping through the windows may offset any solar gain). But complicating a roof by inserting window walls costs roughly ten times more than using skylights because of additional framing, insulation, sheathing, finishing costs, and the higher cost of windows when compared to roofing materials or other daylighting alternatives.

Bale Stacking Cost-Drivers

Framing: Keep framing on either the interior or exterior sill plate—not both. Framing on both sill plates takes extra time to notch, test fit, then twist and pound the bales into place. The insulation will suffer, and the process will take longer if the bale-stacking crew has to cut bales into smaller pieces to fit them between framing. Where interior partition walls intersect a bale wall, temporarily install the stud in contact with the bale wall so it can be removed for the bale stacking, then reinstalled afterward.

Gable walls that require angled bales laid flat slow bale stacking, and the resulting wall's thermal performance may suffer. Consider using stud construction within the triangle of the gable.

Sloped Ceilings over the Bale Wall: Bales laid flat aren't easily shaped into wedges; the result can be a less dense wall that compromises insulation and may be difficult to plaster without first covering with lath. To avoid this, begin the ceiling plane at the inside surface of the bale wall or use a combination of custom-shaped bales and straw-clay to create a dense insulation that will also hold interior plaster.

Tall Walls: The extra effort to place bales above reach of people on the floor makes those portions of the wall more expensive. There's also a structural limit to a bale wall's free-standing height

2-35. Difficult-to-fill spaces.

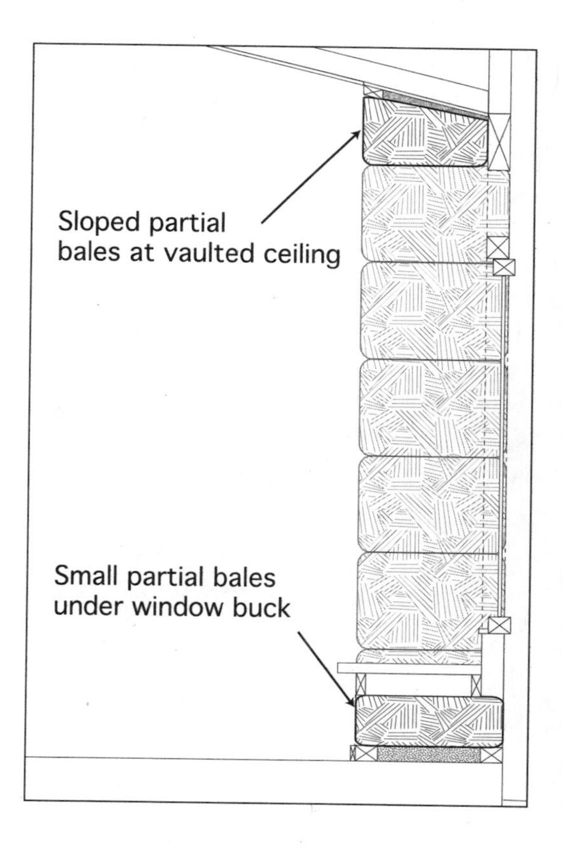

beyond which it requires an intermediate floor or other stabilization. See Chapter 3, Structural Design Considerations.

Window and Door Reveals: Hand-shaping bales or using straw-clay to create desired door and window reveal angles and curves is traditional...and time consuming.

In residential construction, there may be some sequencing conflicts among framers, electricians, plumbers, and the bale-stacking crew. Careful planning and coordination of the trades prevents problems and delays that add expense. See Chapter 4, Electrical, Plumbing, Ducts, and Flues in Bale Walls.

Plaster Preparation Cost-Drivers

Plaster prep includes blocking for lath and mesh. Straw bale walls that require lath and mesh must have places for attachment. Posts and beams are obvious attachment surfaces, but the plans may not show where additional blocking is needed. Framers can install this blocking prior to the bale stack so long as it's not in the way during the bale stack. The bale stacker or lathing crew can also install this blocking if they have framing tools for that task.

Inadequate Plaster Stops: Resolve all questions about how plasters meet wood or other surfaces prior to plastering. See Chapter 6, Plastering Straw Bale Walls.

Posts Set Partially in the Bales: Because most building timbers are green, they're going to shrink, and wood will continue to shrink and swell with seasonal humidity changes. Some plasters—clay in particular—will also shrink. A gap forms as wood shrinks away from the plaster edge. Air seal them even if these gaps are aesthetically acceptable.

Plaster Cost-Drivers

Plaster Amount: Because the uneven bale surface requires more plaster to fill out, straw bale walls use more plaster than conventional walls. Plasters on conventional walls are usually only ⅞" thick. The total plaster thickness on a straw bale wall is typically between 1" and 1½". Consider using straw-clay to fill gaps between

2-36. Frieze boards.

bales and level the wall surface prior to the scratch coat, or order between 15% and 40% more material.

Tall Walls: The extra effort to plaster walls above reach of people on the floor makes those portions of the wall more expensive. Crews use ladders and benches to reach upper wall portions. In many instances, scaffolding is safer and works better, but it may need to be left in place for a month or longer. These costs add up when the scaffolding is rented.

Complex Eaves: Wall surfaces under a sloping eave take longer to complete when plaster must be troweled between exposed rafter tails. Consider closing in the soffit, setting wood trim between the rafter tails, or dropping the frieze blocking or trim down the wall surface a few more inches to facilitate plaster application. See Figure 2-36.

Amount of Edge Detailing: Plaster application moves quickly across an uninterrupted wall surface, but slows considerably as a crew works the edges and reveals of doorways or windows; it comes to a near standstill when they must detail narrow wall surface edges between closely set wall openings and junctions with other finish materials, like exposed posts and beams. See Figure 2-37.

Colors: It's possible to order "ready-mix" clay and lime finish plasters—though selection may be limited and freight costs high. When having custom plaster finishes designed for a project, recognize that it takes time to mix and prepare a palette of suitable finishes. While pigments for natural finishes are capable of producing an astonishing range of colors, they lean toward the "earth tones."

2-37. Design choices affect linear footage openings and edge detailing.

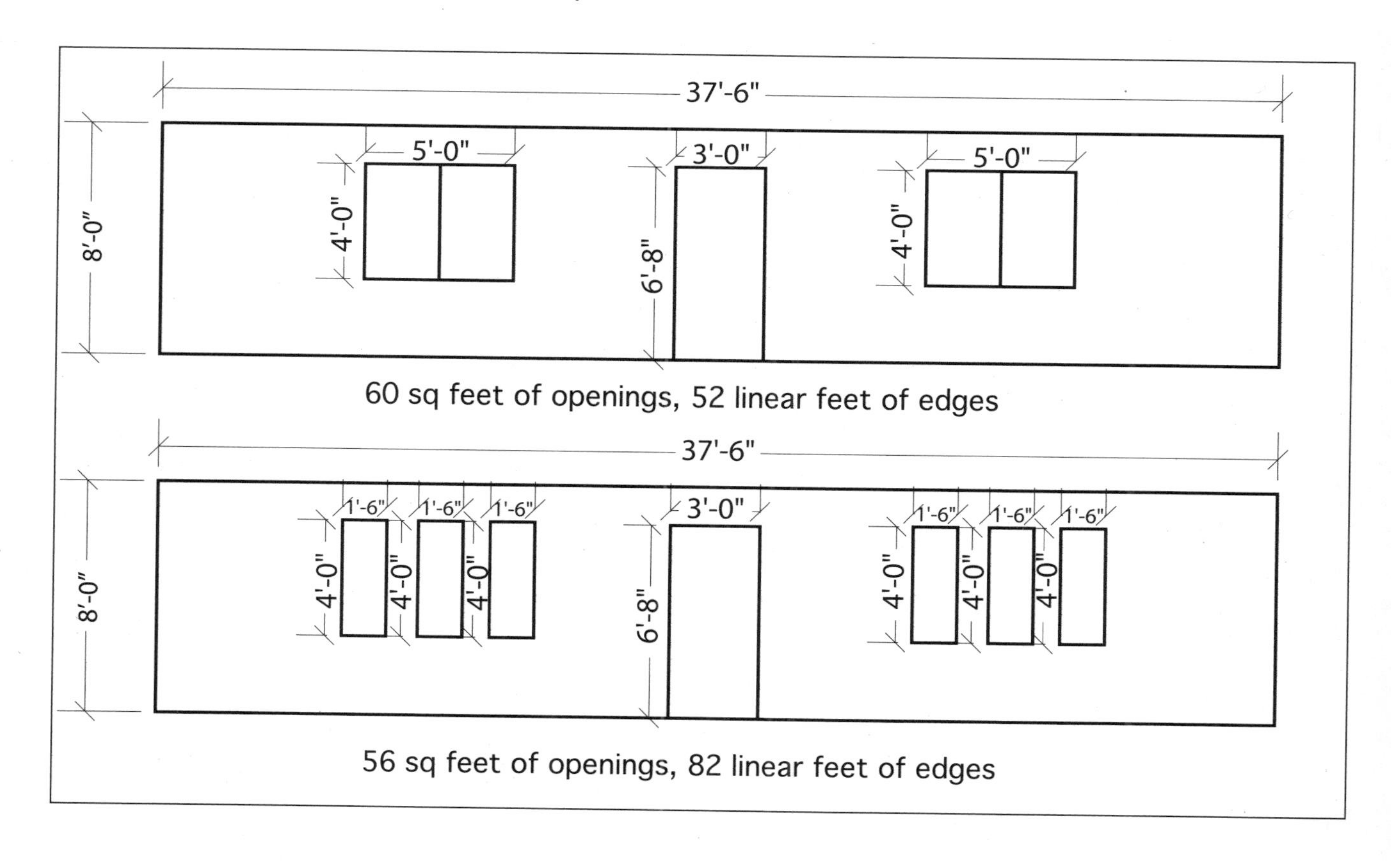

Developing color samples doesn't happen at a computerized color spectrum analyzer often seen at the paint store—it's done manually. While most structures have a single exterior plaster finish color, it's not uncommon for three or four interior finish colors to be used. Be aware that there's a cost for each color that needs to be designed.

Conclusion

Now that we've offered insights into the many detail options of a straw bale wall, outlined a design process, and discussed how design impacts building cost, it's time to move on to structural design considerations.

3

Structural Design Considerations

Note: While this chapter includes technical terms used by structural design professionals, we've tried to make it understandable to all. We have minimized use of these terms and sometimes explained them at first use.

Introduction

Straw bale buildings, like any other building, must support roof and floor loads, withstand wind forces, and if built in a seismically active region, resist earthquakes. Straw bale wall assemblies weigh more than most conventional wall systems used in North America, like wood-frame walls using cellulose, fiberglass, or foam insulation and finished with wood or composite siding or cement plaster on the exterior and gypsum board on the interior. While not as heavy as concrete, concrete block, brick or stone walls, the compressed straw's high volume and dense, thick plaster skins contribute to a significant wall weight that must be considered for both gravity load support and seismic lateral-force resistance (building weight does not add to forces generated by wind).

Loads

Gravity loads act in a downward direction and consist of dead loads (i.e., a building's own weight) and live loads, like occupants and furniture, and snow loads, which arise intermittently and irregularly.

Forces caused by wind or earthquakes act primarily in a horizontal direction.

Wind can also cause uplift forces on roof overhangs, porches, or unenclosed roof structures; occasionally, wind causes downward forces as well. Designs must resist uplift forces, especially in high wind or hurricane regions. Earthquakes can also add downward forces to the static gravity load; however, the building's gravity-load-bearing system, which has a built-in safety factor, comfortably handles these additional loads. Only rare and unpredictable lateral loads from high winds and earthquakes must be resisted by a separate structural system: a lateral-force-resisting system.

3-1. Understanding loads and forces.

Architects, engineers, and builders must take these forces into account when designing any building, including straw bale buildings. Awareness of these loads and forces guides the structural design process from the roof to the foundation.

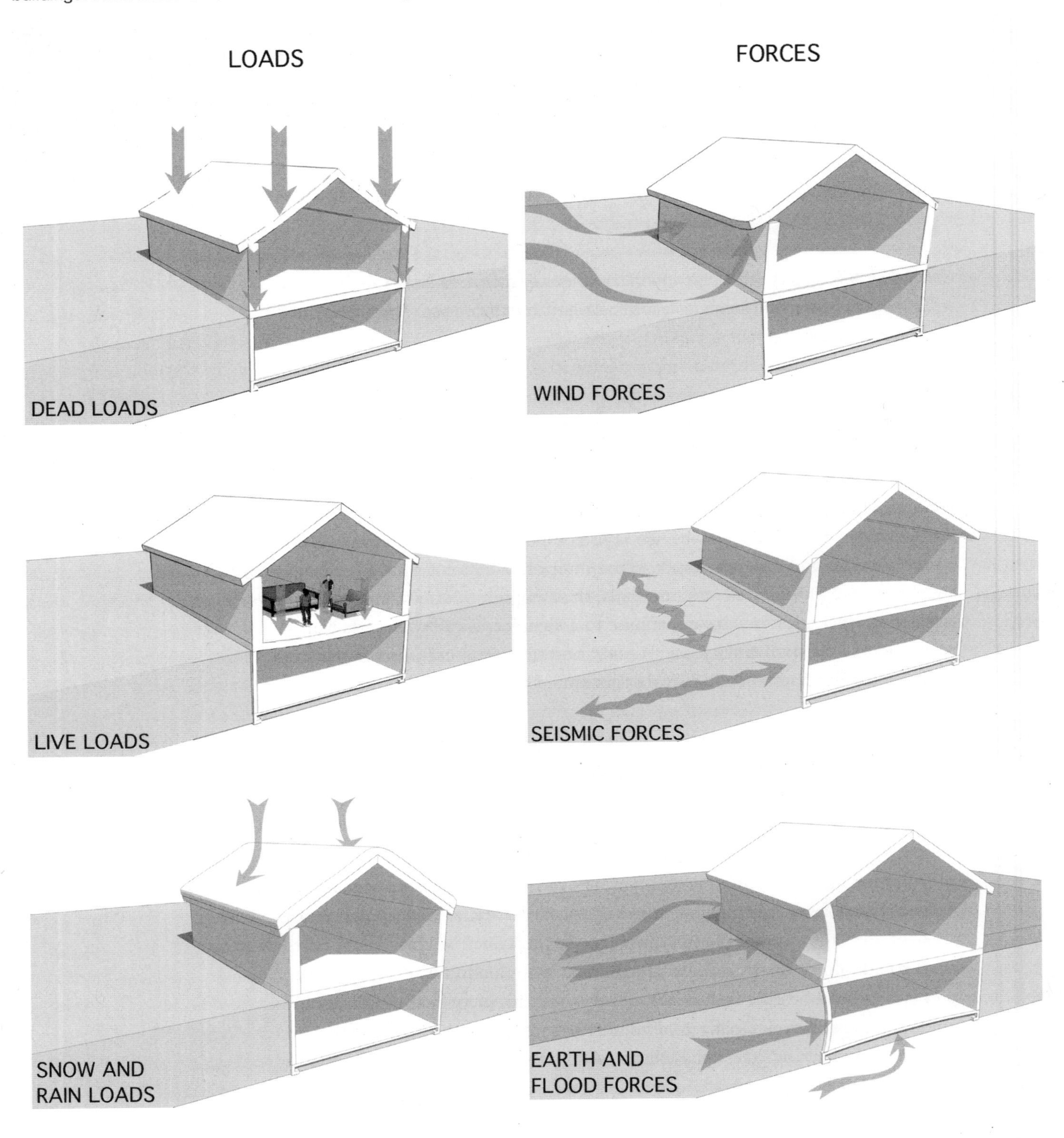

A gravity-load-bearing system carries the weight of the building and its contents to the ground. It does this all day, every day. A lateral-force-resisting system keeps the gravity system in place and functioning by resisting the horizontal forces that come with rare and unpredictable events, like storms and earthquakes.

In post-and-beam straw bale buildings, wood posts and beams carry the gravity loads, and the straw infills between the posts. There are several ways to use wood in this structural system, but we'll refer to all of them as post-and-beam systems.

Load-bearing straw bale buildings use the plastered straw bale walls to carry the gravity loads. These walls contain less wood, usually only sill plates at the bottom, box beams or other roof-bearing assembly at the top, and window and door bucks.

In both post-and-beam and load-bearing straw bale buildings, the properly detailed plastered straw bale walls can resist lateral forces, and conventional shear walls (e.g., wood-frame with plywood) or braced frames can also be used. Because the posts and beams often form part of a shear wall or braced frame, conventional lateral-force-resistance systems more readily integrate with post-and-beam straw bale walls.

The gravity-load system design is quite conventional in post-and-beam straw bale buildings, but rather unusual in load-bearing straw bale buildings. A load-bearing plastered straw bale wall performs like a structural insulated panel (SIP), where the thin but stiff and well-braced skins carry most of the gravity loads. More on this later.

The wall weight makes both post-and-beam and load-bearing lateral system design unusual. This matters for resisting seismic forces. Wind almost always governs lateral structural design in non- or low-seismic areas. However, in moderate- and high-seismic regions, where wind is often the governing design force for conventional wood-frame buildings construction, seismic design forces typically govern the lateral structural design of straw bale buildings.

In addition to affecting seismic design (including connections to the roof system itself, which distributes some of the wall loads), the heavier weight of straw bale walls can affect foundation design.

Wall Assembly Weights

The straw bale wall weight includes the entire wall assembly, but primarily consists of the straw bales and plaster skins.

Straw Bale Weight

Two factors influence a straw bale's weight: the bale's compression and the straw's moisture content. Like framing lumber, the typical equilibrium moisture content for straw bales ranges from 7% to 14%.

Structural designers need to know the bale's actual weight, including its moisture content. This contrasts with the dry density (0% moisture) weight

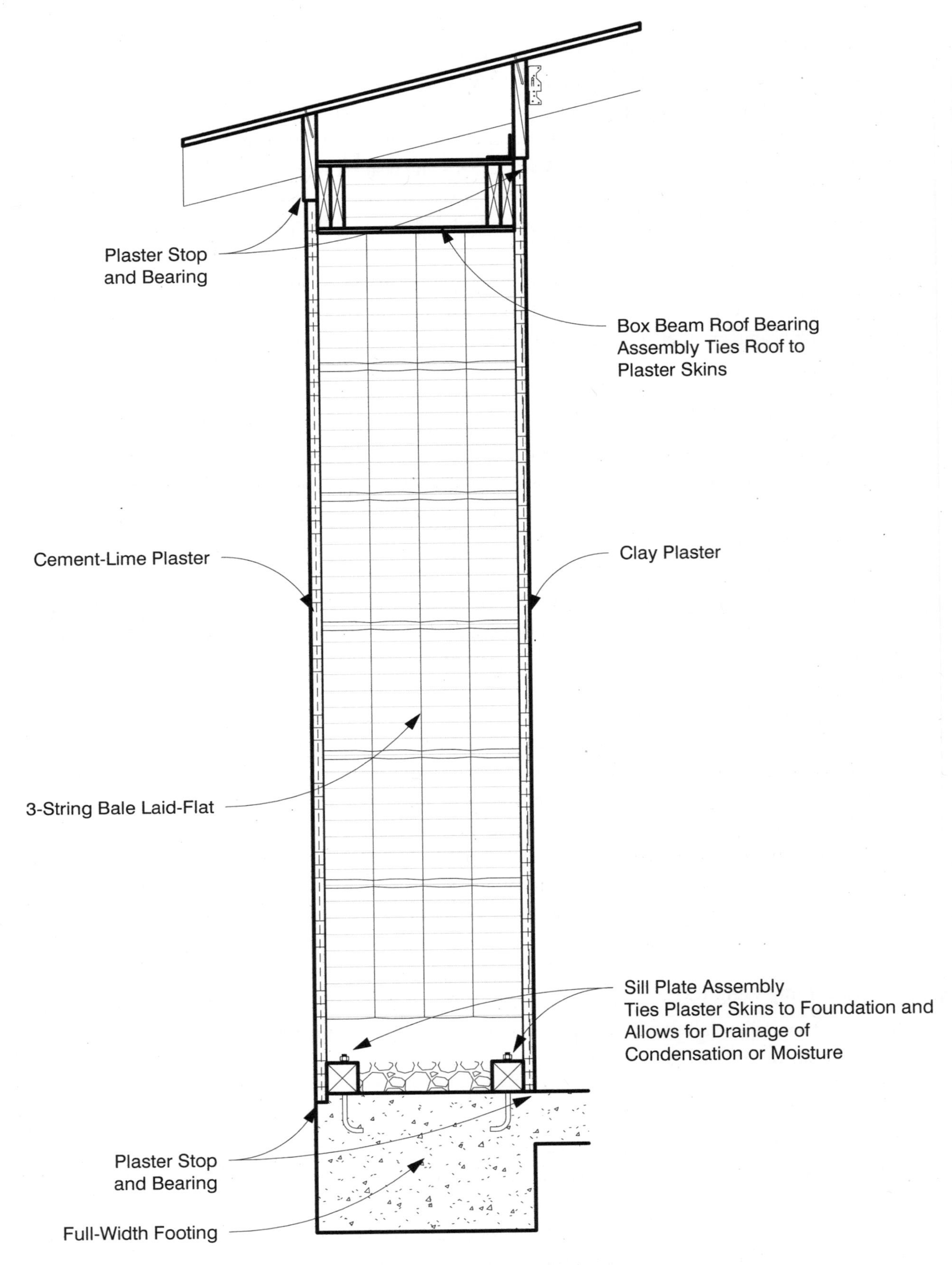

3-2. Straw bale wall section with cement-lime and clay plasters.

referenced in Appendix S that builders use to determine if bales meet the acceptable minimum density. See Chapter 5, Stacking Straw Bale Walls. Since this actual weight is typically not known during the design phase, use a reasonable estimate for bale weight.

The Ecological Building Network (EBNet) testing program found that in-place moist densities for two- and three-string rice and wheat straw bales ranged from 7.2 PCF to 8.5 PCF. To be conservative, use the higher end of that range, or you could use more accurate information supplied by nearby bale suppliers and builders familiar with straw bale construction.

After establishing moist density, multiply it by the wall thickness (bales only, in the chosen orientation) to determine the bale weight in pounds per square foot (PSF) of wall area. For example, 9 pounds per cubic foot (PCF) × 1.92' (three-string bale laid-flat width) = 17.25 PSF of wall area.

Plaster Skins

The plaster on a straw bale wall is often two to three times thicker, and far denser than the interior sheetrock and exterior wood siding typically used in wood-frame construction. The Portland Cement Association plaster/stucco manual lists cement-lime plaster density as 143 PCF, akin to concrete. Appendix S: AS106.3 commentary suggests a 130 PCF value for lime plaster. Soil-cement (clay-cement) for plaster density is similar.

Plaster compression testing data found in the *EBNet Structural Testing of Plasters for Straw Bale Construction* shows clay plaster density ranged from a low of 86 PCF–91 PCF for machine application (at this time, few clay plasters are machine applied) to 105 PCF for plaster labeled as "best for hand application."

The plaster thickness can be controlled by careful wall preparation and application (see Chapter 5, Stacking Straw Bale Walls and Chapter 6, Plastering Straw Bale Walls). Straw bale surfaces are far from uniform, particularly when bales are installed laid flat. Designing for conservative plaster thickness is good practice and provides a higher safety factor for structural calculations. If planning for a 1½" thick clay plaster, account for 1¾", and while Appendix S calls for a minimum thickness of ⅞" for cement-lime and lime plasters, we recommend accounting for 1¼" when considering wall weights.

For 1 square foot of plaster face area:

- 1.75" layer of clay or clay-lime plaster: 1.75" × (1'/12") × 105 PCF = 15.3 PSF
- 1.25" layer lime or soil-cement plaster: 1.25" × (1'/12") × 130 PCF = 13.5 PSF
- 1.25" layer cement-lime plaster: 1.25" × (1'/12") × 143 PCF = 14.9 PSF

Wood Plates, Beams, Bucks

Wood is denser than straw, but less dense than plaster. The wood used in sill plates and box beams add to the wall weight. The wood used in headers and window bucks is often neglected in calculations because their weight is more than offset by the straw bale wall weight removed at the door and window openings.

Douglas fir density, for example, ranges from 32–40 PCF. For an 8-foot wall section, framing can range from 2 PSF for load-bearing to 5 PSF for heavy timber post-and-beam.

Steel and Other Mesh

Steel mesh, when used, is the densest material in a wall assembly, but it only contributes marginally to the wall's weight. (See Straw Bale Shear Walls later in this chapter.) Accounting for two layers (one on each side of the wall), overlaps, and staples, use a conservative weight of a half-pound per square foot. Fiberglass, plastic, and bio-lath (hemp fiber or jute netting, etc.) generally weigh less than steel, and they may have a higher strength-to-weight ratio as well.

Sill Plate Fill

The densest commonly used sill plate fill material is gravel (110 PCF). For three-string bales laid flat with 4×4 sill plates, this adds 43 pounds per linear foot (PLF) to the wall weight, or 5.4 PSF for an 8' tall wall. Other materials used for sill plate fill include pumice (40 PCF), perlite (10 PCF), and rigid insulation (3 PCF).

3-3. Weight per square foot of wall area.

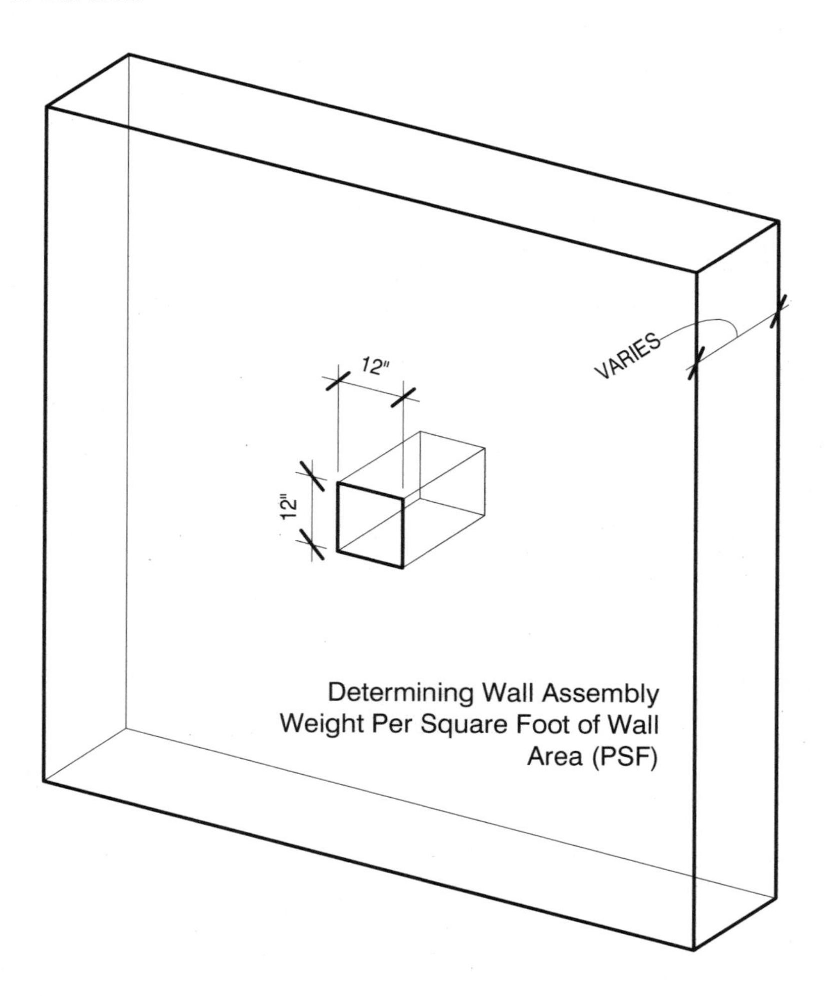

Framing

Wood or steel framing weights can be precisely calculated, loosely estimated, or the wall weight increased by some percentage or by a weight per linear foot. Generally, 3 PSF is a reasonable estimate for wood, steel mesh, and accessory elements.

Total Wall Weight

To summarize, the straw bale wall weight used in structural calculations includes the straw bales, plaster, mesh (when used), wood sill plates, and wood roof-bearing elements (box beam, or posts and beams), sill plate fill, and any accessory elements. Wall assemblies made with two-string and three-string straw bales typically range from the 45 to 55 PSF of wall area, depending on the specific construction and wall thickness.

Examples of Plastered Straw Bale Wall Weight Calculations

EXAMPLE A

Three-string bales laid-flat wall with cement-lime plaster on each side.

1 square foot of wall surface:

Outside cement-lime plaster = 1.25" ÷ 12" × 143 PCF = 14.9 PSF

Inside cement-lime plaster = 1.25" ÷ 12" × 143 PCF = 14.9 PSF

Bale weight = 9 PCF × 23" ÷ 12" = 17.25 PSF

Wood, steel mesh, and accessories = 3 PSF (estimated)

Sill plate fill = 5.4 PSF

Total wall weight = 55.5 PSF

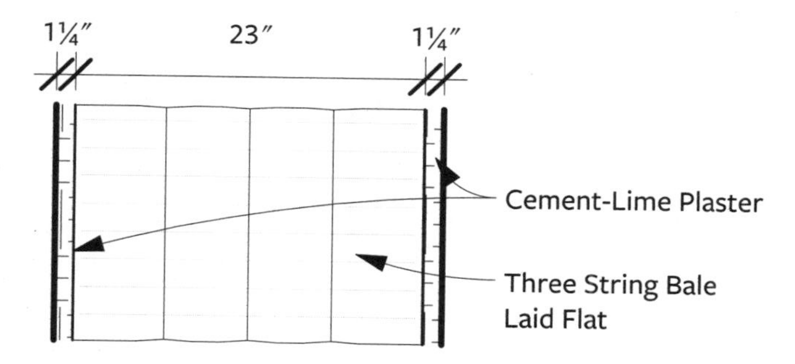

3-4. Three-string bale laid flat with cement-lime plaster both sides.

Appendix S limits straw bale wall assemblies to 60 PSF because its prescriptive "braced wall panel" tables use that as a reasonable maximum. It should be noted that this means the code only accounts for up to 2" of clay plaster on both sides for a wall with three-string bales laid flat; any more and the wall is too heavy for prescriptive standards. Consult an engineer if you intend to build with clay plaster thicker than 2" (or other plasters thicker than 1.5").

EXAMPLE B

Three-string bale on-edge wall with lime plaster on each side.

1 square foot of wall surface:

Outside lime plaster = 1.25" ÷ 12" × 130 PCF = 13.5 PSF

Inside lime plaster = 1.25" ÷ 12" × 130 PCF = 13.5 PSF

Bale weight = 9 PCF × 15" ÷ 12" = 11.25 PSF

Wood, mesh, and accessories = 3 PSF (estimated)

Sill plate fill = 5.4 PSF

Total wall weight = 46.7 PSF

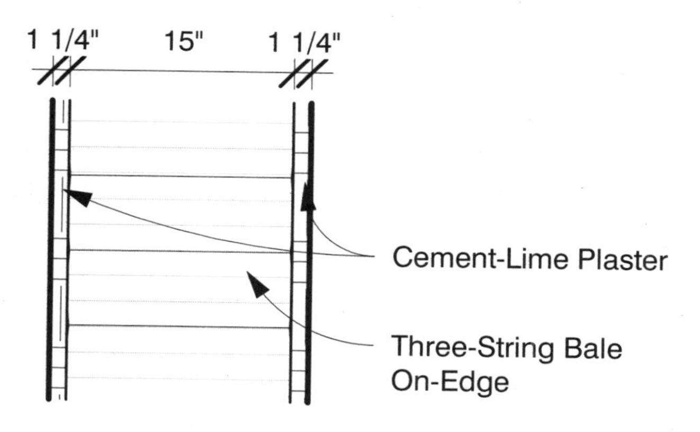

3-5. Three-string bale on-edge with lime plaster both sides.

EXAMPLE C
Two-string bale wall with clay plaster on each side.
1 square foot of wall surface:
Outside clay plaster = 1.75" ÷ 12"×105 PCF = 15.3 PSF
Inside clay plaster = 1.75" ÷ 12"×105 PCF = 15.3 PSF
Bale weight = 8 PCF×18" ÷ 12" = 12 PSF
Wood, steel mesh, and accessories = 3 PSF (estimated)
Sill plate fill = 5.4 PSF
Total wall weight =51 PSF

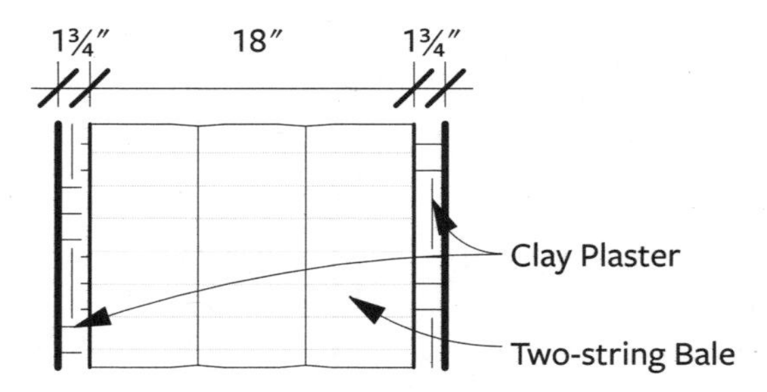

3-6. Two-string bale laid flat with clay plaster both sides.

The parts add up! When compared with light wood-framed structures, a straw bale wall is two to five times heavier. For example, a conventional 2×6 framed wall with fiberglass insulation, sheetrock interior, and plywood sheathing exterior with cement fiberboard siding weighs 11 PSF. A double-stud wall 8" thick with blown-in cellulose insulation and the same siding options weighs 12 PSF. Compare that to the 45–56 PSF calculated just above, and you can see that straw bale walls need special consideration when it comes to structural design.

Foundations

The structure's foundation needs to carry and transfer wall weight, the roof loads carried by the walls or posts, and lateral forces to the ground. Typically, the footings supporting a straw bale wall are wider than in conventional building, especially when the bales rest on a slab-on-grade. The slab edge may be thickened to support the extra wall weight. For a straw bale shear wall, the footing is typically thickened uniformly to provide even support across the wall thickness, as both skins usually play a structural role.

A post-and-beam system may reduce the footing's width on the bale wall's non-bearing side. Typically, the posts bearing gravity loads align along either the interior or exterior wall face, although posts could be in the center, or completely outside the wall on either the interior or exterior.

Straw bale walls can also be built on raised floors—over crawl spaces, basements, or over first-floor straw bale walls or other walls in two-story structures.

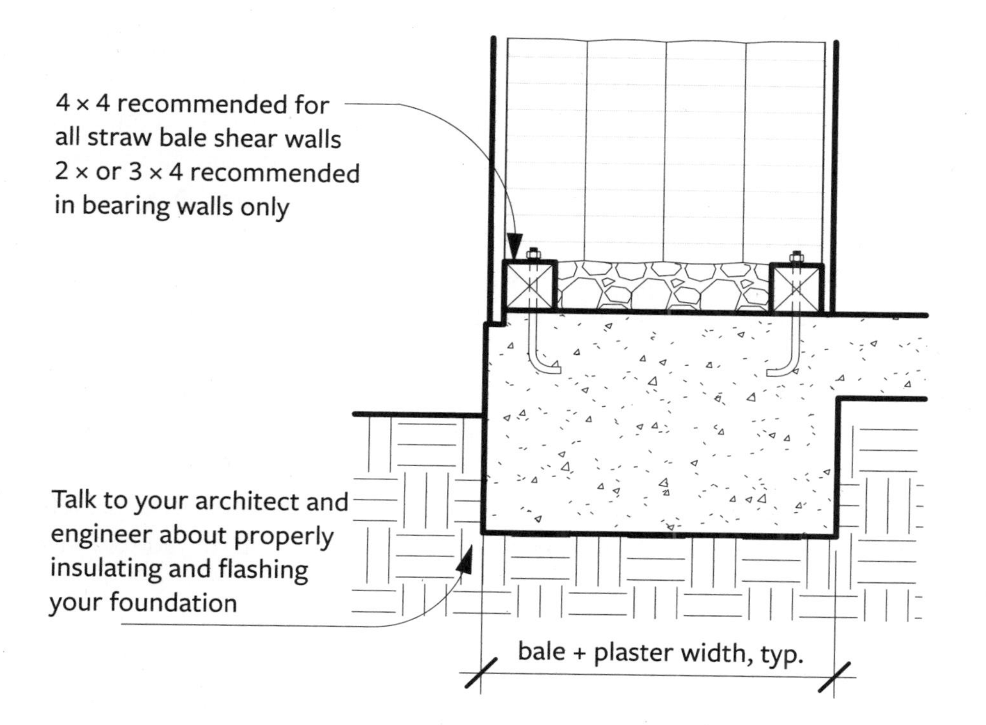

3-7. Thickened slab footing.

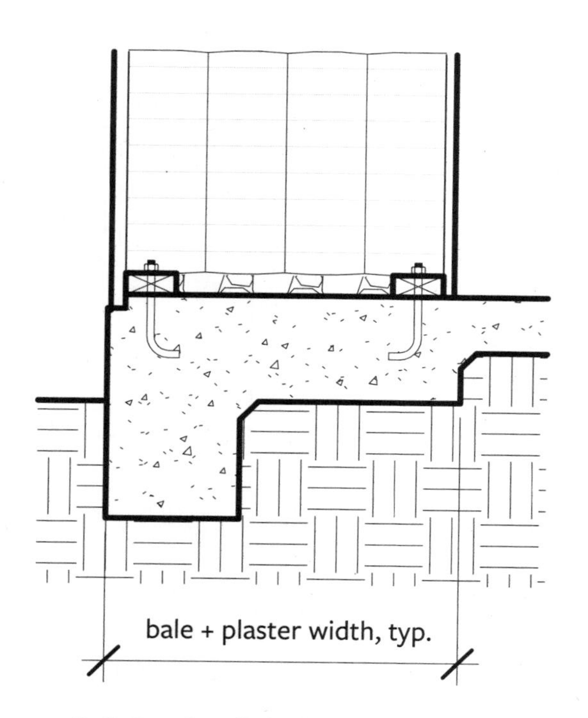

3-8. Exterior sill thickened slab footing.

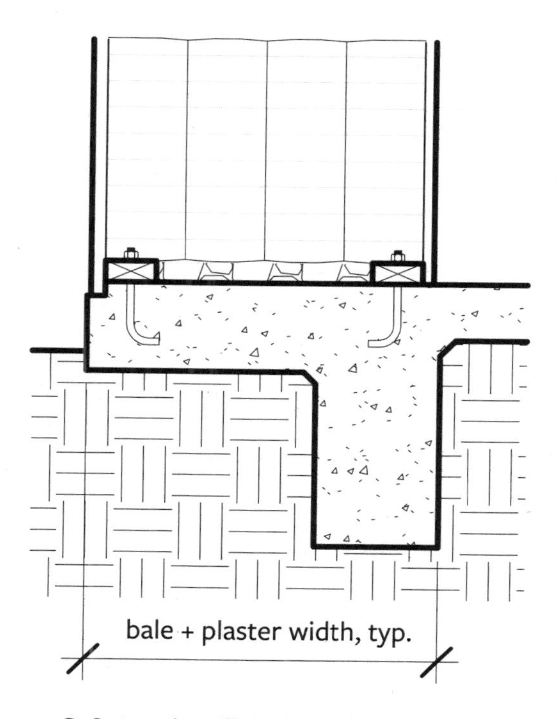

3-9. Interior sill thickened slab footing.

(This last situation is not addressed in this book. Consult with an architect or engineer for two-story straw bale designs.) The foundation wall can be of reinforced concrete, concrete masonry units (CMU), or insulated concrete forms (ICF), with many floor-to-foundation configurations possible. For example, the floor joists either hang from or rest on the sill plate. As with conventional construction, the floor joists must have the required minimum clearance from the soil, typically 18". Check your local building codes.

In both floor joists hanging and resting designs, the footing width and depth depends on soil type, building weight, and the lateral forces the building needs to withstand, as well as the local frost depth.

When a plastered straw bale wall carries gravity loads or resists wind or earthquake forces, its plaster skins carry most of these loads because they are far stiffer than the bales. The plaster does most of the structural "work" in transferring the loads to the foundation, with the bales primarily bracing the skins to prevent buckling. Designs must provide a direct compressive surface for the bottom edge of the plaster onto the foundation.

Appendix S: AS106.10 prescribes several options, the simplest being a cast-in-place plaster shelf in the foundation or slab-on-grade. Alternative methods requiring an approved engineered design include an adequately blocked wood-framed floor or an appropriately anchored steel angle. As with all foundation systems, address thermal isolation and moisture protection, which can vary by region and local climate.

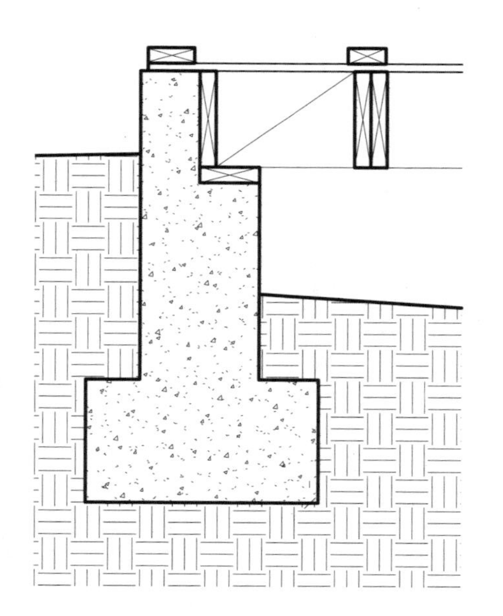

3-10a. Raised floor with floor joists bearing on a depressed sill plate.

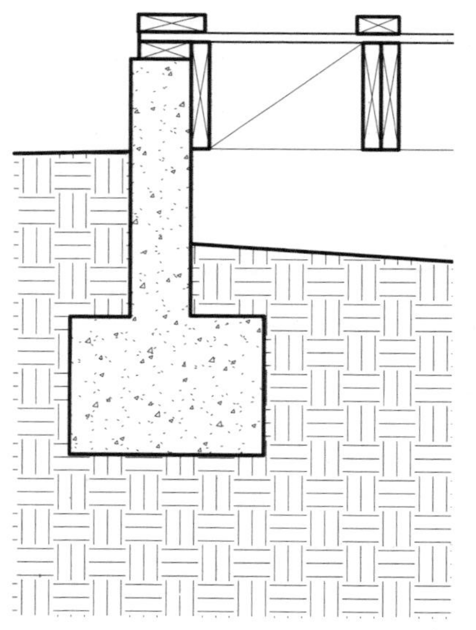

3-10b. Raised floor with floor joists hung from a top sill plate.

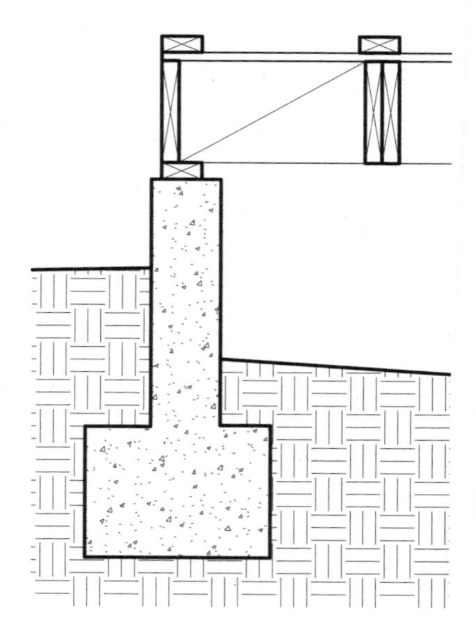

3-10c. Raised floor with floor joists bearing on a top sill plate.

Structural Systems

Think of straw bale buildings fitting into one of four categories, depending on the combination of their structural systems. In the first three, straw bale walls play a structural role, as load-bearing walls and/or shear walls: (1) load-bearing with straw bale shear walls, (2) post-and-beam with straw bale shear walls, (3) load-bearing with a conventional lateral system, and (4) post-and-beam with a conventional lateral system, with straw bale infill. In this last category, straw bale walls are not structural.

As described earlier, straw bale construction uses two types of gravity-load supporting systems. In a *load-bearing* structure—like those first built in Nebraska— the roof is placed on top of the already-built straw bale walls. The combination of straw bales and plaster (applied directly to the bales) support the roof load. *Post-and-beam* structures—often called *non-load-bearing*—are the most commonly built straw bale buildings in the United States. The straw bale walls infill a structural frame, and do not support any gravity loads other than their own weight.

Both *load-bearing* and *post-and-beam* structures must also resist lateral forces, and they do this with structural elements of some sort—including conventional wood-framed and plywood shear walls, prefabricated steel shear walls or braced frames, or the mesh-reinforced plastered straw bale walls themselves.

Depending on the direction from which a lateral force acts upon a wall, it can be described as *in-plane* or *out-of-plane*. In-plane forces act *parallel* to the wall surface. Out-of-plane forces act *perpendicular* to the wall surface. All walls (even non-structural walls) must be able to withstand the out-of-plane wind and/or seismic forces for that location and building design.

Both load-bearing and post-and-beam systems can employ a specially detailed plastered straw bale shear wall that derives its strength from the composite action of the bales, the mesh and its connection to framing, and the plaster itself. See Figure 3-21: Straw Bale Shear Wall Detail.

Straw bale shear walls are commonly an integral part of a load-bearing straw bale building; however, they must be detailed and constructed properly to transfer the lateral forces to the foundation. Conventional lateral-load systems compete for the same load-bearing wall space as straw bale shear walls, and are generally not used in load-bearing straw bale buildings. Post-and-beam structures can also use a straw bale shear wall system, and/or conventional lateral-force-resisting systems such as prefabricated metal or wood shear walls, conventional wood-framed shear walls, prefabricated and custom metal frames, strap assemblies, and more.

Load-bearing systems use the straw bale wall assembly for structure. This system minimizes the use of wood posts and the need to notch bales, and is best suited for small, simple, one-story structures. Load-bearing straw bale buildings require a specific construction sequence: first bale walls, then roof, (allowing time for the stacked bales to compress), then plaster. Because rain can damage

3-11. Load-bearing wall.

The load-bearing wall assembly is the original, least lumber-intensive, and simplest expression of building with straw bales.

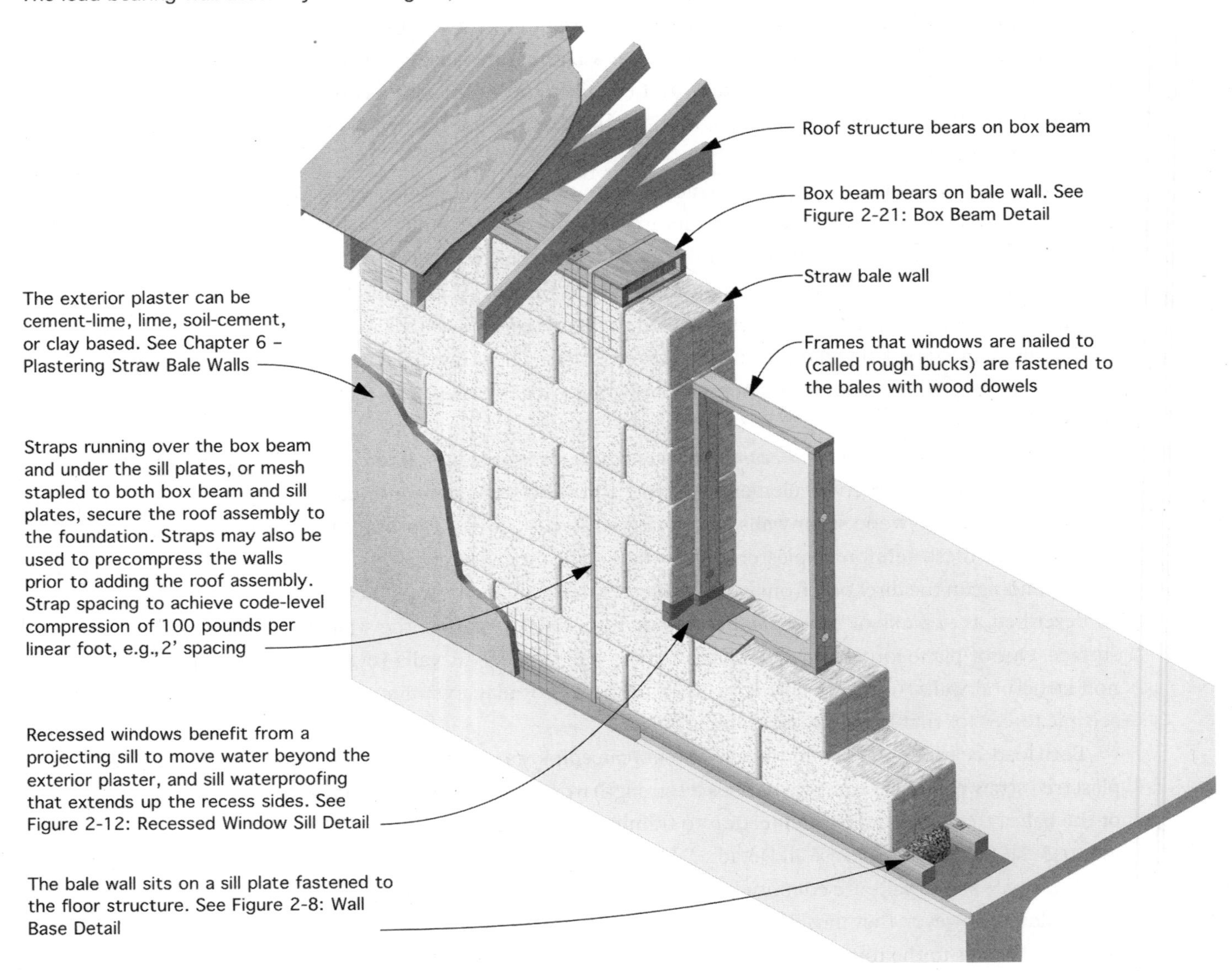

NOTES:

- Threaded rods running from the foundation through the box beam—either through the bales or paired along the sides of the bales—have also been used to hold the roof assembly to the foundation.

- IRC Code Appendix S requires cement plasters to be at least 1/6th lime for minimum acceptable vapor permeability.

- Under certain circumstances, plastered straw bale walls may be used as load-bearing walls in one-story buildings to support vertical loads. See Chapter 7 — Straw Bale Construction & Building Codes (code section: RB473-13, Appendix S — AS106.12).

- Precompression of load-bearing walls must not be less than 100 PLF. See Chapter 7 — Straw Bale Construction & Building Codes (code sections: RB473-13, Appendix S — AS106.12.1).

- Internal pins or external battens. Tightly tied paired wood or bamboo rods on opposite sides of the wall resist out-of-plane loads. Less often, rebar pins are driven through each bale course to pin it to the course below. See Chapter 7 — Straw Bale Construction & Building Codes (code section: RB473-13, Appendix S — AS105.4.2).

or ruin the bale walls before the roof protects them, this system works best in climates with a dry building season.

The type of plaster determines the maximum combination of dead and live loads that can be applied to the plastered straw bale wall. Appendix S: Table AS106.12 gives this in pounds per linear foot (PLF) of wall. The maximum vertical load for walls with cement, cement-lime, and soil-cement plasters is 800 PLF; for lime plaster it is 500 PLF, and for clay plaster it is 400 PLF. Typically, a box beam is the structural element that both supports the roof loads and delivers them into the plaster skins. The box beam spans openings in the wall while also transferring the roof diaphragm shear loads to the plaster skins.

Appendix S: AS106.12.3 offers guidelines for box beam construction, including limits for spanning wall openings. It calls for a minimum of 2×6 lumber on each edge, 15⁄32" sheathing (plywood or OSB) top and bottom, strapping at lumber splices, and blocking at sheathing joints. The assembly must be the width of the straw bale wall and must transfer loads to the plaster skins by continuous direct bearing. Alternatively, the box beam and load transfer can be an approved engineered design.

Trusses, or other roof framing, and box beam design need to be carefully coordinated. Truss diagrams and structural details need to show bearing points on the box beam to ensure that the assembly can transfer the loads adequately to the plaster skins, especially relative to window and door openings.

In a load-bearing assembly, the plaster skins support the roof loads because the skins are much stiffer than the bales. The straw bales brace the plaster skins and prevent them from buckling.

Maximum Wall Height for Straw Bale Walls

Appendix S: Table AS105.4 limits straw bale walls height based mostly on the wall's thickness and method of out-of-plane resistance—for example, reinforced plaster, internal or external "pins," or wood framing. These maximum heights range from five to eight times the bale thickness (in feet). This limit applies to both load-bearing and post-and-beam straw bale walls.

3-12. Box beam with blocking.

Blocking

As shown in Figure 3-12, blocking attached to the box beam transfers the roof loads to the plaster skins. See also Appendix S: Figure AS105.1(3). The box beam connects to the plaster skins by direct bearing and/or through the plaster's reinforcing mesh. All this happens while the box beam's rigidity distributes loads to both skins and across wall openings. Note that the attached structural blocking also doubles as finish trim.

The stacked bale walls in a *load-bearing* structure must be precompressed prior to plastering to prevent subsequent settlement from cracking the plaster. Appendix S: AS106.12.1 requires a minimum precompression of 100 pounds per

3-13. Post-and-beam wall.

In the most widely used system in western North America, larger dimension lumber—posts and beams—carry roof loads and support windows and doors. This permits greater spacing, with bales notched as needed. Nearly any structural shear wall system can be used in post-and-beam buildings.

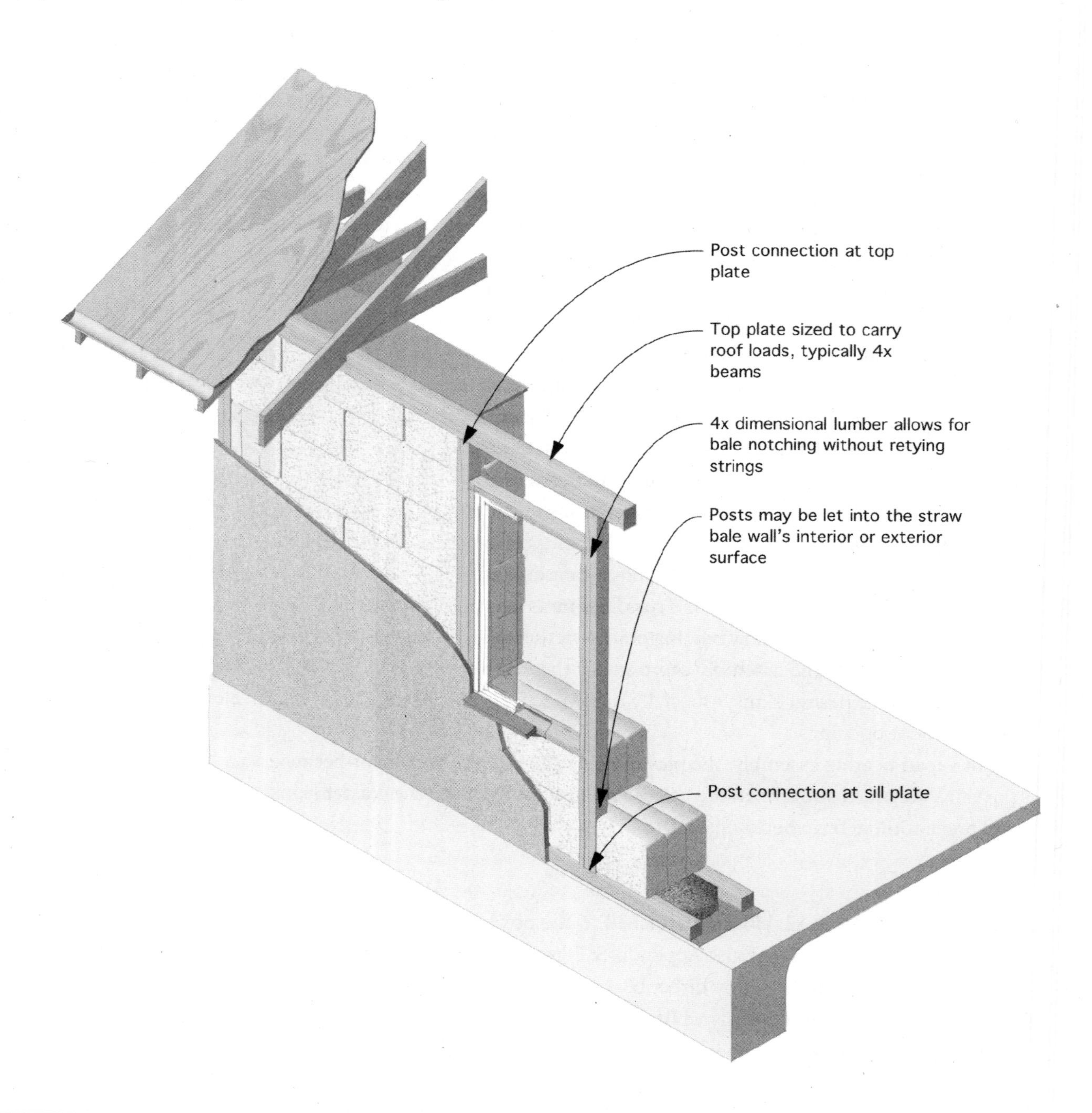

NOTES:

- Posts can be 4x or larger.

- Posts can be used in conjunction with bamboo or 2x battens for out-of-plane loading, see Figure 3-17: External Pins.

lineal foot of wall. Various methods of precompression have proven effective, including:

- ½" to ⅝" nylon package strapping with wire-formed buckles installed under the sill plates and over the box beam. The buckles permit incremental tightening.
- Threaded rod anchored to the foundation, extending through the box beam with couplings as needed. A washer and nut on top of the box beam allows incremental tightening.
- Build the roof assembly on the bale walls, then leave it to compress the unplastered walls for 1–2 weeks. This method often requires releveling the roof assembly by stuffing wood, straw, or straw-clay under the box beam before plastering. The plaster application "locks" the wall height and prevents the wall assembly from settling further.

Post-and-Beam

In a post-and-beam structure, posts and beams carry the building's gravity loads. These posts and beams can range in size from lightweight 2× lumber to large timbers, engineered lumber such as LVLs or I-joists, or structural steel. Assemblies where the straw bales do not bear the gravity loads and simply infill the frame have historically been referred to as "non-load-bearing." However, if detailed properly, these straw bale walls can function as shear walls.

Engineers, architects, and builders often have several reasons for preferring a post-and-beam system to a load-bearing system, including:

- A post-and-beam structure can be roofed prior to stacking the straw bale walls, providing them with immediate protection from weather.
- It provides predetermined wall heights.
- As the bales support only their own weight, there's no need to precompress walls—only to stomp in place with a foot or mallet as the bale courses are laid, although some builders still precompress the wall to reduce settling that might create a gap at the top of the wall and possibly reduce energy efficiency.
- The straw bales can serve simply as insulation and finish substrate. The wood or steel frames handle the gravity loads, while conventional lateral elements more easily address the lateral forces. This opens a project to more conventional structural design, which allows the permitting agency to evaluate the structure more easily.
- Structural design of load-bearing straw bale walls requires strict adherence to plaster detailing which most builders and code officials are unfamiliar with. Limiting the structure to the frame removes the risk of using unskilled labor to install structural walls and plasters (which could happen when the project relies on community work parties).
- Combining post-and-beam construction with conventional lateral elements eliminates the need for temporary bracing.

3-14. Exterior frame.

A post-and-beam frame outside the bale walls separates the structural function from insulation and thermal mass. It supports the roof while the larger overhangs provide exceptional weather protection, and if wide enough, create wall-length porches. This approach lends itself to more design flexibility with curving walls.

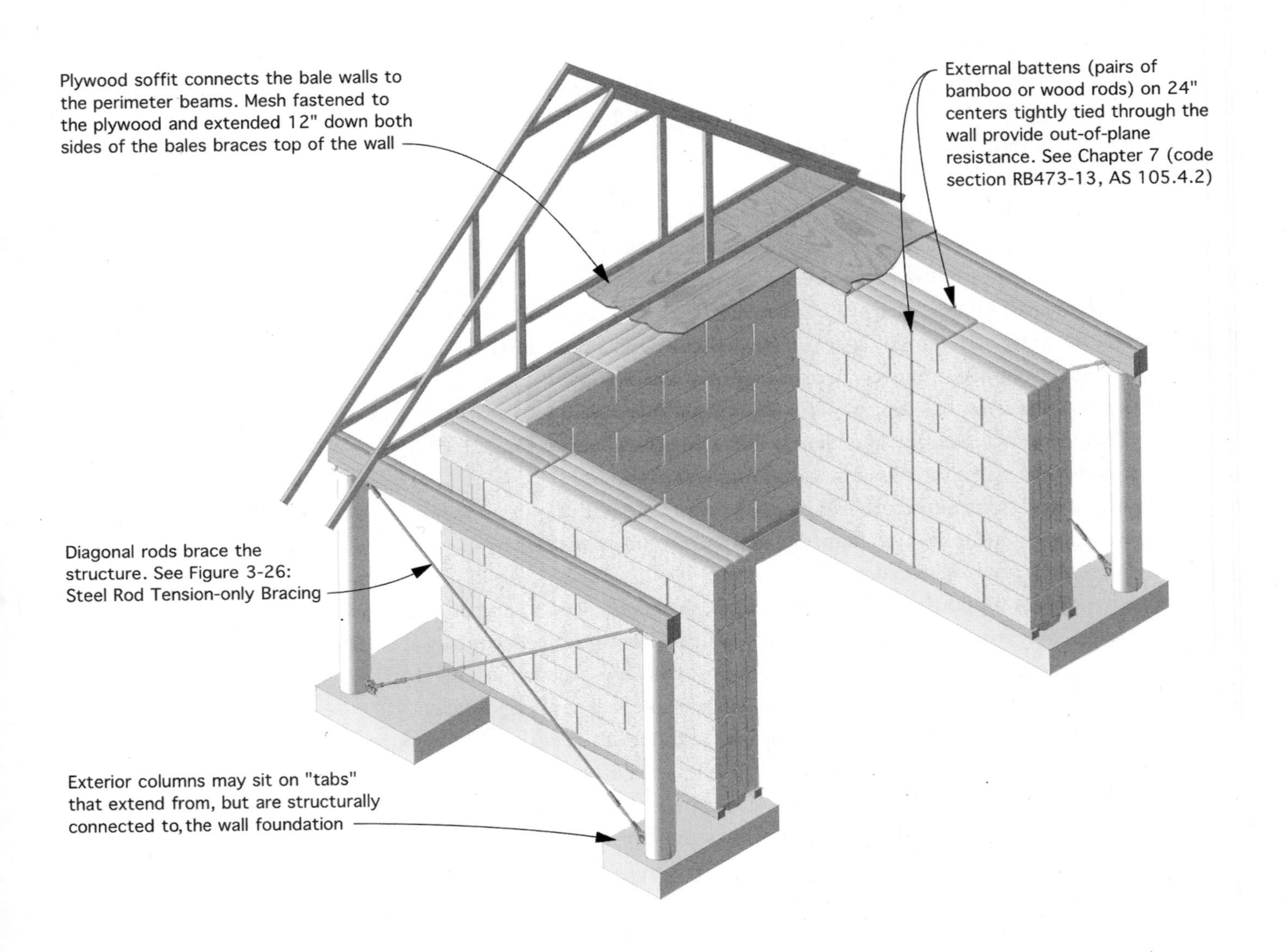

NOTES:

- In seismic areas, end-connection capacity must be carefully engineered.

- Adding a layer of drywall or plaster to the top of the bale walls functions as a fire-stop.

In most post-and-beam systems, the posts and beams are located at the outside face of the bale wall, with bales notched to surround frame members. The finished wall shows no evidence of the structural system.

The type of post-and-beam structure depends mostly on bale orientation. Most commonly, straw bales are laid flat in a running bond with the bales notched around the posts. Normally, the posts are held to a 4× depth so the bale strings will not need to be cut and retied. Keeping the post-and-beam framing and any conventional lateral elements aligned will help simplify the structural design.

Posts and beams can also be completely separate from the straw bale walls—either on the interior or exterior. This requires greater attention to gravity load paths and the lateral-force-resisting system but allows greater design flexibility, e.g., curved walls.

The Redundancy of Straw Bale Walls in a Post-and-Beam System

Non-structural infill straw bale walls have stiffness and strength that creates a secondary structure. The engineer should be aware of this condition, so forces that may be transmitted into the bales will not create unsafe conditions. It is critical to incorporate all wall weights into seismic force calculations and detail the shear path from straw to the frame, even if the walls are not intended to resist the accumulated force.

It's also possible to design a post-and-beam system with 2× "posts." This design integrates the 2× framing with the plastered straw bale walls providing lateral support to prevent the slender posts from buckling. Locating the posts at the bale module can reduce bale notching. A variation uses 2× studs on spacing compatible with an on-end (vertical) bale width, e.g., 23" wide three-string bales squeeze between studs set at 24" centers. Exterior and interior finishes must be vapor-permeable—no OSB, waterproof paints, or wallpaper.

One straw bale wall assembly utilizes I-joists or laminated veneer lumber (LVLs) as vertical members, usually paired with on-edge bale walls. In the United States, manufacturers do not have code approval for using I-joists as studs, therefore a design professional should design the system to ensure correct detailing.

Another configuration uses narrow prefabricated wood shear walls oriented out-of-plane. Standard 18" widths can be field trimmed to gable ends or nonstandard heights. This provides unmatched out-of-plane stiffness and allows for taller walls. This system can provide a great deal of stability and strength for larger wall systems or systems where there's a large distance between either interior or exterior braced wall lines. However, while these walls do double-duty as shear walls, they are expensive. They are also subject to stress concentrations and require careful mesh detailing to prevent plaster cracking.

Lateral-Force-Resisting Systems

All buildings need to resist lateral forces from wind, and many must also resist lateral forces from earthquakes. This section shows and explains methods to resist these forces, carrying them from the roof to the foundation. These lateral forces can act upon the building in-plane (parallel to the wall) or out-of-plane (perpendicular to the wall).

3-15. Encased stud wall.

A variation on the post-and-beam system, a stud wall using smaller-dimension lumber supports the roof, windows, and doors, with bales set around framing flush to the exterior face of the studs.

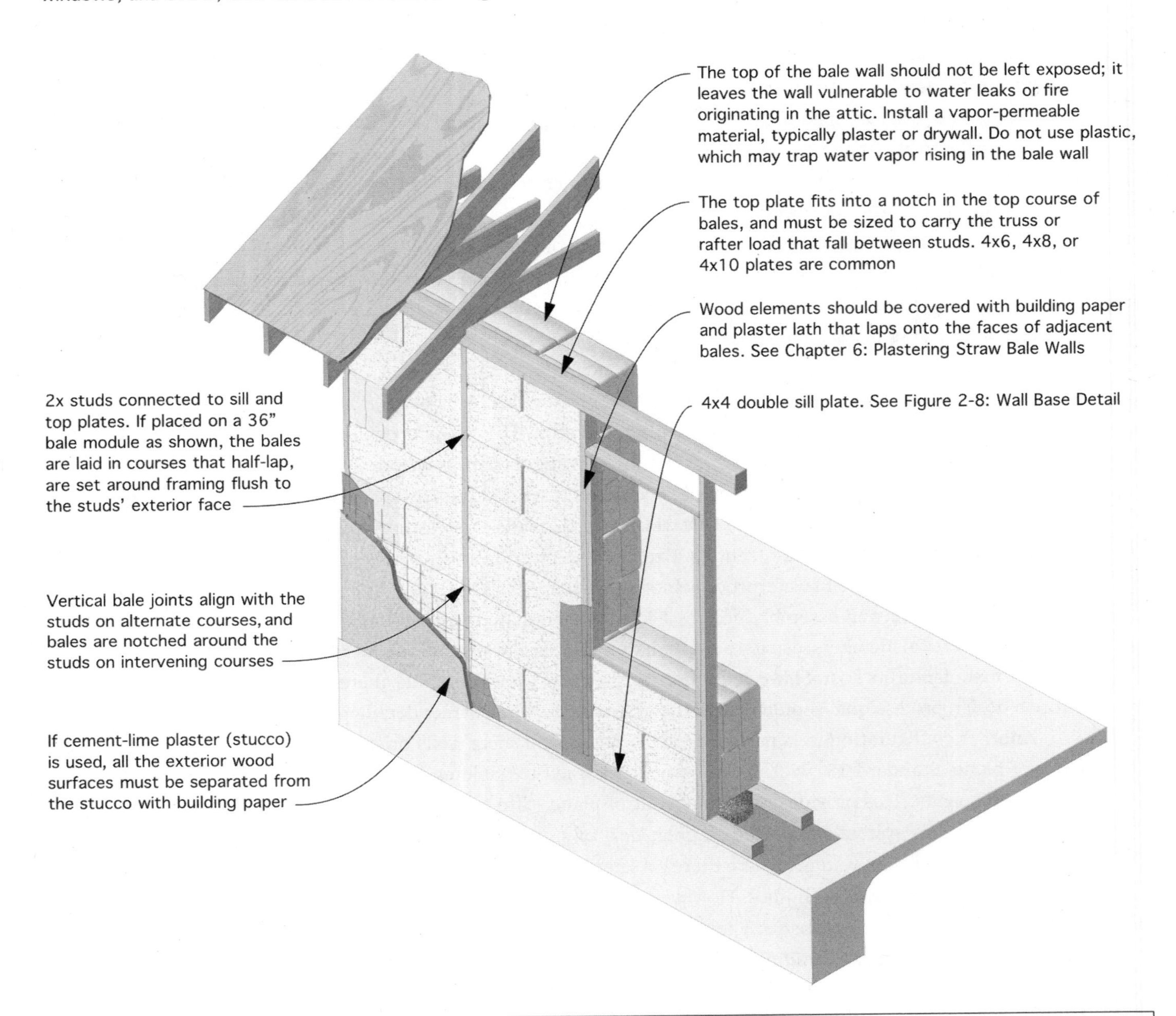

NOTES:

- Modifying bales on-site to create "module-size" bales introduces the possibility that retied bales won't have their original compressive strength, or the compressive strength required by code. Finding a bale supplier to produce module-size bales may be challenging, and this approach may raise the bale cost.

- The stud walls can be nailed up on the ground and tilted up, which takes less time than raising a post-and-beam wall.

- The 1-1/2" stud width tucks into the head joints between bales and may not require careful notching.

3-15a. Encased stud wall—vertical bales (bales on end).

A variation on the post-and-beam system, a stud wall using smaller-dimension lumber supports the roof (standard 2'-0" o.c. spacing), windows, and doors. 23" wide bales are squeezed between studs, flush to the exterior face of the studs.

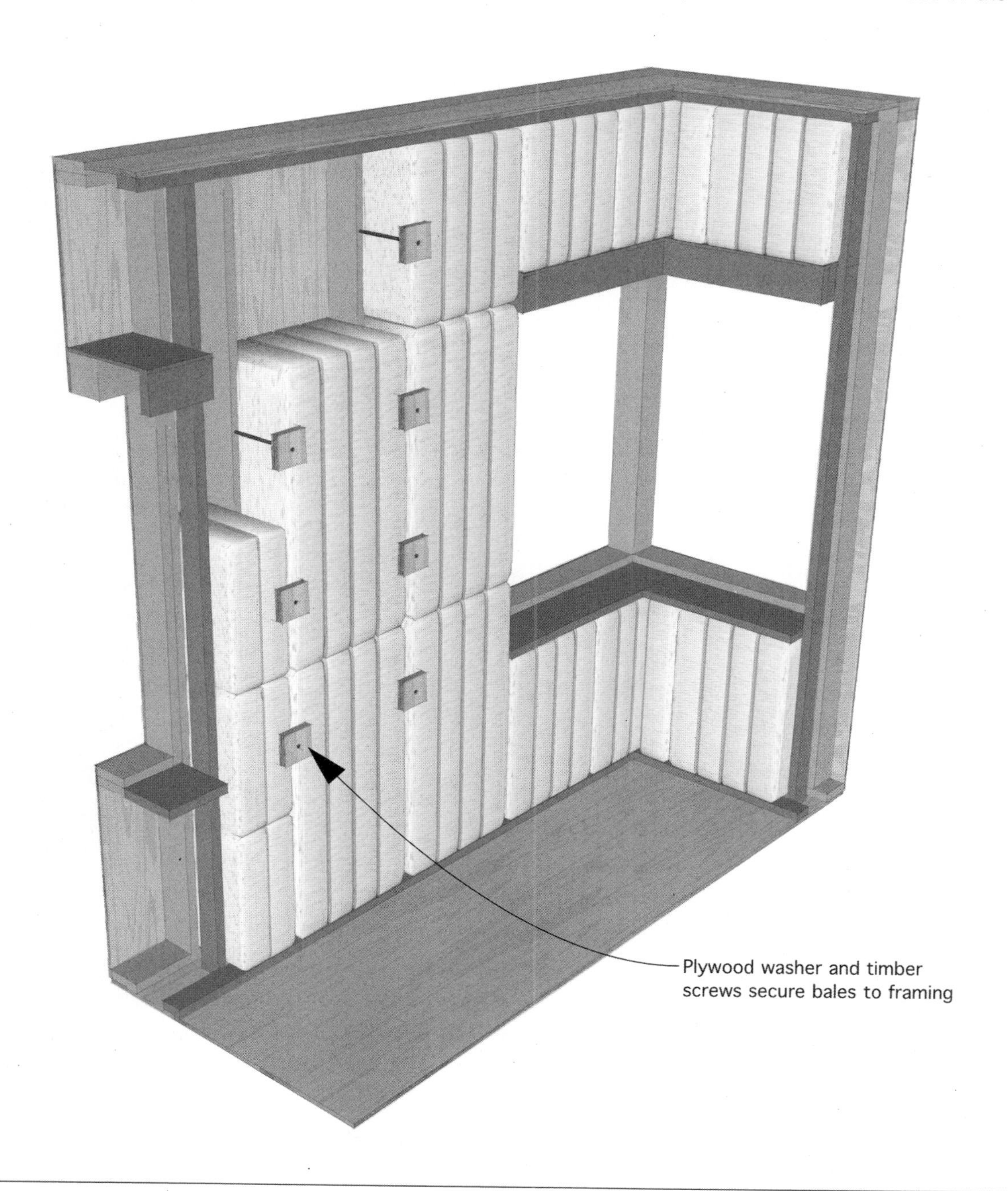

NOTES:

- Plywood and interior finishes must be vapor-permeable (no OSB, waterproof paints / wallpaper, etc).

- Space studs slightly narrower than the bale width, In this example, set exactly 2'-0" o.c. to receive 23" wide bales. Take double and off-module stud locations into account (shear walls, door and window frames, etc.).

3-16. I-joist wall.

Engineered wood joists (called I-joists) as deep as the wall thickness set vertically as posts to support roof, doors, and windows, and as shown, a second floor. Usually paired with bales-on-edge, which results in 15" or 16" straw bale walls, covered by 1 1/2" plaster skins on both sides.

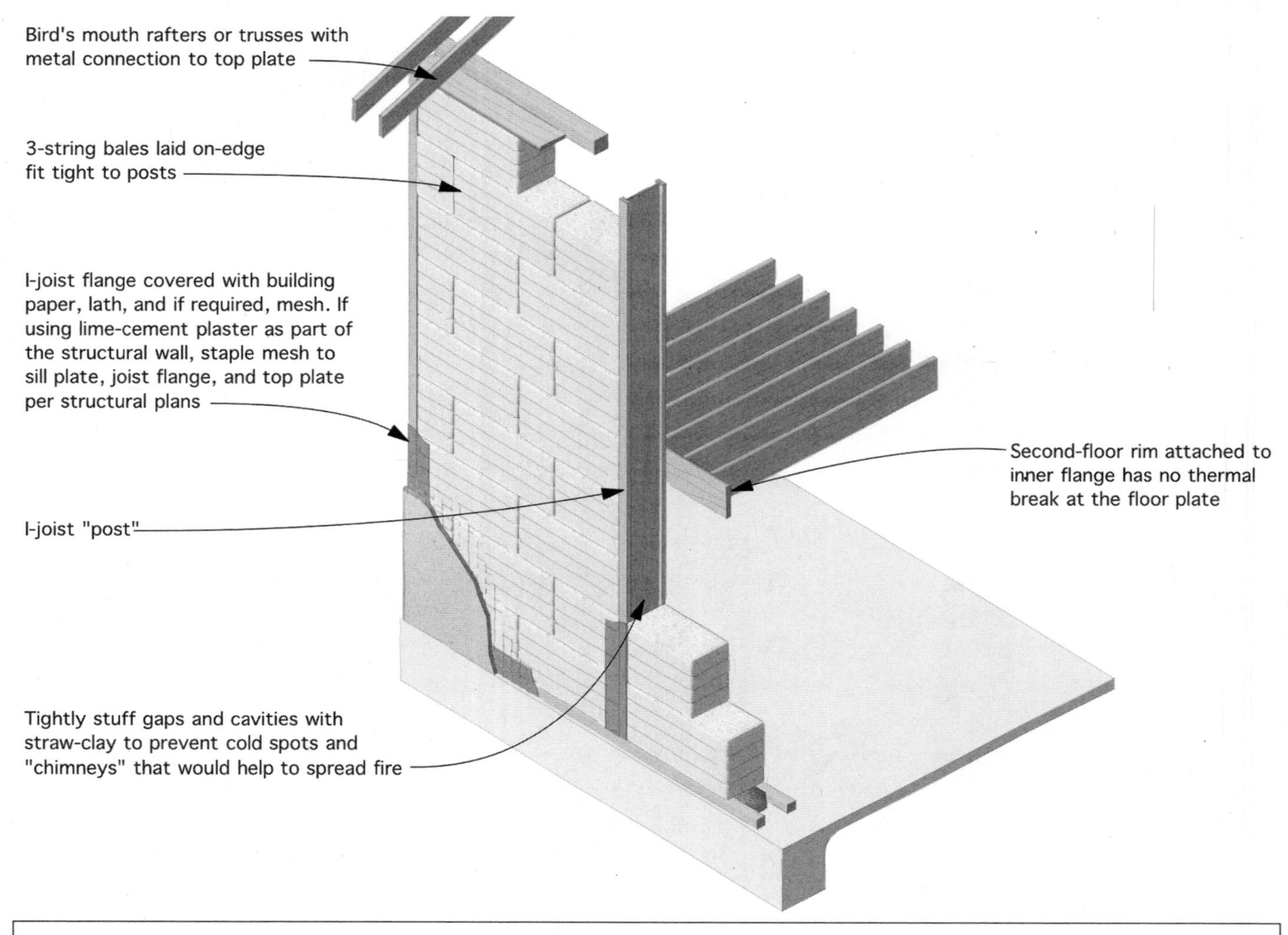

NOTES:

- On-edge bales have the highest R-value of any bale orientation — R-1.85 per inch, compared with R-1.55 per inch for bales laid flat (code section: Appendix S — AS108.1).

- No posts to notch around makes setting bales faster and easier, especially if we space the I-joists in 1/2 bale increments, up to 10'-0."

- Flange attachment makes both interior and exterior skins structural.

- Strings are on the surface, making them more vulnerable to getting cut (not usually an issue once bales are in place).

- Requires foresight to create flared window and door reveals.

- Requires engineered lumber which comes at a cost; could do with solid lumber but with a loss of material efficiency and increased thermal bridging (otherwise the 1/2" thick web of the I-joist is thin enough to not represent a detrimental thermal bridge).

- Often paired with a straw bale shear wall. Many plasters can be used including pise, lime, lime-cement, and clay. Conventional lateral-force–resisting systems can also be used.

- Thoughtfully placed posts address out-of-plane loads.

Out-of-Plane Systems

All straw bale walls, even when non-structural, need to resist out-of-plane forces from wind and/or earthquakes. Appendix S prescribes a number of methods, though others are possible. Section AS105.4 and its table describe the most common methods:

- external or internal *pins* or battens through-tied
- mesh-reinforced plaster on both sides
- tying the bales to wall studs (or posts)

External pins made of wood, bamboo, or steel rebar are placed vertically on opposite sides of the unplastered straw bale wall and through-tied, usually with baling twine. They greatly stiffen the wall. Internal rebar pins embedded in the foundation and running up through the bales were once common, and even required. They are still allowed, but have fallen out of favor because of installation difficulties and possible condensation problems. The term *pin* is accurate for internal pins, but it has migrated for use with external pins, where some say the word *batten* is more appropriate. Many believe bamboo to be the superior pin material because of its high strength-to-weight ratio and a thermal expansion akin to the bales. Sometimes external pins can be paired with posts on the opposite side. See Figures 3-17 and 3-18: Internal and External Pin Details, and Chapter 5, Stacking Straw Bale Walls.

Mesh-reinforced plaster can also be used as the out-of-plane system. Appendix S does not require through-tying the mesh because the reinforced plasters on opposite sides are sufficiently connected via the straw bales themselves. However, through-tying the mesh provides even better connection and out-of-plane strength, and it provides stability during construction before plastering.

Regularly spaced posts in post-and-beam assemblies provide the necessary out-of-plane force resistance. Appendix S uses the term *stud*, and specifically, 2×6 or 2×4 studs with various spacing options, but other equivalent framing members can be used, such as 4×4 posts. Attaching the bales to the framing is vital. Appendix S gives flexibility here by saying: "Bales shall be attached by an approved method." The commentary to Appendix S identifies wood stakes or plywood gussets as common means of attachment, and notes that this method should be used only with an engineered design unless a local successful history exists.

In-Plane Systems

Many designers use the building's plastered straw bale walls as shear walls—whether load-bearing or not—to resist in-plane lateral loads. Properly detailed straw bale shear walls are very strong and resilient, and they efficiently use the straw bale walls as both enclosure and structure. But there are pros and cons, and limits to their use. A variety of methods commonly used in conventional construction can also be employed in straw bale buildings to resist in-plane forces. These include conventional wood-framed plywood shear walls, prefabricated wood or steel shear walls, prefabricated steel-braced frames, steel tension-only straps or cables, exterior buttresses, and steel moment frames.

3-17. External pins.

Pins (sometimes called battens) paired on opposite sides of the wall and tied tightly together help resist out-of-plane forces.

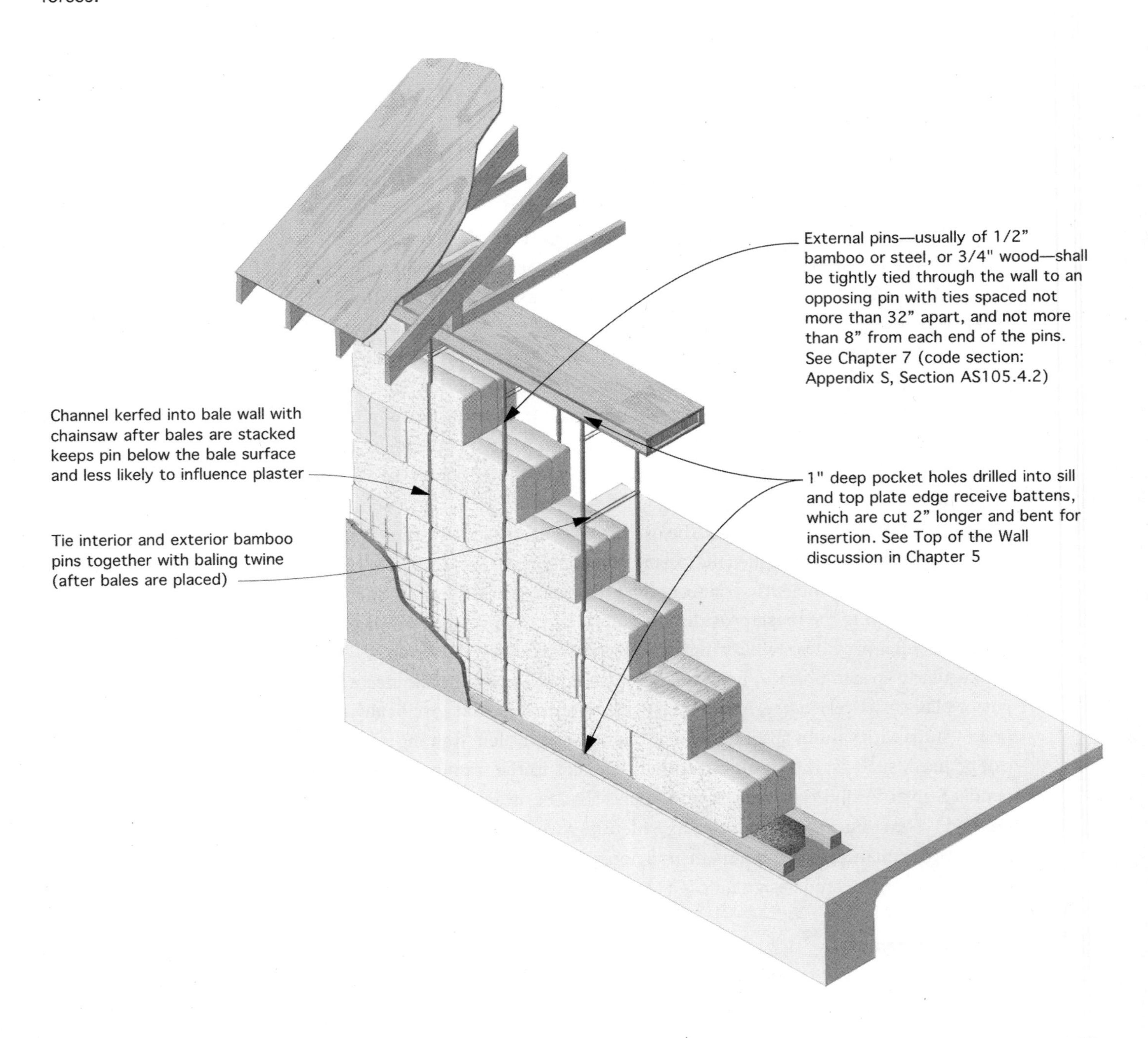

NOTES:

- Particularly useful in load-bearing and post-and-beam wall systems with widely spaced posts; pins help to stiffen the wall.

- Can eliminate the need for wire mesh.

3-18. Internal pins.

Internal pins made of rebar were in common use to resist lateral forces early in the straw bale building revival. They were embedded in the foundation, and bales were impaled over them. Pins extend from the foundation through to the top plate. Although more efficient methods have evolved, internal pinning may have application in some projects. See requirements for internal pinning in Appendix S: AS105.4.2(3).

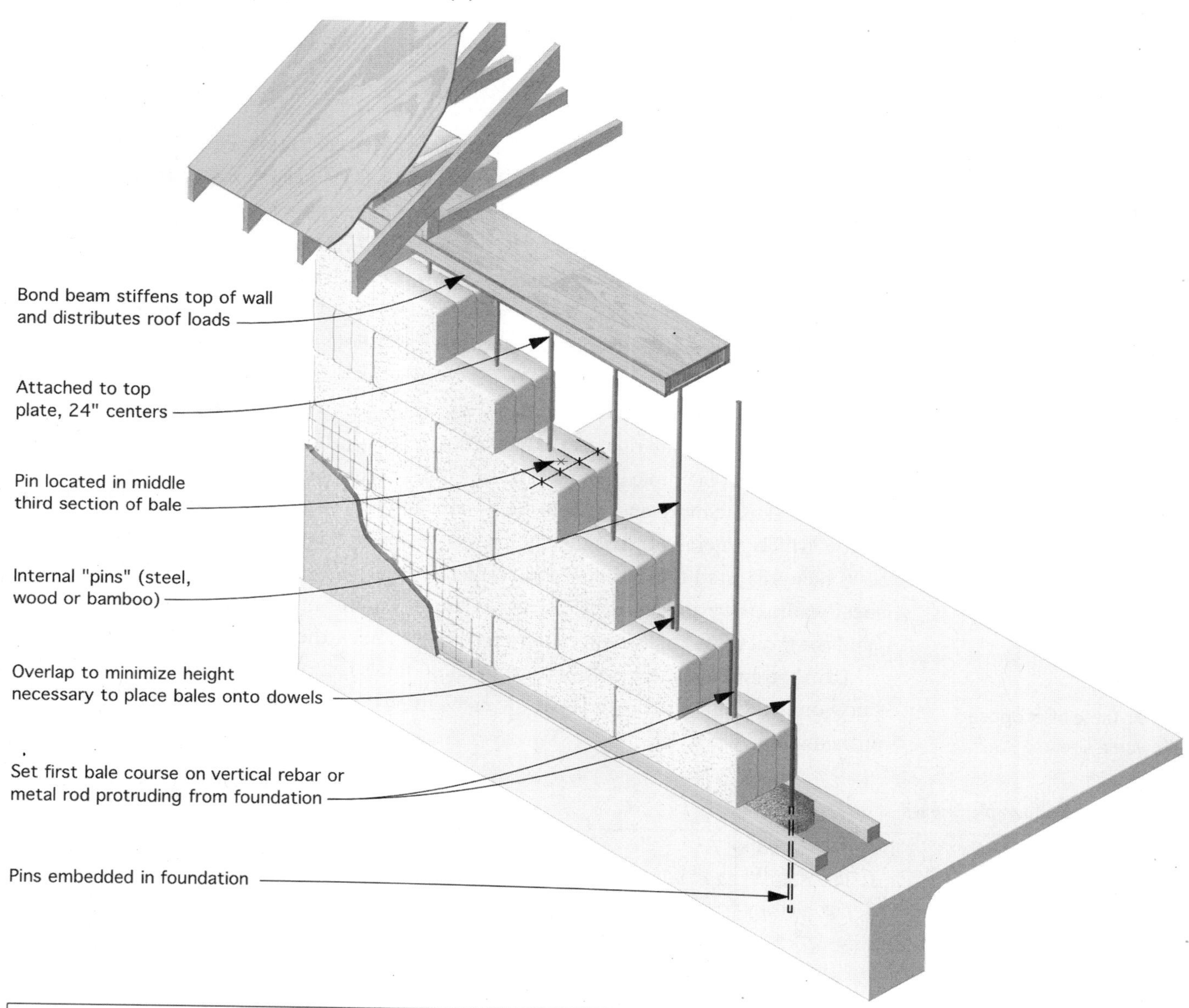

NOTES:

- Largely a discontinued practice due to the difficulty of impaling the bales over the pins and concerns for condensation forming around the rebar.

- Internal pinning with bamboo or fiberglass rebar between several bale courses, if not the entire wall, is sometimes used to align bales during bale raisings when other methods aren't available.

- Particularly useful in load-bearing and post-and-beam wall systems with widely spaced posts; pins help to stiffen the wall (external pins are more widely used, see Figure 3-17: External Pins Detail).

- Can eliminate the need for wire mesh.

Shear Wall Terminology

We have used the common terminology *shear wall* throughout the text to refer to a wall designed and constructed to resist lateral seismic and wind forces parallel to the wall plane. Shear walls are also called "braced wall panels" in Appendix S and throughout the IRC.

Many straw bale buildings employ more than one system. They may combine interior wood-framed shear walls with exterior plastered straw bale shear walls (see below). The project engineer will calculate how much shear is required (the building's "base shear") and how much each of those systems will contribute to resisting that shear. Because some systems are more flexible than others, the engineer will also evaluate the systems for compatibility, and provide details accordingly.

Straw Bale Shear Walls

With proper design, detailing, and construction, a plastered straw bale wall assembly can resist in-plane wind and seismic forces; it's a shear wall, or in the terminology of Appendix S, a *braced wall panel.*

In 2003, the Ecological Building Network and CASBA conducted tests to demonstrate the load capacity of plastered straw bale wall assemblies. The table below, adapted from the 2012 *Structure Magazine* article "Strawbale Construction," shows proposed allowable shear values for various assemblies based on that testing.

Interestingly, the maximum force in these tests occurred at 7.5% drift (deflection of the top of wall relative to the bottom of wall) whereas typical plywood shear walls reach their maximum force at 1.4% drift. This means that a straw bale wall will dissipate more energy before ultimate failure. It is more ductile, or less brittle, which generally translates to higher performance and greater safety. The tests demonstrated the stability and resiliency of properly detailed straw bale shear walls. However, this drift can also result in significant damage to the wall (and the structure) that would require repair after an earthquake.

Using straw bale shear walls as part of a building's lateral-force-resisting design—one that can be approved by the local building department—can be accomplished in two ways:

3-19. Table of proposed allowable shears.

Development of Allowable Shears

Plaster Type	Plaster Thickness (min)	Plaster Reinforcement Mesh	Shear Strength[1] (kips)[2]	Factor of Safety	Proposed Allowable Shear (plf)[3]
Clay	1.5"	2"×2" high-density polypropylene	3.05	2.78	140
Clay	1.5"	2"×2"×14 ga.	4.10	2.78	180
Soil-cement	1"	2"×2" x 14 ga.	16.26	3.87	520
Lime	⅞"	17 ga. woven wire	10.18	3.87	330
Lime	⅞"	2"×2"×14 ga.	13.97	3.87	450
Cement-lime	⅞"	17 ga. woven wire	11.71	3.87	380
Cement-lime	⅞"	2"×2"×14 ga.	16.07	3.87	520
Cement	⅞"	2"×2"×14 ga.	16.70	3.87	540
Cement	1.5"	2"×2"×14 ga.	17.45	3.22	680

[1] for an 8-foot-long wall with a 1:1 aspect ratio
[2] 1 kip = 1000 pounds of force
[3] pounds per linear foot
Source: Adapted from *Structure Magazine*, September 2012

1. With an engineered design (by a licensed design professional) using the proposed allowable shear values shown in the *Structure Magazine* article (see table in Figure 3-19) (assuming those values and the design are accepted by the local building official) (see page 93 regarding an appropriate R factor) or
2. By using the prescriptive tables in Appendix S (see Examples, below) (assuming Appendix S has been adopted by the local jurisdiction or is accepted by the local building official).

Appendix S gives prescriptive minimum wall lengths for straw bale shear walls depending on wind and seismic forces. As with any construction system, you or your engineer can determine your project's ultimate design wind speed and seismic design category using the IBC, IRC, or ASCE 7. Consult ground motion parameter maps in the IBC or ASCE 7. Obtain prescriptive bracing requirements

3-20a. Table AS106.13(2) from the 2018 International Residential Code (IRC) Reproduced with permission of the International Code Council.

TABLE AS106.13(2)
BRACING REQUIREMENTS FOR STRAWBALE-BRACED WALL PANELS BASED ON WIND SPEED

• EXPOSURE CATEGORY B[d] • 25-FOOT MEAN ROOF HEIGHT • 10-FOOT EAVE-TO-RIDGE HEIGHT[d] • 10-FOOT WALL HEIGHT[d] • 2 BRACED WALL LINES[d]			MINIMUM TOTAL LENGTH (FEET) OF STRAWBALE BRACED WALL PANELS REQUIRED ALONG EACH BRACED WALL LINE[a, b, c, d]		
Ultimate design wind speed (mph)	**Story location**	**Braced wall line spacing (feet)**	**Strawbale-braced wall panel[e] A2, A3**	**Strawbale-braced wall panel[e] C1, C2, D1**	**Strawbale-braced wall panel[e] B, D2, E1, E2**
≤ 110	One-story building	10	6.4	3.8	3.0
		20	8.5	5.1	4.0
		30	10.2	6.1	4.8
		40	13.3	6.9	5.5
		50	16.3	7.7	6.1
		60	19.4	8.3	6.6
≤ 115	One-story building	10	6.4	3.8	3.0
		20	8.5	5.1	4.0
		30	11.2	6.4	5.1
		40	14.3	7.2	5.7
		50	18.4	8.1	6.5
		60	21.4	8.8	7.0
≤ 120	One-story building	10	7.1	4.3	3.4
		20	9.0	5.4	4.3
		30	12.2	6.6	5.3
		40	16.3	7.7	6.1
		50	19.4	8.3	6.6
		60	23.5	9.2	7.3
≤ 130	One-story building	10	7.1	4.3	3.4
		20	10.2	6.1	4.8
		30	14.3	7.2	5.7
		40	18.4	8.1	6.5
		50	22.4	9.0	7.1
		60	26.5	9.8	7.8
≤ 140	One-story building	10	7.8	4.7	3.7
		20	11.2	6.4	5.1
		30	16.3	7.7	6.1
		40	21.4	8.8	7.0
		50	26.5	9.8	7.8
		60	30.6	11.0	8.3

For SI: 1 inch = 25.4 mm, 1 foot = 305 mm, 1 mile per hour = 0.447 m/s.

a. Linear interpolation shall be permitted.

b. All braced wall panels shall be without openings and shall have an aspect ratio (H:L) ≤ 2:1.

c. Tabulated minimum total lengths are for braced wall lines using single-braced wall panels with an aspect ratio (H:L) ≤ 2:1, or using multiple braced wall panels with aspect ratios (H:L) ≤ 1:1. For braced wall lines using two or more braced wall panels with an aspect ratio (H:L) > 1:1, the minimum total length shall be multiplied by the largest aspect ratio (H:L) of braced wall panels in that line.

d. Subject to applicable wind adjustment factors associated with "All methods" in Table R602.10.3(2)

e. Strawbale braced panel types indicated shall comply with Sections AS106.13.1 through AS106.13.3 and with Table AS106.13(1).

TABLE AS106.13(3)
BRACING REQUIREMENTS FOR STRAWBALE-BRACED WALL PANELS BASED ON SEISMIC DESIGN CATEGORY

• SOIL CLASS D[f] • WALL HEIGHT = 10 FEET[d] • 15 PSF ROOF-CEILING DEAD LOAD[d] • BRACED WALL LINE SPACING ≤ 25 FEET[d]			MINIMUM TOTAL LENGTH (FEET) OF STRAWBALE-BRACED WALL PANELS REQUIRED ALONG EACH BRACED WALL LINE[a, b, c, d]	
Seismic Design Category	**Story location**	**Braced wall line length (feet)**	**Strawbale-braced wall panel[e] A2, C1, C2, D1**	**Strawbale-braced wall panel[e] B, D2, E1, E2**
C	One-story building	10	5.7	4.6
		20	8.0	6.5
		30	9.8	7.9
		40	12.9	9.1
		50	16.1	10.4
D_0	One-story building	10	6.0	4.8
		20	8.5	6.8
		30	10.9	8.4
		40	14.5	9.7
		50	18.1	11.7
D_1	One-story building	10	6.3	5.1
		20	9.0	7.2
		30	12.1	8.8
		40	16.1	10.4
		50	20.1	13.0
D_2	One-story building	10	7.1	5.7
		20	10.1	8.1
		30	15.1	9.9
		40	20.1	13.0
		50	25.1	16.3

For SI: 1 inch = 25.4 mm, 1 foot = 305 mm, 1 pound per square foot = 0.0479 kPa.
a. Linear interpolation shall be permitted.
b. Braced wall panels shall be without openings and shall have an aspect ratio (H:L) ≤ 2:1.
c. Tabulated minimum total lengths are for braced wall lines using single braced wall panels with an aspect ratio (H:L) ≤ 2:1, or using multiple braced wall panels with aspect ratios (H:L) ≤ 1:1. For braced wall lines using two or more braced wall panels with an aspect ratio (H:L) > 1:1, the minimum total length shall be multiplied by the largest aspect ratio (H:L) of braced wall panels in that line.
d. Subject to applicable seismic adjustment factors associated with "All methods" in Table R602.10.3(4), except "Wall dead load."
e. Strawbale braced wall panel types indicated shall comply with Sections AS106.13.1 through AS106.13.3 and Table AS106.13(1).
f. Wall bracing lengths are based on a soil site class "D." Interpolation of bracing lengths between S_{ds} values associated with the seismic design categories is allowable where a site-specific S_{ds} value is determined in accordance with Section 1613.3 of the *International Building Code*.

3-20b. Table AS106.13(3) from the 2018 International Residential Code (IRC). Reproduced with permission of the International Code Council.

for your project from Appendix S: Table AS106.13(2) based on wind speed (Figure 3-20a) and Table AS106.13(3) (Figure 3-20b) based on seismic design category. Consult both tables; the table with the greater minimum wall length will govern your project. These wall lengths are subject to the adjustment factors in the footnotes of each table.

Using Appendix S for a Building's Lateral Force-Resisting Design

EXAMPLE 1

A 20'×20' rectangular structure, with an ultimate design wind speed of 110 mph, in Seismic Design Category D1, with simple roof construction and no interior shear walls.

Consulting Table AS106.13(2) for wind, the building needs 8.5' total of type A2 shear walls along each wall line. Consulting Table AS106.13(3) for seismic, the building needs 9.0' total of type A2 shear walls along each wall line. In this case, the greater seismic requirement governs. These shear walls must be continuous top to bottom, with no windows. Table AS106.1.13(1) shows that A2 walls are plastered with 1.5" of clay plaster, 2×4 sill plates, anchors at 32" o.c. and 2×2 high-density polypropylene mesh, stapled to the top and bottom plates at 4" o.c. Linear interpolation is permitted, so for a 25'×25' structure (longer walls), 10.6' would be required, that is, (9.0' + 12.1')/2.

EXAMPLE 2

A 30'×30' rectangular structure, with an ultimate design wind speed of 130 mph, in Seismic Design Category C, with simple roof construction and no interior shear walls.

Consulting Table AS106.13(2) for wind, the building needs 14.3' total of type A2 shear walls along each wall line. Consulting Table AS106.13(3) for seismic, the building needs 10.2' total of type A2 shear walls along each wall line. In this case, the greater wind requirement governs. One way to decrease the required length is to use a stronger wall type. In this example, a C2 wall, with lime plaster and 2×2 welded wire mesh, requires a total length of only 7.2'.

Footnote "b" on both tables states that braced wall panels cannot exceed a height-to-length ratio of 2:1. Therefore, the minimum length for a 9' tall shear panel is 4.5'. As given in footnote "c," if more than one panel is needed to meet the required length, the maximum height-to-length ratio is 1:1. So, in Example 2, if two walls were needed, each wall would need to be at least 9' long.

For post-and-beam systems not using the straw bale walls as shear walls, the prescriptive lateral design process in the IRC is the same as for conventional framing, except that the required braced wall panel lengths in R602.10 must be increased by 60% to account for the straw bale wall weight, per AS105.2.4.

The IRC and Appendix S offer no clear way to mix straw bale shear walls and conventional shear walls for a building's prescriptive lateral system. Consult a licensed design professional if you want to do this.

Notes on Designing with and Employing Straw Bale Shear Walls

In engineered seismic design, the Response Modification Coefficient (R factor) greatly affects the total design shear load (the base shear) that must be resisted. The type of lateral system used dictates the R factor, which varies from 1 to 8. The lower the number, the higher the base shear will be. Most conventional systems have well-established R factors, as listed in ASCE 7 and other references. For buildings using straw bale shear walls, engineers in California have typically used an R factor of 4.

In buildings with conventional lateral systems, ignore any lateral resistance contributions of the plastered straw bale walls. However, load paths should be checked due to the straw bale weight, as well as their inherent in-plane shear capacity. For example, the unaccounted-for shear capacity of the straw bale wall could damage a raised wood floor system if not properly detailed.

While under construction, a structure designed with straw bale shear walls should use an engineered bracing system to resist anticipated wind and construction impact loads until the plaster is applied and cured. Depending on the structure's design, this can create a considerable challenge because all the trades must work around the bracing, from installation of posts and beams until the plaster reaches final design strength, which is typically 28 days.

Installing reinforcing mesh on a straw bale shear wall takes considerable time and attention to detail. Mesh folds and laps must be carefully made; the mesh

3-21. Straw bale shear wall.

A shear wall using an approved mesh embedded in a plaster with the needed compressive strength is an elegant solution that combines wall protection and structural functionality.

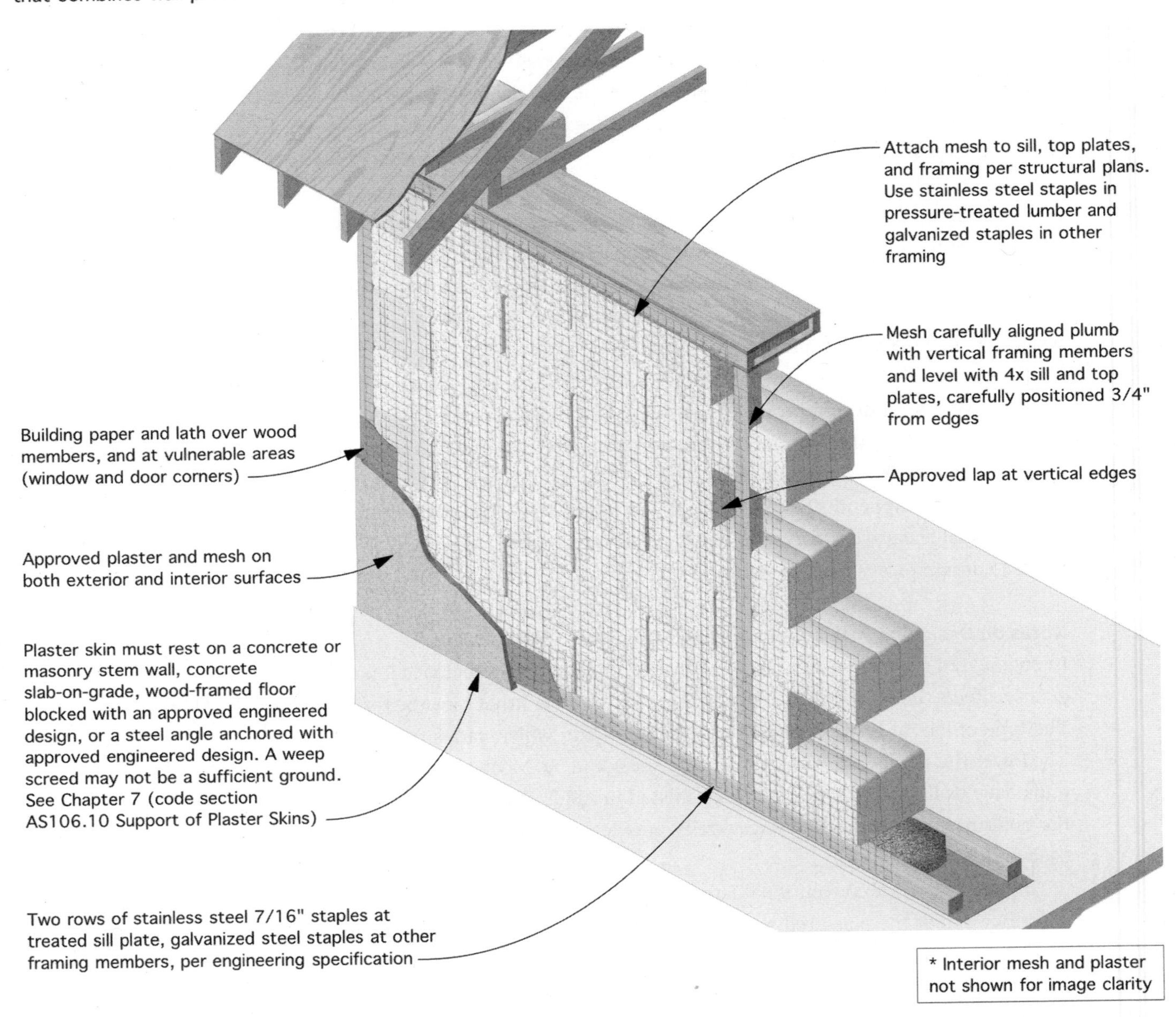

NOTES:
- Neatly folded top and bottom laps facilitate accurate stapling.

- Stretch mesh tightly from top to bottom before stapling.

- Building paper obscures the sill and top plates; mark building paper to ensure staples secure the mesh "centered" on the plates and posts.

- Work the scratch or "discovery" coat through the mesh into the bales to embed it.

must be tightened across the bale wall surface, then accurately attached to framing that is often hidden behind building paper. See mesh discussion in Chapter 6, Plastering Straw Bale Walls.

Straw bale shear walls require specific details which are not needed for the entire building. For example, narrow wall sections or walls with windows typically do not perform a shear function. So 2× plates can be installed at these walls, with 4×4 plates used only at shear walls. This requires the first course of bales to be notched around the 4×4s so the bale courses remain level. High-strength mesh and attachment need only be used as plaster reinforcement at the required shear walls. If the structure is not using straw bale shear walls, then details for sill plates, mesh (if any) and its attachment, and the plaster support at the bottom edge can be simpler. In practice, however, it's often easier and less costly to specify and install a single, more robust system all around, even if it exceeds what is required in places. Weigh the additional time spent with custom details and varying installations against the saved material, labor, and embodied energy costs.

Conventional Lateral-Force-Resisting Systems

Use conventional lateral-force-resisting systems in straw bale buildings when:

- the shear capacity of the desired plaster is insufficient for the building's design forces
- architectural design does not provide enough exterior wall length to use straw bale shear walls
- the building department is unfamiliar or uncomfortable with straw bale shear walls
- conventional materials and systems are familiar to engineers, contractors, plan checkers, and building code officials
- cost and construction sequence present concerns

Any lateral-force-resisting system has trade-offs. The following descriptions show alternatives to straw bale shear walls, including notes about their advantages and disadvantages. Design all structural systems per the applicable building code sections. Carefully detail changes in plaster substrate from straw bale to wood or metal surfaces to minimize cracking and prevent moisture damage. See lathing discussion in Chapter 6, Plastering Straw Bale Walls.

Conventional Wood-Framed Shear Walls

A straw bale building's lateral-force-resisting system can incorporate conventional wood-framed shear walls on both interior and exterior walls. Common passive solar design uses large window areas on the south face, which often requires conventional shear walls on the south wall because they provide more bracing strength per lineal foot. Careful design can also use interior partitions as conventional shear walls to decrease the force on the exterior walls. Keep an

3-22. Plywood shear wall.

This familiar shear wall system can be inexpensively made from readily available materials, and is most often located at building corners.

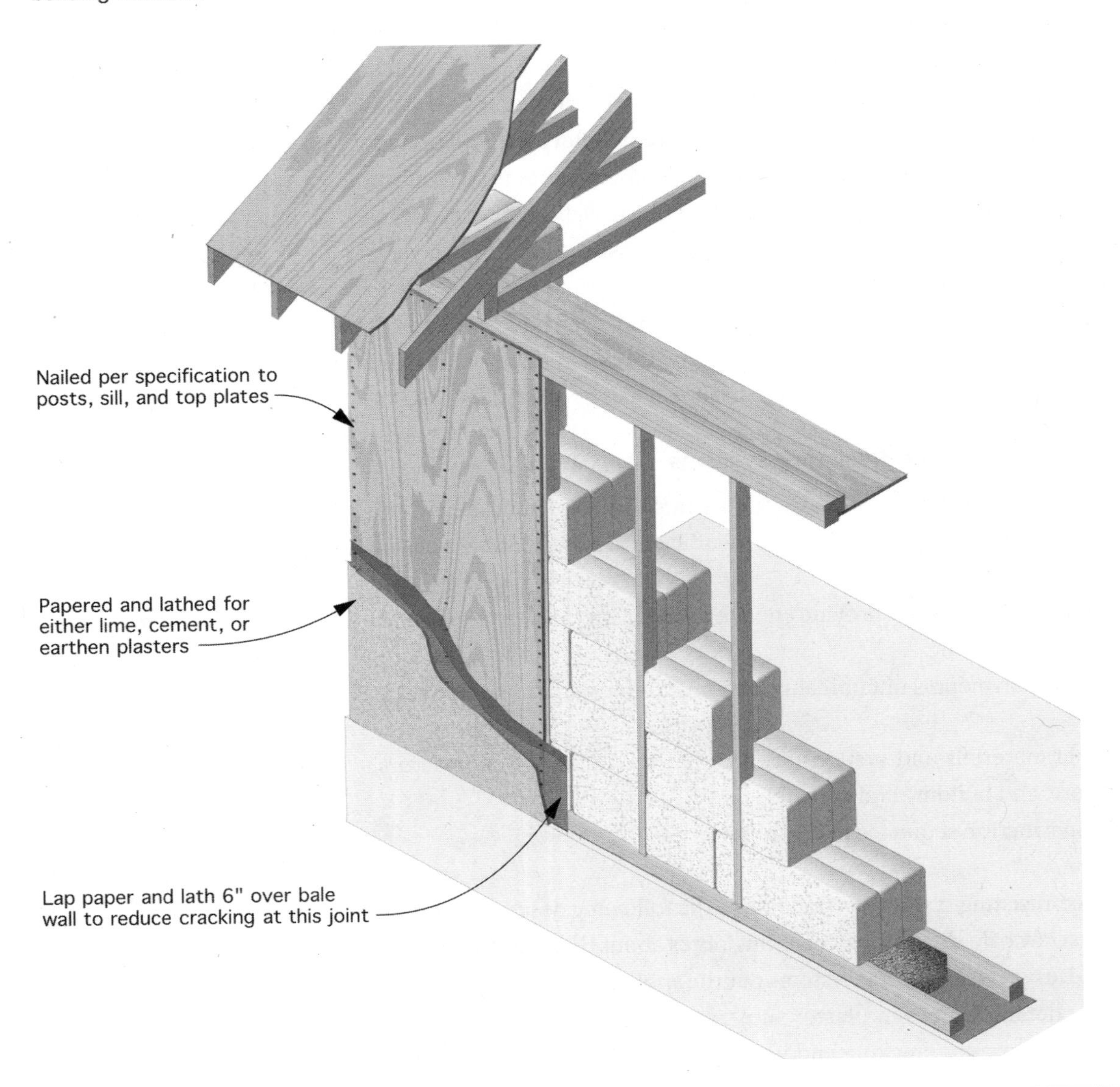

NOTES:

- Plywood has a low vapor permeability — keep plywood shear wall lengths short so moisture in the bales can move around them.

- Seal the top and bottom of the wall to prevent air movement.

- To install a fire and insect barrier and reduce air movement through the bales abutting the plywood panel, temporarily brace the wall, stack the bales, and spray slip or apply a first coat of plaster. Install the plywood when the slip or plaster has dried.

- This shear wall system can work with external pinning as well; drill holes in plywood and loop baling string from exterior plywood panel to interior pins prior to applying building paper and lath.

exterior wood shear wall as short as possible when it has bales notched around and stacked behind it—the glued construction of wood sheathing has a very low vapor permeance; the short lengths help avoid trapping moisture in the straw bale wall.

In order to maintain the bale wall's fire resistance, some builders "butter" the bale surface with clay slip or plaster prior to stacking them against the plywood. Batts of 1" thick low-density mineral wool have also been used for this purpose; they have the advantage of not introducing moisture inside the wall assembly. The top and bottom of plywood shear walls should be carefully detailed to prevent air movement vertically along the surface of the bale wall where it abuts the wood shear wall.

Prefabricated Shear Walls and Braced Frames

Several manufacturers make prefabricated wood and steel shear walls and steel-braced frames—commonly available through local lumber yards—for wood- or steel-stud framed buildings. Post-and-beam straw bale buildings can be designed so these prefabricated units resist lateral forces. Like conventional wood-framed shear walls, notch the straw bales around them to provide a flush wall surface. The "Simpson Strong-Wall" in wood and metal and the "Hardy Frame Panel" and "Brace Frame" work well due to availability of different heights and the 3.5" panel thickness, which allows for bale notching without having to cut and retie the bale strings. Prefabricated wood shear walls can be cut to height in the field and are easier to paper and lath than steel shear walls, though the airspace between the shear walls and the bales must be carefully detailed (see the discussion above about wood-framed shear walls).

Due to concentrated forces, prefabricated systems require larger footings with more reinforcement, but they install quickly and can eliminate the need for temporary wall bracing throughout construction. They also simplify design detailing because the manufacturers provide prescriptive footing details. The prefabricated units have accepted shear values, and the lateral-forces design becomes a simple matter of how many of which type to deploy on each braced wall line. They require careful detailing to transfer the lateral loads from the roof to the shear walls or braced frames, often with beams in line with the top of the shear walls or frames, and steel straps to carry tension across the beam splices.

Locate prefabricated shear walls and braced frames at least a bale width from building corners to simplify bale stacking and tying, base plate connections, and the foundation. Custom-made structural steel-braced frames can be designed to satisfy project requirements. Custom steel frames typically use square tubes or angles for the diagonal member in the frames. A qualified engineer should design them. Each configuration has varied options and well-documented design provisions. A certified welder or a certified steel fabrication shop must fabricate these braced frames.

3-23. Prefabricated steel or wood-braced frames.

Prefabricated steel or wood-braced frames are often used to handle lateral forces in post-and-beam wall systems. Available in a variety of heights, their 3 1/2" width facilitates bale notching with no need to retie strings.

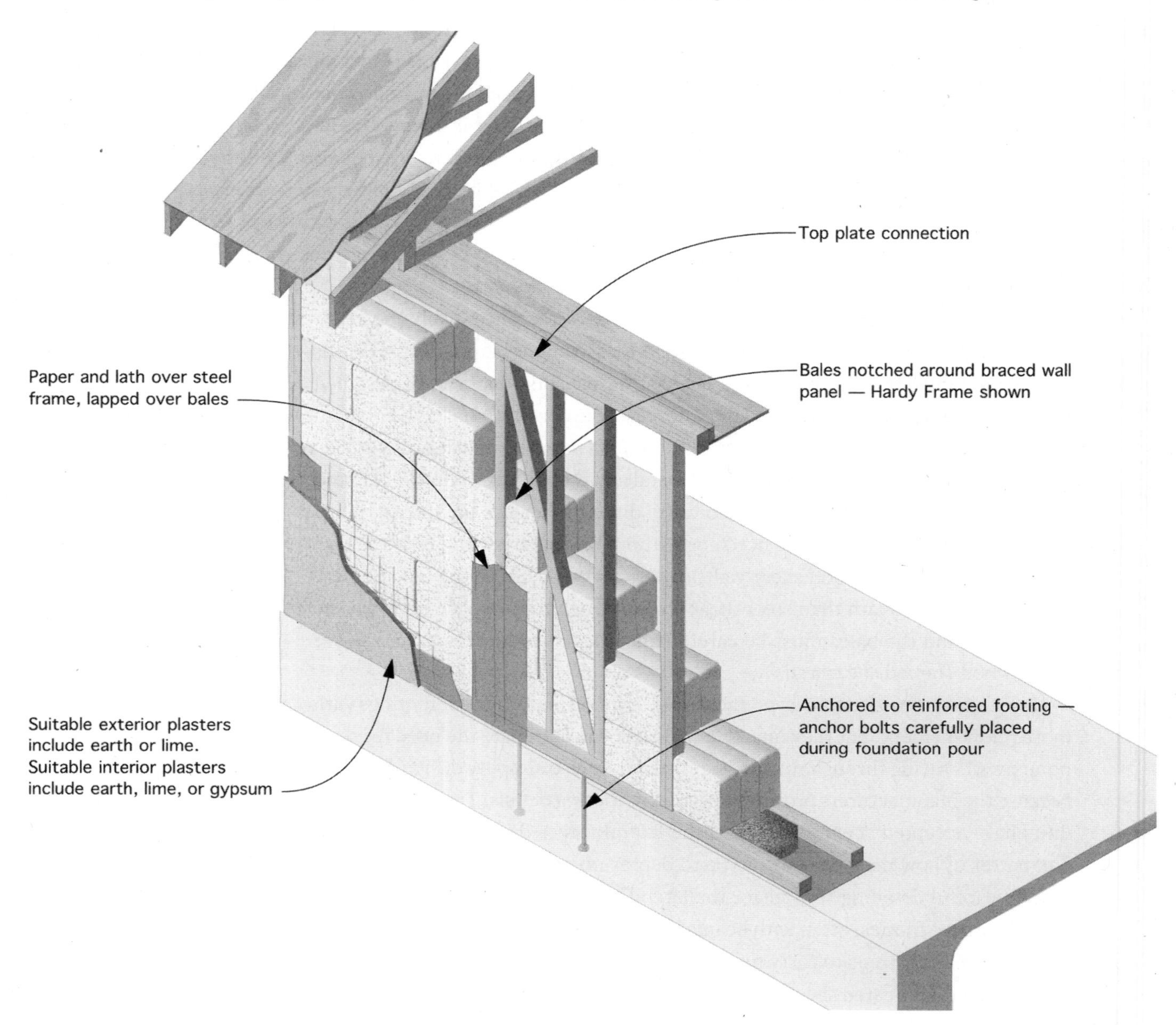

NOTES:
- Footing may need to be several inches wider under brace frames.

- Keeping frames a bale width (18" or 24") from corners allows one to avoid having to notch the butt-end of bales where they interlock at corners. This makes stacking much easier.

- Accepted shear values simplify lateral design.

- Building paper and lath attaches more easily to wood-braced frames. This option also offers some insulation value and doesn't pose a condensation problem as steel-braced frames might.

3-24. Interior shear wall.

This option combines shear with an interior partition wall using commonly available bulding materials.

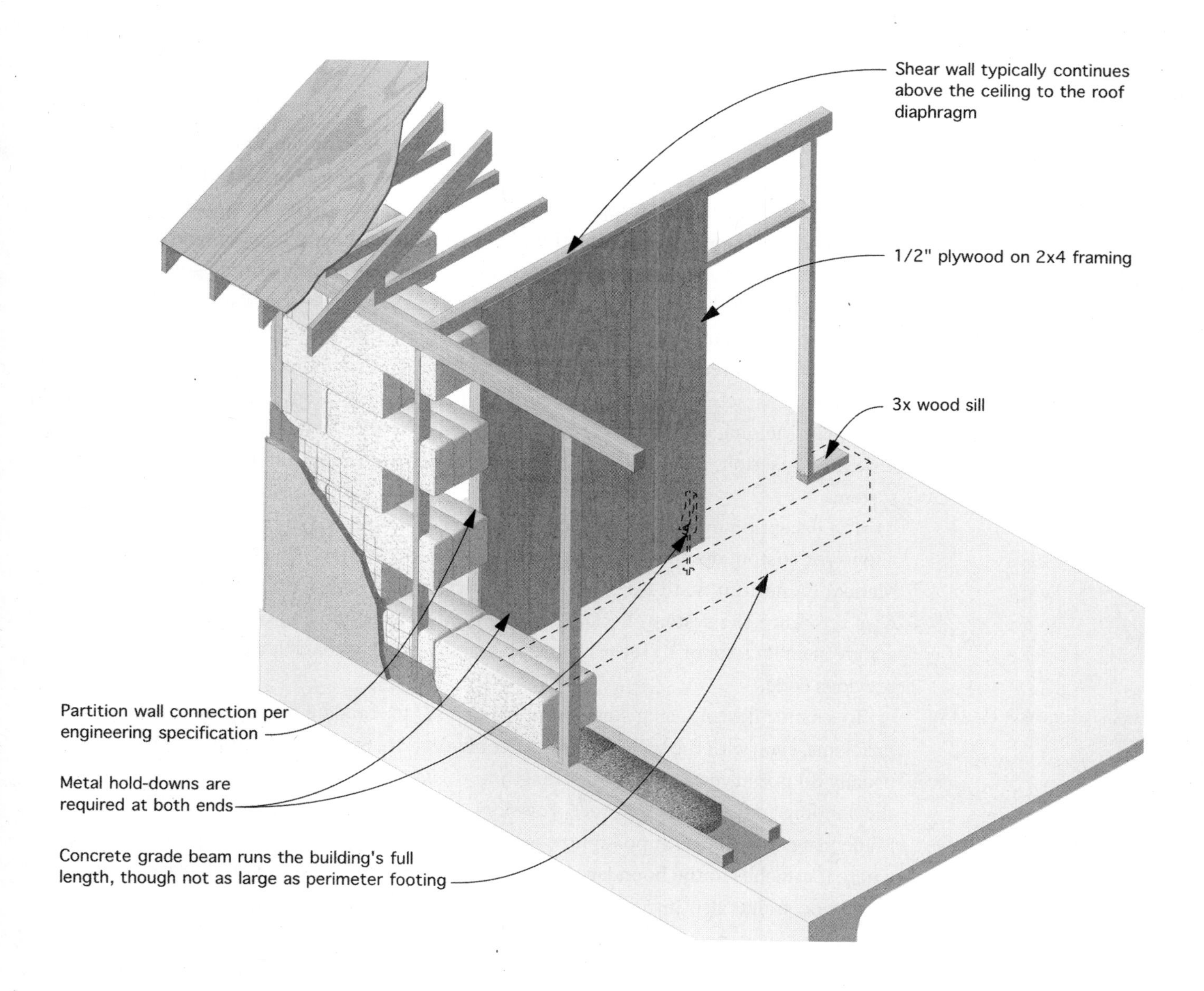

NOTES:
- Rule of thumb: for a one-story house ≥ 16 feet of shear wall is required in each direction. The shear walls should be fairly equally distributed to all sides of the building.

- Interior shear walls can generally be set in, up to 12' from the perimeter.

- A plaster finish over plywood requires proper surface and joint preparation to reduce cracks and ensure adhesion.

- Sheetrock over the plywood is also an option, but may require custom-made door jamb depths.

Interior Shear Walls

You can use interior walls to reduce the required strength or length of exterior straw bale shear walls. Interior shear walls are typically conventional wood-framed, but can be of any type—even a straw bale shear wall. Interior shear walls require a foundation even if the wall is not load-bearing. As with structural walls of any type, pay attention to proper wall connections at the top and base.

Steel Tension-Only Bracing

In tension-only bracing systems, metal strapping or threaded or unthreaded steel rod create a network of triangles that resist the lateral loads. The diagonal of these triangles—the steel strapping or rods—resists the lateral loads in tension only, and only in one in-plane direction. They connect at the top to a horizontal structural member or a vertical member that is in compression when the diagonal is in tension. At the bottom, they connect directly or indirectly to the foundation.

Once common in seismic design, steel tension-only bracing has fallen from favor in seismic areas due to complex connection design, undesirable energy-dissipation characteristics, over-strength load requirements, and construction costs. The yielding of a tension rod X-brace upon load reversal likely induces a slamming action, resulting in greater-than-anticipated forces applied to the system.

The exception in ASCE 7-16 Section 14.1.2.2.1, referenced in the 2018 IBC, allows for custom design of steel systems under AISC 360 – Steel Construction Manual (American Institute of Steel Construction), though it does not otherwise reference the system. Because of this requirement, tension-only systems are given an R factor of 1 (see earlier explanation) which is much lower than the previous code.

In seismic design, using steel strapping or rods for bracing presents several problems, even with the still-allowed "X" configuration. Steel strapping and rods usually do not have the required strength in this configuration when following the building code's structural steel provisions. Even if the steel strap or rod itself is adequate, it is almost impossible to provide adequate end-connection capacity where it attaches to the boundary elements. Straps or rods attached on the face of a wood frame also introduce load path eccentricity. Finally, the code indicates that this system may be used for light-framed walls—very different from straw bale walls. Therefore, we discourage using steel strapping or rods as a seismic bracing system for straw bale walls, though they can be a suitable and cost-effective solution for resisting wind loads.

Steel Moment Frames

Moment frames—steel frames without diagonal bracing—can be used where the diagonal member of a braced frame or wall panel would interfere with door or window openings.

3-25. Steel strap tension-only bracing.

This option can be a cost-effective system to resist wind loads; it has fallen from favor for use in seismic applications.

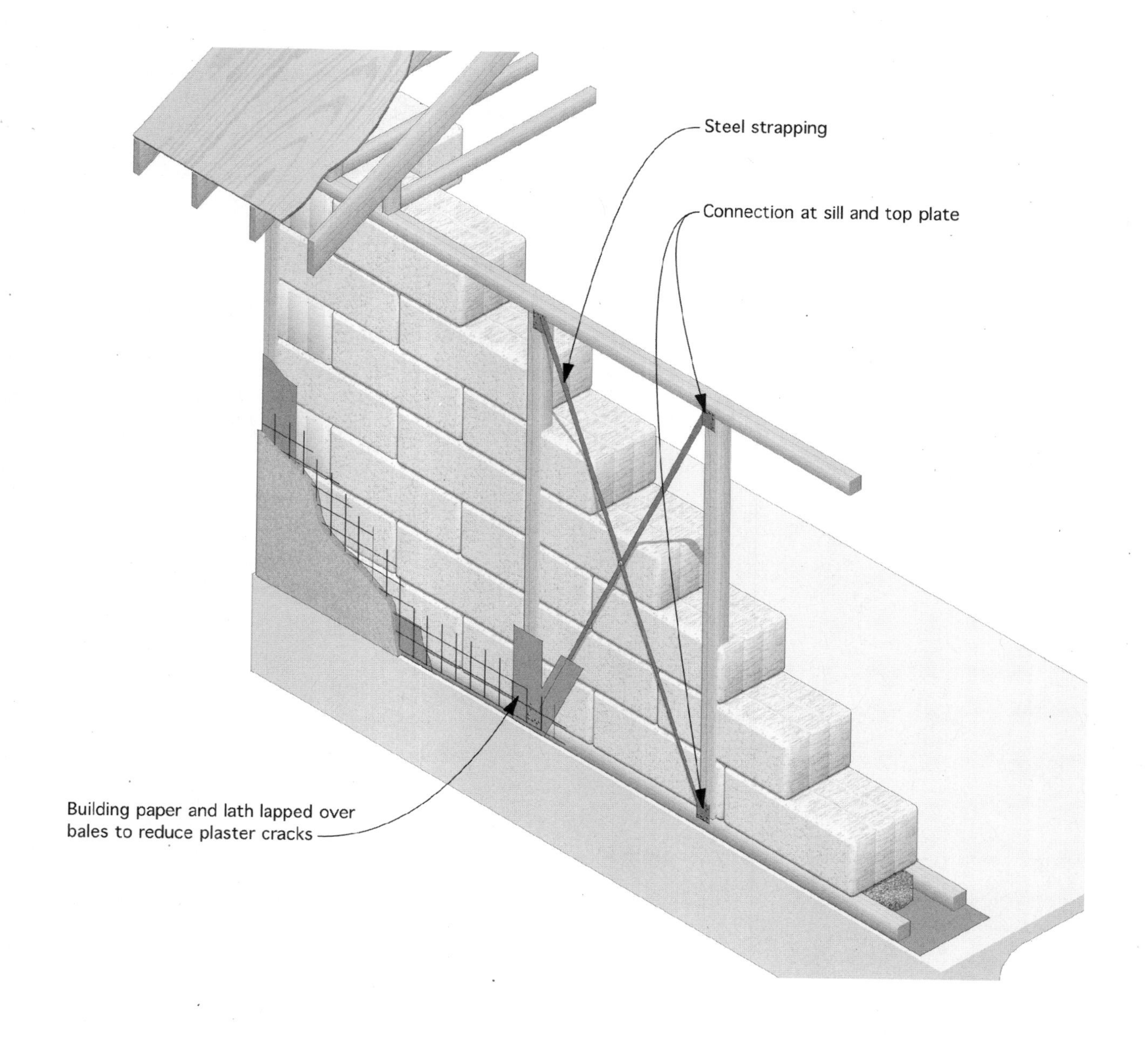

NOTES:

- This system does not perform well in high seismic areas prone to multiple seismic events due to cyclic loading.
- End connection capacity improved with custom-made plate steel brackets bolted to framing.
- Straps attached on the face of a wood frame introduce load path eccentricity.

3-26. Steel rod tension-only bracing.

Placing structural elements to the interior or exterior of the straw bale wall allows for straight or curved walls, eliminates the need for notching around posts, and facilitates plaster preparation. This option may not be suitable in seismic design.

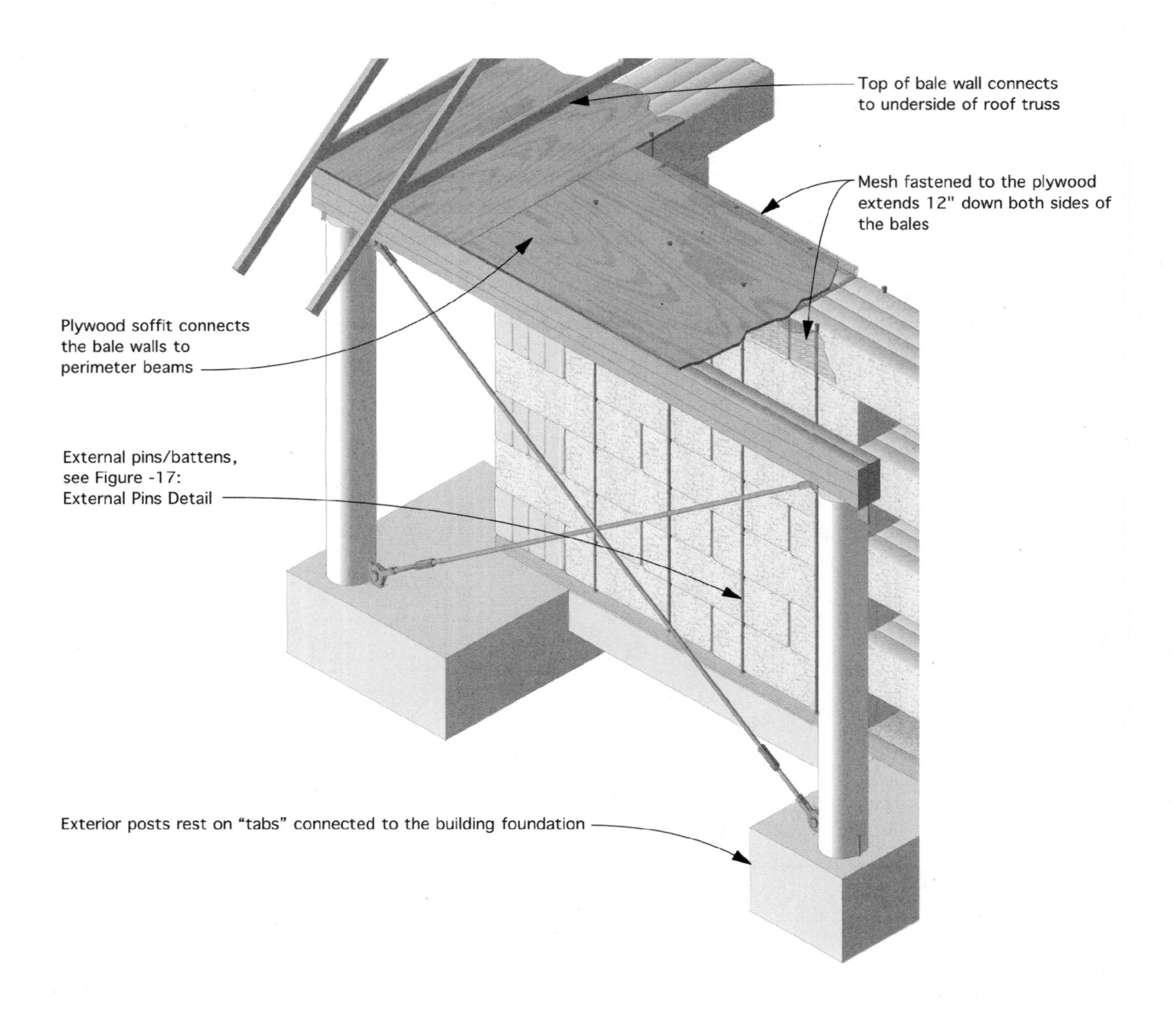

NOTES:
- In seismic areas, end-connection capacity must be carefully engineered; may be expensive to build.

- Well-placed exterior bracing helps define a porch or patio's outside edge, directs pedestrian movement near the building, and doesn't interfere with window and door function.

3-27. Moment frame.

Moment frame welded steel tubing can be fabricated in any size to fit nearly any wall location.

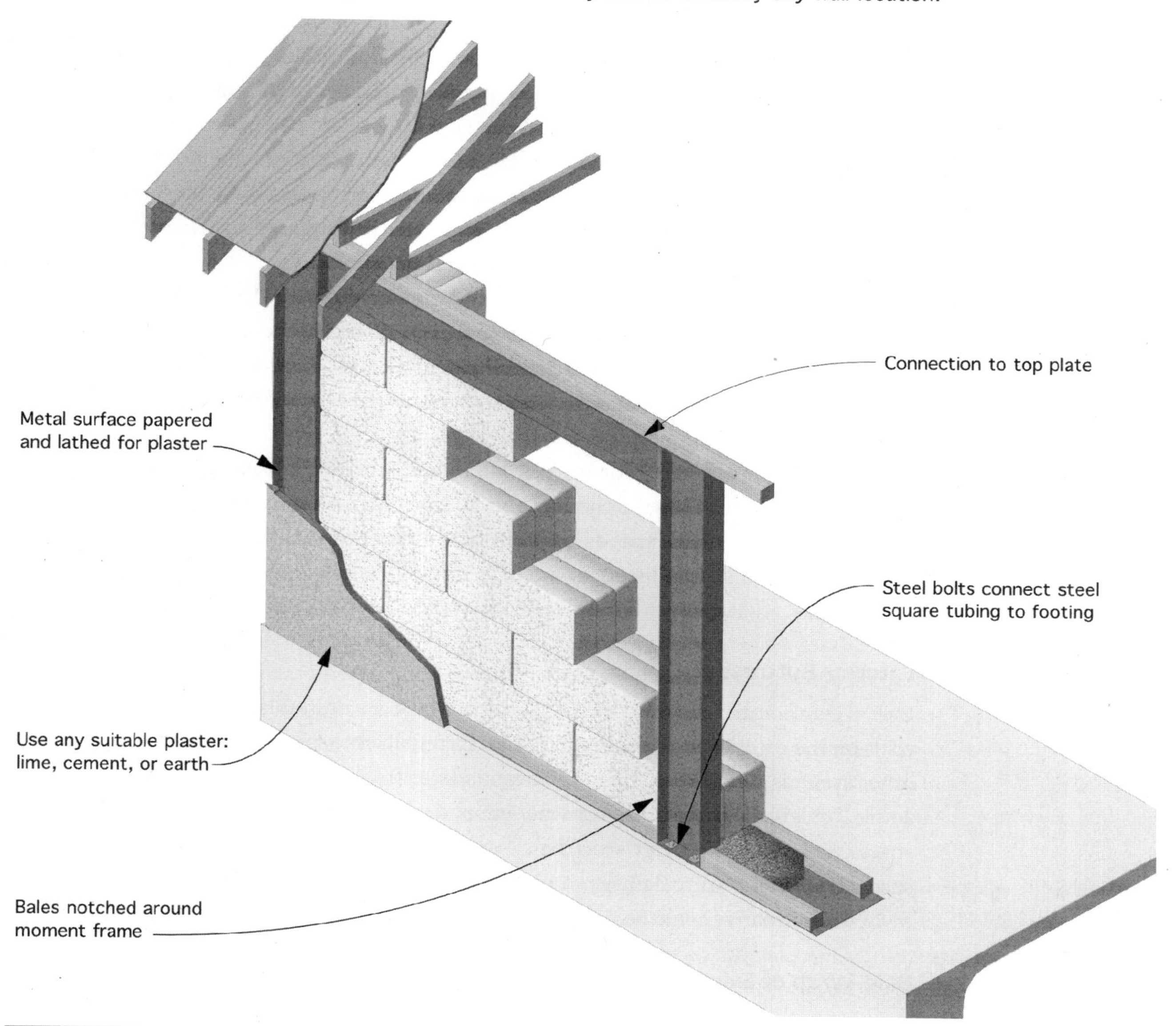

NOTES:
- Must be designed by a qualified engineer and fabricated by a certified welder, or made in a certified steel shop.

- Architectural flexibility—doors and windows can be located within the moment frame itself.

- A viable solution for multistory structures

- Not an inexpensive option but can work when other options aren't feasible.

- Address condensation issues in climates with cold winters.

- Rarely used in straw bale construction due to cost and complexity. They require larger footings and additional reinforcement.

ASCE 7 limits the dead load for the roof and floor, and—most problematically for straw bale walls—it limits the wall dead load to 20 PSF for use of "ordinary" and "intermediate" moment frames. Straw bale walls weigh between 45 and 60 PSF. However, the ASCE 7 limit applies only to the highest seismic design categories D, E, and F. Most of North America is categories A, B, and C.

The terms *ordinary* and *intermediate* are classifications of moment frame design. "Special" moment frames can be used, but they are more complex and difficult to design in a wood-framed straw bale structure. Also, moment frames concentrate the lateral forces at the foundation, requiring larger footings and more reinforcement. Therefore, moment frames are rarely used in straw bale structures.

Simpson Strong-Tie offers a prefabricated moment frame called a "Strong-Frame." This arguably puts moment frames, even those that require "special" detailing, more in reach for conventional residential construction. It remains to be seen whether this will result in more moment frames being used in combination with the insulating value of straw bale walls.

Consider thermal behavior in extreme temperatures when using moment frames. A large steel mass is a thermal conduit for heat loss and a potential condensation source, and it expands and contracts differently than the surrounding plaster and wood. Carefully detail moment frames to reduce condensation and resist cracks.

Exterior Buttresses

Use an exterior masonry or concrete buttress where it's desirable to locate shear walls on the exterior of the bale walls. This system likely uses more cement than other systems, though it offers the advantages of freeing up wall space, for example, for a bank of south-facing windows, and it helps define or divide outdoor space. We highly discourage straw bale exterior buttresses because they are very susceptible to moisture damage. A more environmentally friendly but weather-resistant alternative could be a cement- or lime-stabilized rammed earth buttress.

Bale-Wrap of Existing or Prefabricated Metal Buildings

Commonly called a *bale-wrap*, an existing structure can be wrapped in bales if it has sufficient strength to support not only itself but the additional straw bale wall load. For example, a premanufactured metal building that includes a lateral bracing system can be purchased and erected to support the roof, with straw bale walls wrapped around the structure. "Free-formed" straw bale walls under the rectangular roof can make the structure interesting, but simple shapes and rooflines use a metal building's efficiency to best advantage.

Be sure to address the following concerns with the metal building manufacturer, any of which may raise the building cost.

1. Most moment frames used in metal buildings are classified as *ordinary* or *intermediate*. These moment frames have limits in seismic design categories D,

3-28. Exterior buttress.

Exterior buttress shear walls lie outside of the straw bale wall. This allows for more flexibility with window and door placement, plaster selection, and can create “positive” outdoor spaces, courtyards, or sitting nooks.

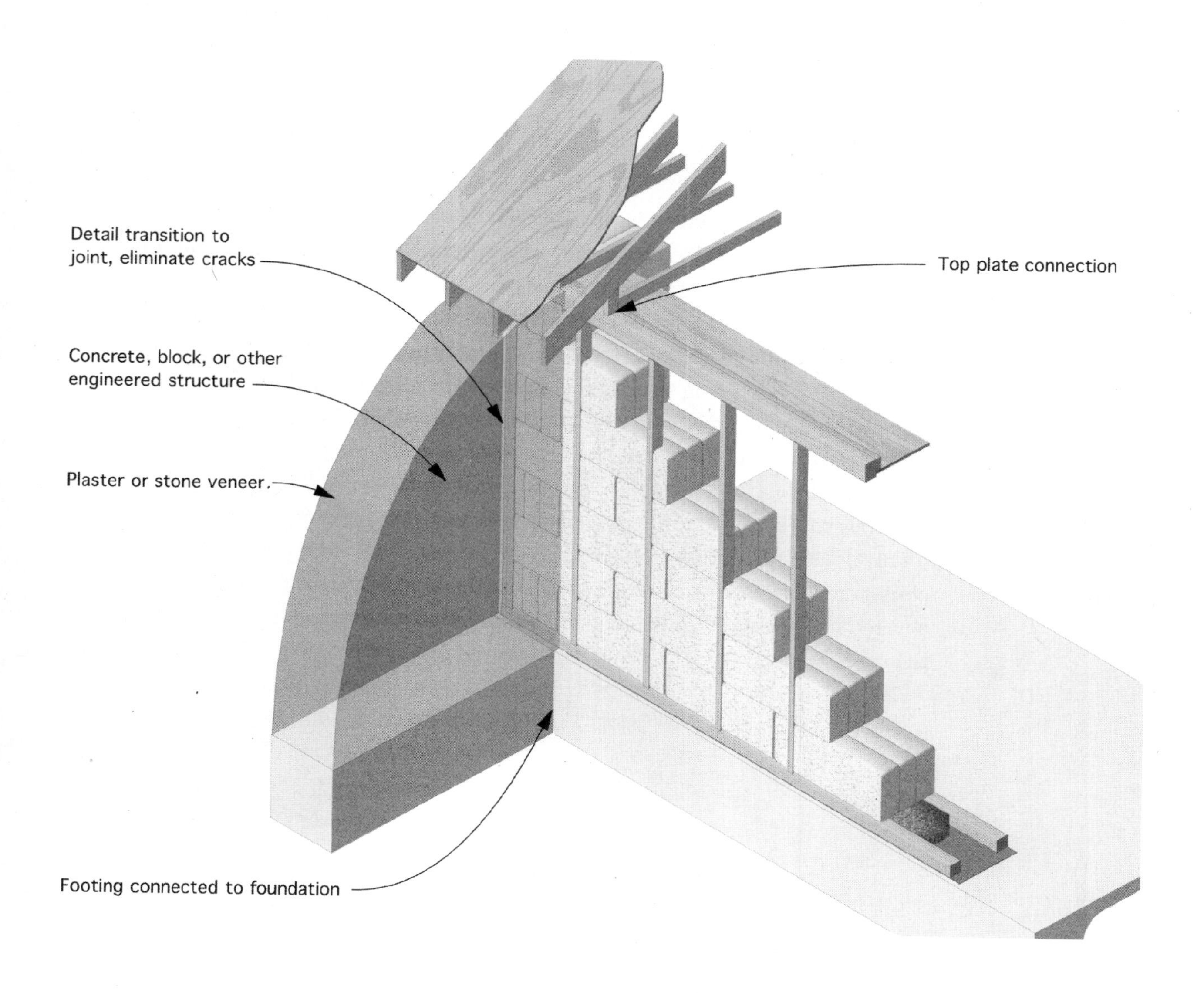

NOTES:

- Carefully detail the buttress-straw bale wall joint to prevent moisture intrusion, especially at the top.
- This is not a common solution, but it has been used to good effect.

E, and F in conjunction with walls as heavy as straw bale walls. See the above note on moment frames.

2. Metal buildings are typically designed to lateral drift limits that do not work well with "typical" house interior finishes. You may need to specify stricter drift limits.
3. The bale wall weight must be included in the foundation design and the metal building's lateral system design.
4. Clarify with the metal building supplier which design professional is responsible for the entire structure. The metal building supplier will typically certify only the metal building portion, not the foundation.
5. The roof overhangs and foundation width must be sized to accommodate the bale-wrap walls.

Conclusion

There are many engineering design solutions for straw bale buildings. Buildings can be load-bearing or post-and-beam and can employ straw bale shear walls or conventional shear walls or lateral bracing elements.

IRC Appendix S has placed straw bale construction closer to the mainstream of construction by providing guidelines applicable throughout the world. It simplifies one-story building structural design and detailing for engineers and non-engineers alike. It allows plan checkers and code officials unfamiliar with straw bale construction to evaluate a design and defer to internationally recognized standards.

Straw bale construction is simply another wall system that can be used structurally or as infill. With Appendix S and the International Code Council in your corner, and the assistance of a qualified engineer or architect for designs not covered in Appendix S, designing and building with straw bales is easier today than ever before.

4

Electrical, Plumbing, Ducts, and Flues in Straw Bale Walls

Electrical Planning

Installing electrical wires, cables, and fixtures in a bale building is similar to wiring a conventional structure: wires and cables need to be routed through the building, and electrical boxes and fixtures secured to walls and ceilings. Many straw bale buildings use conventionally framed interior walls and ceilings so only the exterior straw bale walls present unfamiliar work sequence and techniques.

Electrical work is a highly skilled job that comes with continually changing codes and a high degree of risk, and most electricians prefer to do all of the electrical work, from rough to finish. Yet some tasks, such as running cables between bale courses and installing outlet and switch boxes in the bale walls, may need to be done during the bale raising—when the electrician may not be available. Involve the electrician early in planning and decide which tasks must be handled by the electrician and which can be handled by the builder or homeowner. Develop a work plan that clarifies timelines and the responsibilities of the owner, the general contractor, electrician, and other trades.

As with any building, identify the service size (in amps), how utility power reaches the site, and the meter and panel locations. For some projects, solar or wind power or a backup generator must be considered. Develop a circuit plan and decide on wiring, conduit, box types and sizes, and locations. Consider lighting needs, code-required circuits for appliances or special equipment, floor outlets, switch locations, HVAC and water-heating requirements, communication wires (phone, internet, cable TV), and smoke detectors.

Review the structural system through and around which the wires must pass in bale walls, paying particular attention to the top and bottom of the wall. Consider whether running the wiring through the bale walls or under the slab (or in raised floor framing) will be more efficient, including the impact on the circuit layout and work sequence. Identify appropriate routes to run wires (e.g., along posts, along sill plates, between bales, etc.).

Wires? or Cables?

An electrical wire is an individual conductor, usually run with other wires in conduit. Cables contain two or more insulated wires wrapped in protective jacket—they are usually not run in conduit. When electricians "wire" a house, they most often use electric cables. In this chapter, we're using *wire* to describe both wires and cables, but you should understand the difference.

Attaching the Meter, Panel, and Subpanels

When choosing the meter and main panel location, consider meter reader access, how the meter connects to the grid source (overhead or underground), the meter's connection to the inside subpanel, and visual impact. If the site will include a grid-tied photovoltaic (PV) system, how will it connect to the meter? If the structure will be off-grid PV, where will the batteries and inverter be located, and how will the converted AC power run to the house's main panel?

Ideally, locate the meter and main panel on an exterior stud wall of a garage or outbuilding, or a power pole or post near an access road in an inconspicuous location. Locating the meter panel and any subpanels on a stud wall has advantages, including familiar installation for electricians, panels designed for gypsum board finish, and open stud cavities for wire runs.

If you must attach the meter panel to a weather-exposed straw bale wall, detail it properly using one of the following methods. If it is a load-bearing or shear wall, first check with an engineer or architect.

Surface Mounted: Fasten the meter panel to a piece of plywood supported by two studs running from the exterior sill plate to the beam or box beam. Install the studs before stacking the bales and notch the bales around them, or stack the bales and let the studs into them. Plaster between the studs where the panel will be located, allow the plaster to dry/cure, then attach the plywood. Fasten the meter panel to the plywood. If the plywood will remain exposed, paint or seal it before installing the panel. Some builders skip the plywood and attach the panel directly to the studs. Other builders plaster between the studs, attach plywood to the studs, then mount the panel.

Recessed: Create a 2×4 and plywood box that will hold the meter panel and "float" in a recess in the bale wall. After the bale raising, carve a recess into the bales somewhat larger and deeper than the box and carve channels for wires or conduit that will connect to the box. Plaster the recess, and when it has dried/cured, insert the box, securing it through the wall using threaded rod in four corners (see Figure 2-31: Attaching/Hanging Objects from Straw Bale Walls). Mount the meter panel in the box. Cable or conduit will enter/exit through holes drilled in the 2×4 frame and channels in the straw bale wall.

With both methods, prepare the installation for exterior plastering with appropriate flashing, waterproofing, and lath. Mount PV system elements (inverter, charge controller, and power center) in the same fashion.

Confirm the correct size and location and complete the meter or subpanel prep work before the electrician arrives. The electrician will likely handle the meter installation and prepare for the grid or off-grid service connection. In many areas, a generator or a temporary power hookup may be needed for the construction. Final meter connection or power is usually not allowed until after the final inspection.

Alternatives to Electrical in Bale Walls

Because it can take more time and materials to run wires in bale walls than in studs, first consider and try to maximize alternatives that use conventional installation methods:

- Through wood framing:
  - interior stud walls
  - framed ceiling/roof assemblies
  - floor joists (for designs with raised floors)
- In conduit under the slab
- Along the bale wall base trim, or between the sill plates (some builders use glued PVC conduit in this potentially damp location)
- Behind cabinets or in cabinet bases

If the building has a slab-on-grade floor, it can be more cost-effective to run conduit under the slab/floor, which also offers greater schedule control. This is particularly effective for home runs and for getting around the building when an open ceiling makes running wires above the walls difficult, or when wires must run around numerous doorways. Discuss how the conduit will rise in the interior walls (in the center, away from plumbing) and through the bale wall sill plates (generally at the interior face). Use either PVC or galvanized conduit under slabs. Review the bale and interior sill plate dimensions and the plaster thickness with the electrician to ensure proper riser locations.

To keep electrical wires well into the wall where they can't easily be damaged, conduit stubs should rise about ½" back from the face of the interior plate. Electricians usually stub up 3"–6". The bale-stacking crew will later notch bales around the conduit stubs. Conduit cable then runs to the first appropriate electrical box, and conduit or cables continue from there. Mark the floor or face of the sill plate near conduit risers so they can be more easily located while stacking bales.

Wires and Boxes in Bale Walls: Inspections

Straw bale walls present an inspection sequencing challenge. In conventional wood-frame buildings, the framing inspection is usually simultaneous with the rough plumbing and electrical inspections. That is difficult for post-and-beam straw bale buildings because electrical boxes and wiring and plumbing pipes are often fastened to, channeled in, or run between bales. However, the bales can't be stacked until the structure has passed a framing inspection because the bales would conceal the framing—as well as any electrical or plumbing that has been secured to it. Discuss this issue with the building inspector because some components may need to be inspected out of their familiar order.

Builders handle this sequencing challenge in different ways:

Pre-stack inspection: This option works well for post-and-beam systems. Secure as many outlet and switch boxes and as much electrical cable as possible to the framing or run wire in conduit; attach any water lines and vents to framing before the bales are stacked. After passing inspection, stack the bales around the electrical and plumbing, which, with thoughtful planning and coordination, will not be difficult. A bale crew showing up for the bale raising might find electrical boxes secured to interior sill plates or posts, or supported by rigid or flexible conduit mid-wall. If some boxes or wiring can only be installed during or after stacking bales, then install and have them inspected as described below.

Mid-Stack Inspection: Call for a partial rough electrical inspection to look at everything the bales will cover. Afterward, stack the bales to the point that additional cables or conduit can be run to outlet and switch boxes. Place horizontal runs that will connect outlet boxes on the first bale course and install the boxes. Screw electrical boxes to ½" plywood plates placed horizontally or vertically between the bales, which are notched to receive the box. Position the boxes and secure the plywood plates with long screws driven into adjacent bales. After pulling wire into each box, pack straw-clay around and behind each one. This secures the box against air infiltration and provides a fire-resistant barrier. Alternatively, wrap the box with a self-stick membrane, allowing for knock-outs (partially stamped opening) that will be used. Trim any stray straw poking into the box. Also install any plumbing lines or vents passing through the straw bale walls at this time. Then call to have the inspection completed. The bale stacking resumes after passing inspection. This method inconveniently delays the bale stacking for several days.

Post-Stack Inspection: A third option is to install electrical and plumbing after the bales are stacked, then call for the rough electrical and plumbing inspection. Tuck electrical wires into the gap between bale courses or a channel cut into the bales, and secure boxes to *vampire stakes* (see below and Figure 4-1) driven into notches that have been carved into the bale for each box. Secure wires with galvanized landscape staples or use a loop of baling twine to "sew" them to the exterior bale surface.

Vampire stakes are made from 1× or 2× wood scraps about 12" long and tapered to a point on one end. Barbing the stake with nails or using a thicker stake can increase holding power. The receptacle box can be screwed to the stake's exposed blunt end either before or after the stake is driven. As described earlier, seal the box with self-stick membrane or fill around the box with straw-clay.

Installing and inspecting electrical after stacking bales has some drawbacks:

- The straw bales and loose straw may obscure the cables and conduit. Coil cables and secure them to the bale wall with a landscape staple, and mark conduit locations.
- Carving channels into the bales risks cutting the bale strings. This may not be a problem mid-wall where adjacent bales maintain bale density, but could be a problem nearer windows and doors where the bale may expand and deform the reveal.
- Vampire stakes may twist as they are driven into the wall, making it difficult to mount outlet and switch boxes plumb and flat in the wall plane.

- Inspectors may be unable to see wires tucked too deeply into the bales. Keep them visible.

Many projects mix techniques and methods described in these installation and inspection sequence options, but all have common features:

- Place wires far enough into the wall that they can't be nicked by someone driving a nail to hang a picture. One inch from the bale surface is usually sufficient.
- When laying the bales flat, notching for the boxes and the conduit will not affect the strings. If the bales are laid on-edge, take care to avoid cutting the strings.
- Determine the finish plaster depth, use a level to plumb up from the bottom plate, and set the boxes plumb and at the correct depth. The boxes (sometimes with plaster rings) will act as screeds for the finish plaster. Use a reference jig to maintain consistent outlet or switch box heights.
- Carve recesses for electrical boxes and channels for wires and conduit with a chainsaw, an angle grinder with a Lancelot blade, a hay knife, or even a hammer's claw (see Chapter 5, Stacking Straw Bale Walls—Tools). The first two are power tools and should be used with caution and only by those with experience. Niche and channel depth depend on the box size and finish plaster thickness. For boxes attached to the sill plate, run wires under the first course of bales. Theoretically, this area could become damp, so consider glued PVC conduit.

Material Choices

When choosing materials—NM or UF cable, plastic or metal boxes, rigid or flexible conduit—consider installation ease, cost, flexibility, and durability. Understand the code requirements and limitations of these materials for your project and code jurisdiction. Check with your local building department.

Electrical Cable

Ordinary non-metallic sheathed cable (NM) is usually allowed in bale walls. Appendix S is silent on this and all electrical issues, but the commentary under AS101 mentions the historical use of NM cable without conduit, and might be helpful if a building official insists on conduit in a straw bale wall. Run NM cable between the bales or in notches cut into the surface of the bales. Take care to avoid nicking the wire during mesh installation and plastering. NM cables are susceptible to cuts by staples, screws, nails, bale needles used while through-tying, *Robert pins*, galvanized landscape staples securing mesh, and saw blades used to notch bales. Wire damage discovered after the wall is plastered is difficult to find and repair, and a shorted wire can cause a fire. To help prevent this problem, mark the wire paths on the bale walls with fluorescent spray paint. Then carefully work around those areas while through-tying, securing mesh and lath, or cutting into the bales.

4-1. Electrical boxes.

Attach boxes to wood elements in the wall wherever convenient. Absent that, the boxes require an anchor as the bale faces are not dense enough to brace the boxes against movement.

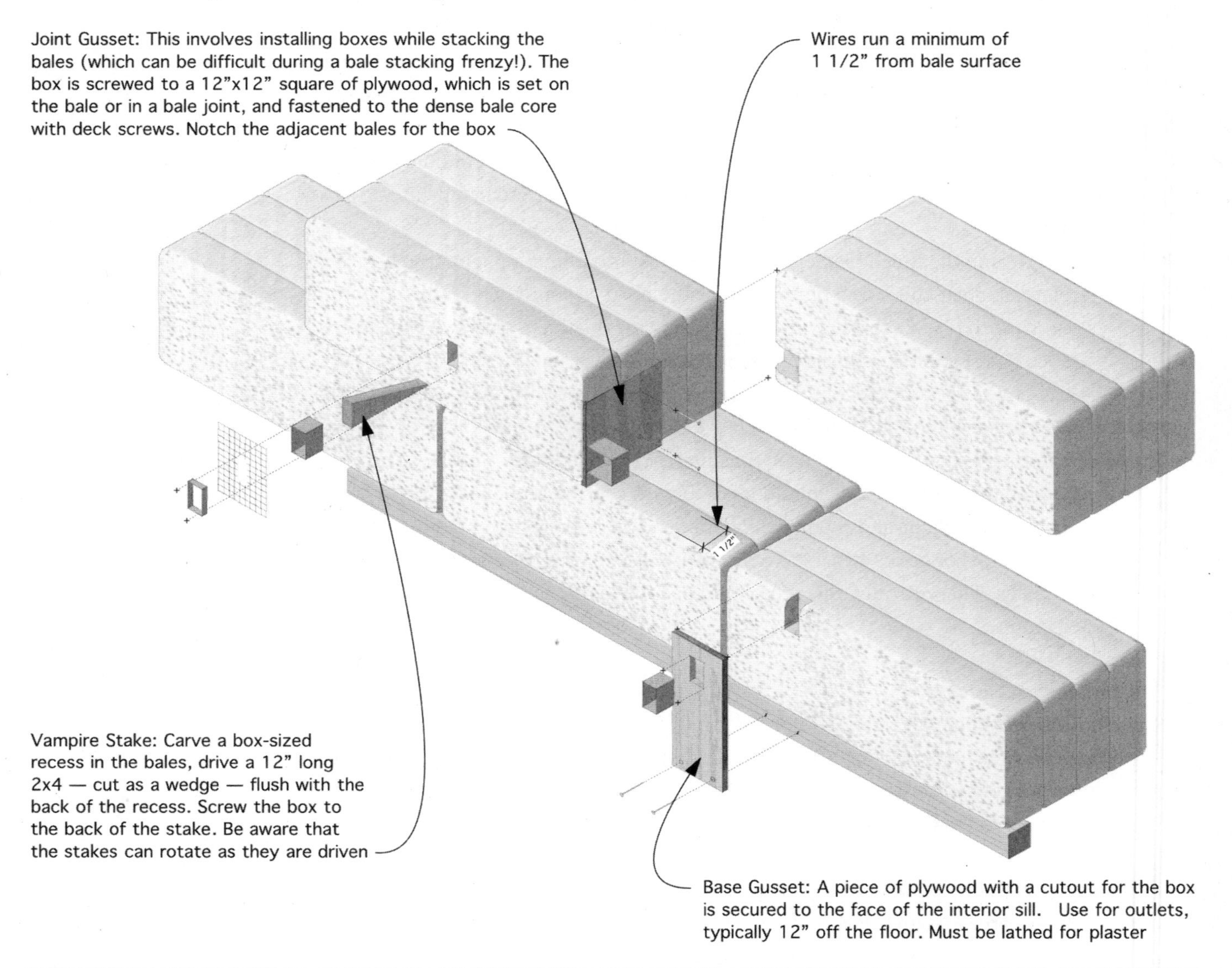

NOTES:

- With all these techniques, remember that the face of the box should project the thickness of the plaster from the bale surface, that any cavities around the box should be filled with straw-clay, and that a lath collar around the box assists plastering.

- Electrical Code may allow running non-metallic cable (NM) or require UF (underground feed) as it is more resistant to damage from accidental nicks and deterioration with lime-cement plasters.

- Using conduit facilitates future electrical upgrades, and may facilitate a complete framing inspection prior to the bale stack.

- Where the finish plaster depth is unknown, consider using adjustable depth boxes or boxes that use mud ring covers, available in several depths.

Some building codes don't allow NM cable to contact plaster because moisture might degrade the cable sheathing. In that case, use cable rated for underground feed (UF). Although more resistant to moisture and inadvertent nicks, UF cable's thicker sheathing takes longer to strip.

Before installing any mesh, electricians should test all circuits by powering the panel or checking with a continuity tester. Test again before plastering to find and correct any shorts.

Plastic or Metal Boxes

Where cost overrides other factors, choose plastic boxes. They work with NM or UF wire but are usually incompatible with conduit. Some plastic boxes may not be robust enough to withstand repeated bashings during bale stacking and plastering—you might need to replace them mid-project. Use the most robust plastic boxes available and make certain they can be mounted as intended: some are designed for nailing to a stud and may not work with mounting methods used for straw bale walls. NEC-rated plastic box extenders are available to compensate for incorrect depth placement. Also, consider adjustable-depth plastic boxes.

Metal boxes offer more secure attachment to bale walls, better endure the abuse of bale stacking and plastering, and can be used with NM, UF, armored cable, or conduit. Only slightly more expensive than plastic boxes, they also offer greater flexibility. Where you need a single-gang box, consider installing a double-gang box (4-square) and a plaster ring that reduces the box opening back to single-gang size. These boxes offer more room for wires and devices, and more easily attach using methods described earlier.

Metal boxes can, like plastic boxes, be set to the plane of finish plaster, or they can be recessed and use metal plaster rings—available in different depths—to form the box opening at the finish plaster surface. This gives greater useable depth to the box and allows potential adjustment if needed. Assume a plaster ring depth and set the metal box back from the plane of finish plaster by the same amount. For example, if you use a ½" plaster ring, set the face of the box ½" in from the finished plaster. Some plastic box extenders can also be used with metal boxes.

Conduit

Conduit allows for future flexibility for changing or adding wires. Exposed and accessible wiring runs require conduit, for example, in a garage or utility room. Use EMT (rigid metal conduit), AC (armored cable, flexible metal conduit), gray rigid PVC, or plastic flex tubing. Run conduit between the bales or in channels cut into the surface of the bales. Plastic flex conduit costs less but punctures more easily. Flexible metal conduit combines flexibility and strength—at a higher price. Gray PVC works well in wet areas, though limited fitting options make it less flexible in tight turns. Consult with an electrician or the building department if NM cable will be run through conduit. A conduit that is too small makes pulling NM cable through it difficult, and might cause the cable to overheat.

4-2. Cold (wet) sleeve in straw bale walls.

Conventional walls often serve to bring water lines into a kitchen or bath area, but keeping plumbing out of bale walls avoids potential damage to the bales from leaks and condensation. Where plumbing through bale walls is unavoidable, careful detailing guards against such possibilities. Cold water lines in particular have the potential to condense enough water on their exterior to cause damage — especially in humid contexts. Running pipes through a properly detailed sleeve in bale walls directs potential leaks or condensation outside.

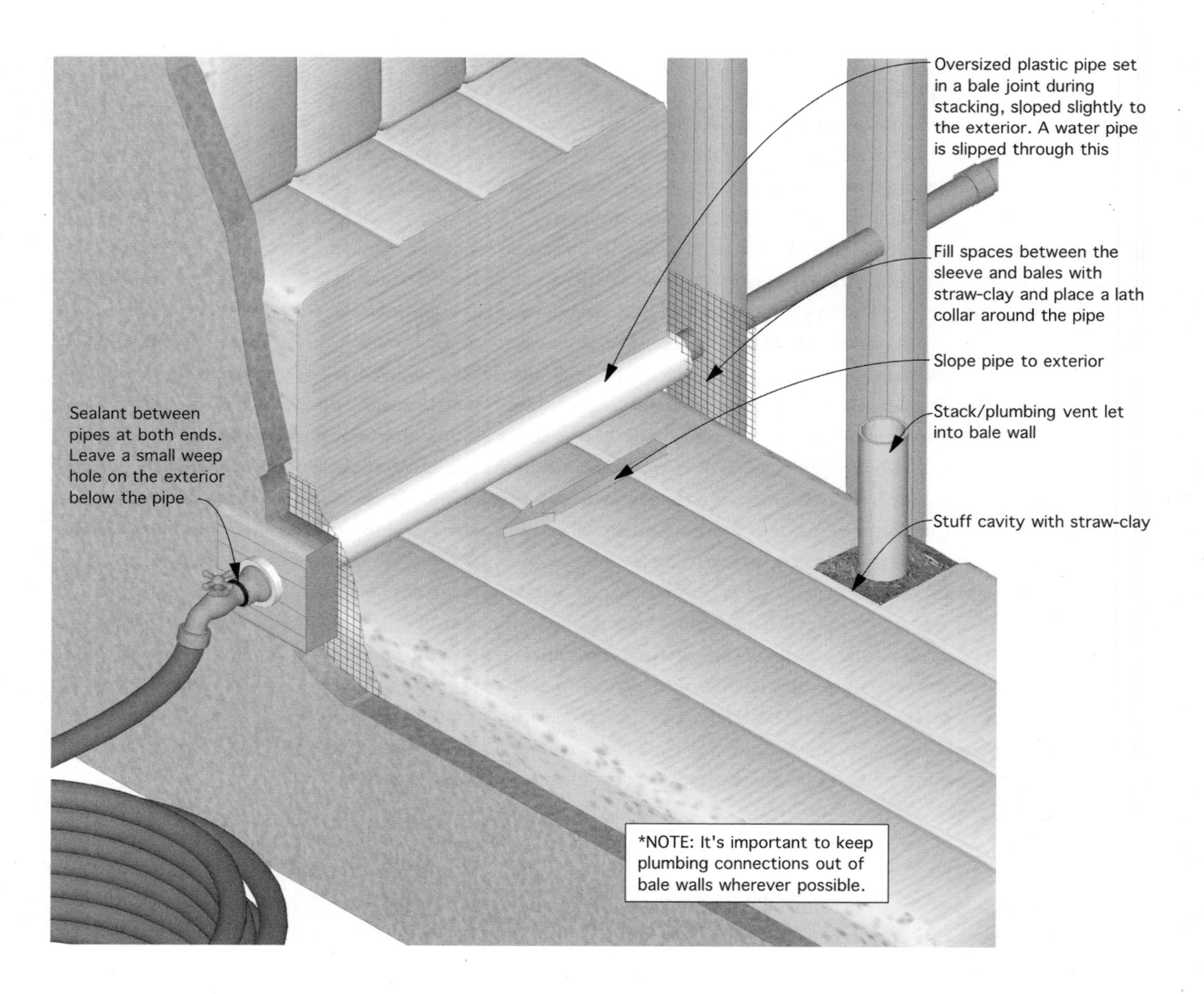

NOTES:
- Option not shown:
  - Plywood chase to house plumbing passing through bales from the sill plate

Plumbing

Planning

Thoughtful design keeps plumbing out of straw bale walls altogether—routing pipes in wood-framed walls, cabinets, or under floors. But when plumbing must run in a straw bale wall, the same options for installation and inspection exist as they do for electrical: pre-, mid-, and post-stack, although load-bearing projects generally have mid- or post-stack inspections. Plumbing supply and waste and vent lines must be planned carefully to avoid complications during construction and installed correctly to avoid leaks.

Keeping supply lines out of straw bale walls reduces the likelihood of water entering the walls from condensation (especially from cold water pipes) or leaking. Waste and vent lines in a bale wall are much less likely to cause water damage, but it's often easier to run them in an interior framed wall than through a box beam often found atop a straw bale wall.

If a supply line must pass through a straw bale wall, sleeve it inside a larger ABS or PVC pipe that slopes and drains to the outside. Pipe insulation around cold water pipes reduces condensation. Where plumbing vents must rise through a straw bale wall, notch bales laid flat around vents. In on-edge wall systems, run vents along posts because the bales can't be notched.

Keep plumbing joints—the most common source of leaks—out of the walls if possible. Continuous and flexible PEX tubing is a good choice for supply lines in straw bale walls. Copper is a particularly poor choice because soldering joints involves open flame in the vicinity of exposed straw.

Kitchens present the most common plumbing challenge because the kitchen sink is often on an exterior wall and has a window above it. Instead of running supply and wastepipes up through the interior plate, consider routing them through the back of cabinets, or move cabinets about 3" from the wall to create a pipe chase—this offers the advantage of a deeper counter top, too. Vent pipes can also run through this chase, rising slightly until they turn vertically inside an adjoining interior framed wall. To avoid these complications, many straw bale structures simply employ a wood-framed exterior wall in the kitchen area. See Figure 2-29: Straw Bale Wall and Conventional Wall System Transitions Detail.

Bathrooms also present challenges, though it is easier to locate toilets, sinks, and showers on an interior framed wall. If you must locate a sink on a bale wall, as with kitchens, you can bring supply pipes through interior walls or cabinets. If locating a shower or tub adjacent to a straw bale wall, avoid running water lines through the bales, and don't aim a shower nozzle at a plastered bale wall unless a full rainscreen has been installed. See Figure 2-24: Rainscreen over Straw Bale Wall Detail. IRC Appendix S requires straw bale walls enclosing a shower or steam room to be protected by a Class I or Class II vapor retarder (≤1.0 Perm). Bathrooms also often employ a wood-framed exterior wall because of the large amount of moisture generated in bathrooms daily. This eliminates the concern of that water vapor being driven through and condensing in a bale wall.

4-3. Air vent in straw bale walls.

Ventilation and exhaust is important for occupant health in all construction; either actively or passively drawing in fresh air/exhausting stale air while resisting air/water intrusion. Check your local and state building code(s) to be sure you address sizing and any additional fire or condensation concerns.

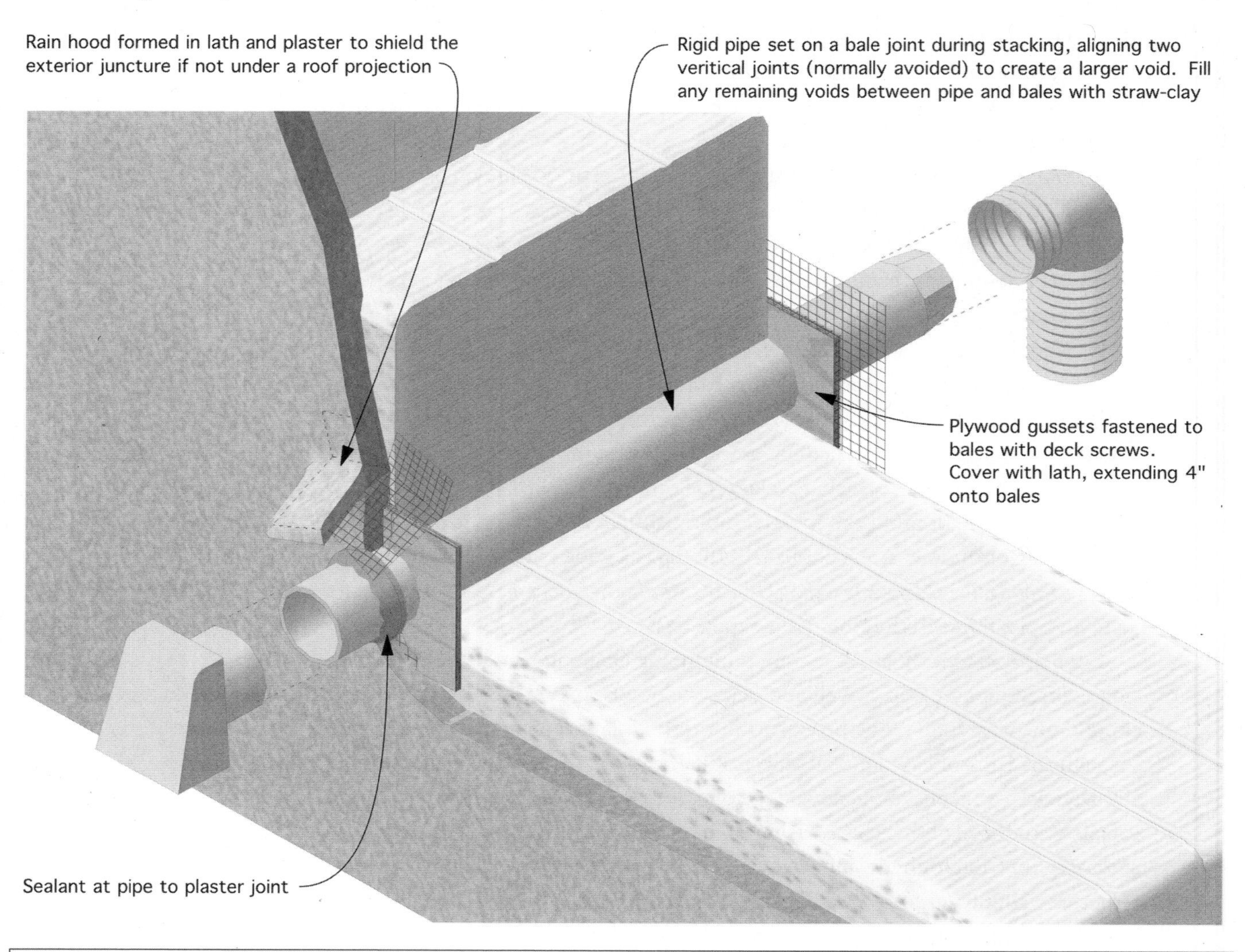

NOTES:
- Try to run outside make-up air vents for exhaust fans or wood-burning stoves through conventional walls.

Supply pipes for garden hose spigots are often run in straw bale walls and should be sloped to the exterior in a sleeve as previously described. In cold areas, use a freeze-proof faucet that locates the valve near the inside of the wall where it's less likely to freeze, burst, and cause damage. Some builders frame a small wood chase through the wall, sloped toward the outside, that provides room for the pipe and pipe insulation and a secure faucet attachment surface. Alternatives include locating hose spigots on stand-alone risers with insulation and freeze-proof faucets, or in below-ground boxes.

Gas lines can also condense water vapor; sleeve these inside a larger pipe where they pass through a straw bale wall.

Exhaust and Intake Ducts and Flues through Straw Bale Walls

Homes have a variety of ducts and flues that often pass through exterior walls from kitchen and bathroom ventilation fans, heating, ventilation, and air conditioning (HVAC) systems, and clothes dryers. Planning for these during design makes for smoother installation. When possible, run these ducts and flues through framed exterior walls. This facilitates installation by subcontractors who may be unavailable during the bale stacking or unfamiliar with straw bale wall construction.

Carefully detail ducts and flues that must run through straw bale walls. Sleeve dryer, intake air, and fan ducts with a large ABS pipe, and insulate with fiberglass or mineral wool to prevent condensation.

Straw bale buildings often require smaller HVAC systems than conventional buildings. In areas with mild climates or with designs employing passive heating and cooling, a simple heating system and operable windows may suffice. There are many choices in HVAC systems today, and all can be integrated in a straw bale structure. To avoid costly mistakes and delays, involve an HVAC contractor during your design phase to learn options and installation requirements for the system you choose.

Energy-efficient forced air systems route ductwork through conditioned spaces in soffits framed down from ceilings or under floors. Supply and return ducts and registers are much more easily located in interior stud walls than in a straw bale wall. Similarly, hydronic floor heat system manifolds are best located on stud walls.

Consider air-to-air, water source, and geothermal heat pumps as options. One type of heat pump—ductless mini-splits—requires a relatively small (approximately 3") hole through an exterior wall for a line that runs between the evaporator inside and the compressor outside. Locate the evaporator and compressor to avoid surface mounting conduit on the plastered straw bale wall, and insert a 3" sleeve in a bale joint during the wall stacking. The lines connecting the two units carry power, refrigerant, and condensate. Run it behind cabinets or through stud walls. If planning to mount a mini-split unit on a straw bale wall, be sure to embed a ledger during the bale stacking (see Figure 2-31: Attaching/Hanging

4-4. Hot sleeve in straw bale walls.

Sometimes stove pipes and other hot exhaust vents must pass through a straw bale wall; keep them away from combustible materials.

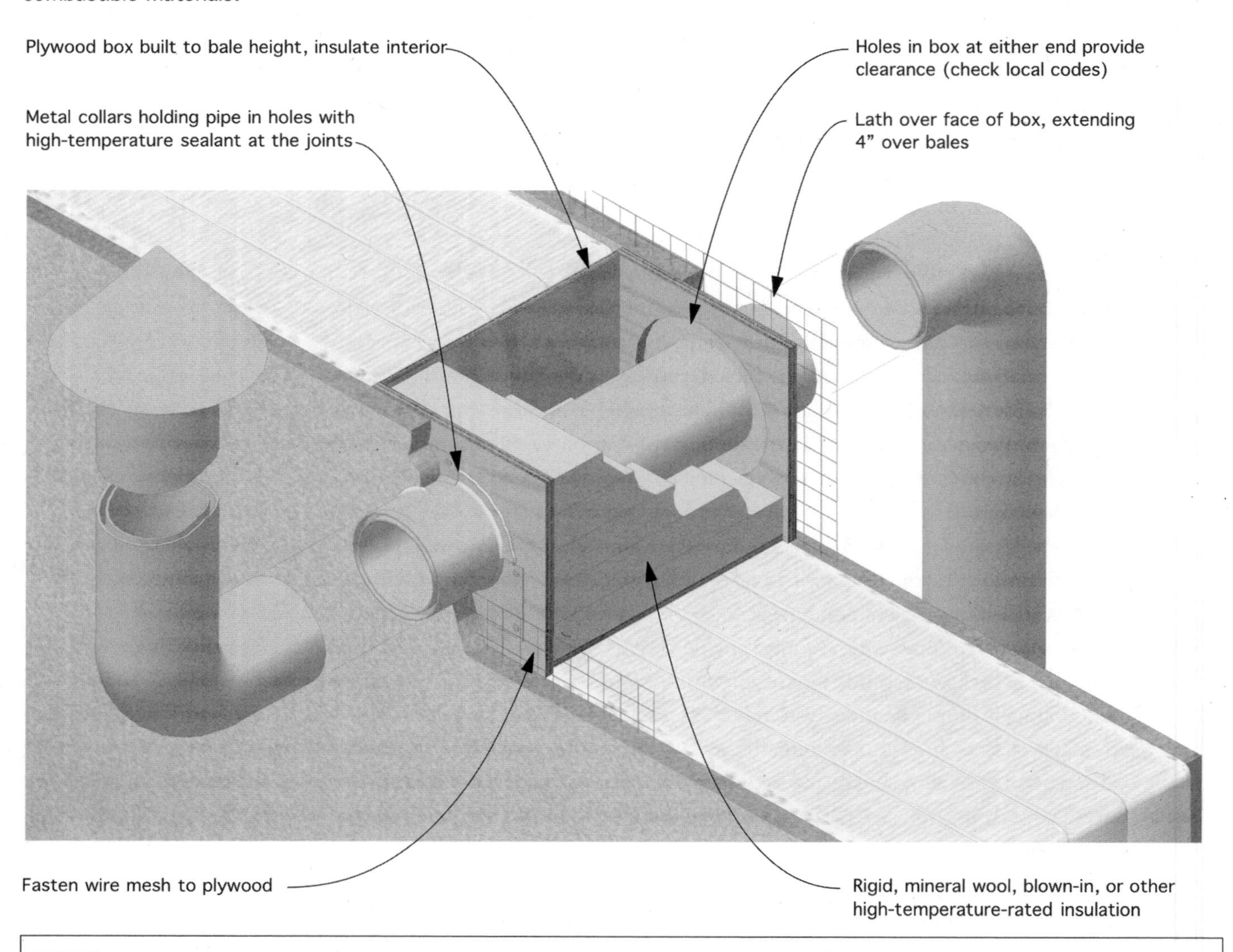

NOTES:

- A plastered straw bale wall assembly is fire-resistant, not fireproof. Follow wood-burning stove and stove exhaust pipe manufacturer guidelines for clearance from combustible materials.

Objects from Straw Bale Walls) to hold the unit's weight. Speak with the manufacturer or installation contractor to learn about heat pump system installation requirements.

Heat Recovery Ventilators (HRVs) and Energy Recovery Ventilators (ERVs) also require intake and exhaust ducts. If passing through a straw bale wall, handle them in much the same way: place an appropriately sized PVC pipe sleeve or chase in the wall during the bale stacking. Check the manufacturer's installation specifications for the correct size.

Flues and Fireplaces in a Straw Bale Wall

Woodstove, gas or wood manufactured fireplace, or traditional masonry fireplace installation in a straw bale building is similar to installation in a conventional building: observe required clearances to combustibles for the appliance or fireplace, and flue, and finish the wall with non-combustible materials. Manufacturers publish clearance specifications. Follow them. See also Appendix S: AS107.3 Clearance to fireplaces and chimneys.

As with conventional structures, it's generally much simpler to run a woodstove flue (stovepipe) or water heater flue up through the roof than out through a wall. If it must exit through a straw bale wall, build a non-combustible chase sized for manufacturer's recommended clearances. Most builders fabricate this chase from plywood faced with cement board. During the bale stacking, leave a space in the wall somewhat larger than the chase. Before inserting the chase, plaster all exposed straw bale surfaces and edges.

Direct-vent gas wall heaters do not have vents long enough for straw bale walls. Either relocate the heater to an exterior wood-framed wall or recess the straw wall area so the heater fits with the required clearances. The through-wall vent for an on-demand water heater may pose the same challenge. Consider choosing space- or water-heating units that vent through the roof.

Instead of setting a fireplace into a straw bale wall, consider framing around it with metal studs and non-combustible insulation. Where a fireplace must be set into or at a straw bale wall, follow all code requirements, including IRC Appendix S. Section AS107.3 requires that straw bale surfaces adjacent to fireplaces or chimneys receive a minimum ⅜" thick plaster, and that the required clearance to combustibles stated in the code or in manufacturer's instructions be maintained to the plaster surface. The construction sequence would be at least one plaster coat (two or three is better) over the straw bales, then the required air gap, then the firebox and chimney.

5

Stacking Straw Bale Walls

Straw bales are the defining part of a complete wall system that is both similar to and different from conventional construction. Those familiar with post-and-beam or stud framing systems may be unfamiliar with straw bales as significant structural elements, or as insulation, or how materials are sequenced and handled. Here we offer guidance about:

- obtaining and handling straw bales
- tools for a bale raising
- site preparation
- stacking bales
- shaping and placing bales to fit framing
- trimming straw bale walls

Obtaining and Handling Bales

Bale Take-Offs

Most builders use the building plans to "take-off," or estimate, the number of bales the project requires. Two methods described below—wall elevation counts and wall calculations—start with a known bale height and length. If the building has gable ends or a sloping wall made of straw bales, be sure to account for these areas. With both methods, it's wise to add 10% to allow for damaged bales, mistakes during the bale stack, stuffing voids with straw or straw-clay, chopped straw in the plasters, use as scaffolding, and unanticipated changes.

Using the wall elevation count method, draw scaled-to-size bales on side views (called *elevations*) of every bale wall, then count the bales, being careful to not double-count at corners. Bales are usually stacked in a running bond, with offset vertical joints at each course, and full bales are used as much as possible at corners, doors, and windows. This method previews where you'll need full and partial bales, which helps you develop a time estimate for stacking bales. Placing full bales takes just minutes, but shaping and placing customized partial bales takes much longer.

Using the wall calculation method, add up the total linear feet of bale walls, divide that number by the typical length of the bales you expect to use, and multiply by the number of bale courses. Determine the number of bale courses by subtracting sill plate height (and box beam if used) from the wall height, and divide by the "height" of the bale, depending on its orientation (laid-flat or on-edge). Subtract for windows and doors. However, if you know that windows and doors account for approximately 10% of your total wall space, ignoring them gives the recommended 10% extra.

Uses for Extra Straw Bales

Leftover bales can be used as jobsite scaffolding, steps, seats, workbenches, and more. Straw flakes and loose straw have many uses, too: mixed with clay slip to stuff gaps or to form curves and other sculpted surfaces, or packed dry into cavities, around windows, in gaps, and at rake wall (a wall with an angled top edge that extends to the underside of a sloped roof or ceiling) locations. Finally, loose straw—both long and chopped—can be used for clay plasters.

Best to order extra bales. If you end up short during the bale raising, your only option may be to purchase bales from a local feed store, which will likely be in different dimensions and/or not building quality—and they will probably cost more.

When the project is complete, use extra bales and loose straw for erosion control and mulch. You might be able to sell extra bales to other builders, local farmers, or ranchers, or you could donate them to a nearby school or public garden.

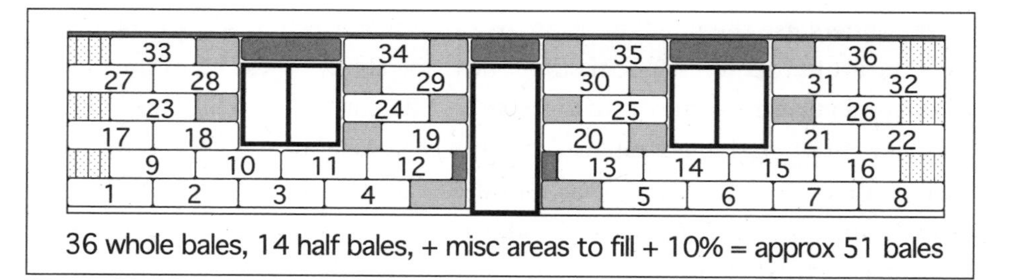

5-1. Elevation count method of estimating bales.

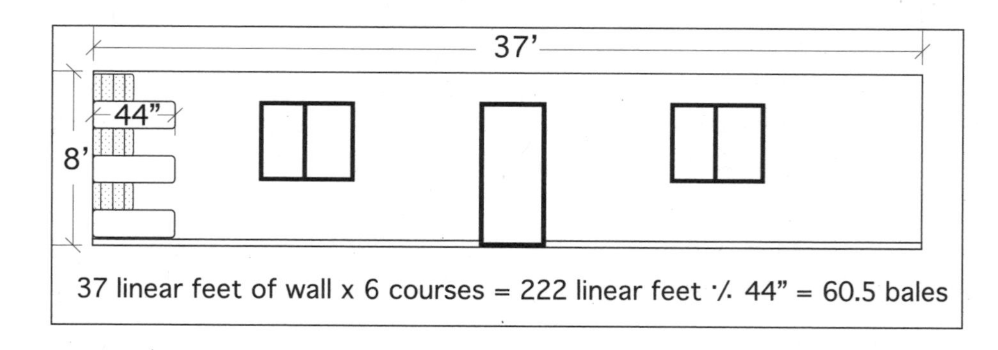

5-2. Linear footage method of estimating bales.

Sourcing Bales

Find out where other builders in your area have obtained straw bales, and if they were satisfied. Contact a natural building organization in your area. Inquire at local farmers markets, the county Department of Agriculture office, state Grain Growers Associations, feed stores, and farmers boards. A university Agricultural Extension office may know of grain growers in your area. Be sure to communicate the quality of bales you need, in terms of moisture content and density.

Ordering Bales

Chapter 1 discussed the time of year straw is baled—be sure to order bales well before you need them. Learn from your bale supplier when you need to place an order. Some will have bales stored from last year's harvest and can support a number of projects from their inventory. Other suppliers can only deliver bales directly from the field, shortly after harvest. Knowing bale availability helps with construction planning. Some builders or owners reserve bales as soon as the design is completed. Others when they have permits and funding. Some suppliers may require a deposit or full purchase to reserve bales, whereas others won't. Find out who will arrange for delivery—you or the supplier. Receiving bales immediately prior to the bale raising is ideal—they can often be unloaded and staged in convenient locations around the building without concern for long-term dry storage.

Load Sizes

Because bale suppliers often arrange for bale transportation, talk with them about delivery truck size. You will need to know both how many bales the truck can carry, and if the truck can deliver them to your building site. In many areas where hay and grain crops are grown, a "squeeze"—a specialized forklift—loads blocks of 64 layered bales (8 layers of 8 bales) onto a truck. Bales stacked in blocks are more stable during transport. Specialized hay trucks and trailers are designed with deck heights at or under 4'; a block of 15" tall three-string bales sitting on them is just under 14' high—the regulation maximum load height for a truck. Regular flatbed trucks and trailers often have 51" high decks and are configured differently, which limits what they can carry. Depending on size, a truck can carry as few as 100 and as many as 500 bales.

Order what safely fits on the truck, closest to the number of bales you need. If your project needs more, you may need two trucks. If you need fewer bales than the truck can carry, see if you can coordinate with other building projects or others in your area who can use the bales to share the delivery costs.

Receiving and Unloading Bales

Site access—road width and condition, height of obstructions (bridges, utility lines, trees), and turn-around at the site—must also be considered. Ideally, the delivery truck can get to the building site where a squeeze will efficiently offload the bales, then turn around and leave. However, if your access road is steep, has tight turns, or poor traction, the truck may not be able to reach your site, and you'll need to arrange for a nearby staging area to offload, and then you'll have to complete the delivery with a smaller vehicle.

Approach a local feed store about staging a bale delivery—they're accustomed to large truck deliveries. You may even be able to piggyback a partial load of straw bales on one of their hay bale shipments, and possibly rent their squeeze. Also consider asking a nearby lumber yard, farmer, or rancher about staging the delivery. Depending on the distance from the staging area to the building site, a squeeze or a bobtail truck can deliver blocks of bales, or you might use one or more pickup trucks.

Handling individual bales takes time, and trucking companies charge for time. If loading and unloading pickup trucks by hand, have eight to ten people available. Make sure the trucker and truck company are aware of this and approve the extra time. Develop an unloading plan and follow it. Do not hurl the bales from the truck because they can deform past the point of use. Instead, place a few bales on the ground alongside the truck and "drop" the others onto them. The bale strings should be tight enough that you can just slip your fingers under them to pick up. Many builders use bale hooks to pick up bales to avoid deforming them. Use wheelbarrows, hand trucks (dollies), and/or sturdy garden carts with large wheels to quickly move the bales to a stack at the building site. If building a post-and-beam structure and the roof is already on, stage some bales inside the building, and some along the outside. If you can stack piles of bales adjacent to the building, you can build a ramp from plywood and 2× lumber to slide bales from the stack to the building floor. Several layers of bales can be easily moved in this way before the ramp isn't steep enough for the bales to slide.

Evaluating Bales during Delivery

Building-quality bales are dry and dense. IRC Appendix S requires random tests for both moisture content and density. At least 5% and no fewer than 10 project bales must be tested and have no greater than 20% moisture content (see AS103.4); at least 2%, and no fewer than five project bales, must be tested for "dry density" no less than 6.5 lbs./cu. ft. (see AS103.5). You'll need a scale and a moisture meter for straw to determine dry density. You may be able to borrow or rent a moisture meter from a local feed store or hay or grain farmer, or you can purchase one. Appendix S and its commentary describes the dry density evaluation process. While Appendix S allows moisture content as high as 20%, most builders look for moisture content lower than 15%. Problems start when moisture levels rise above 20%, so use dry bales—and keep them dry! (*Note:* Because plasters are applied wet, the moisture content of plastered bales can temporarily exceed 20%, especially near the bale surface. Straw has the ability to manage this temporary increase in moisture without damage.)

Bale quality control doesn't end when the truck is unloaded—it continues through the bale raising. Although you may have checked for bale moisture con-

1. Weigh the bale.
2. Measure its moisture content.
3. Calculate its volume in cubic feet by multiplying its length x width x height. 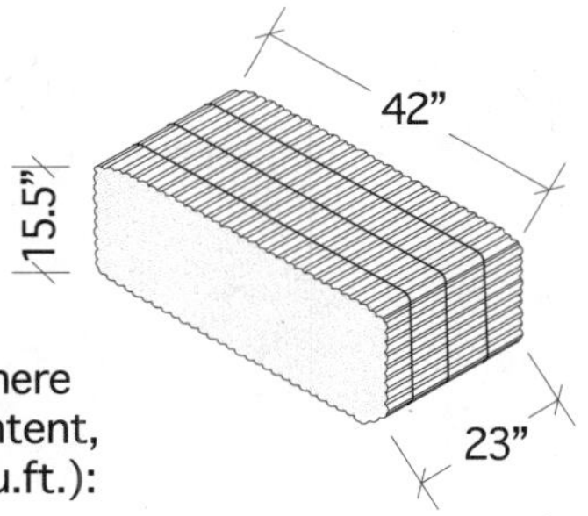
4. Enter the results in the following equation, where A = total weight (lbs), B = percent moisture content, C = Volume (cu.ft.) and D = Dry Density (lbs./cu.ft.):

$[A - (B \times A)] \div C = D$

Our example:
A = Total weight = 80 lbs.
B = Percent moisture = 14.5%
C = Volume 23" x 15.5" x 42" = 14,973 cu. in. = 8.66 cu. ft.
D = Dry density

$[A - (B \times A)] \div C = D$

B x A = 14.5% x 80 lbs. = 0.145 x 80 = 11.6 lbs. of moisture
A - 11.6 = 80 - 11.6 = 68.4 lbs. dry weight
68.4 ÷ C = 68.4 ÷ 8.66 cu ft = 7.898 lbs. per cu. ft.

5-3. Bale density.

tent and density before or at delivery, the building inspector can require it to be demonstrated again—before, during, or after the bales have been stacked. During the frenzy of unloading the delivery truck, it's possible that unsuitable bales escaped the crew's attention. During the bale stack, the crew should be alert for bales that feel heavier than the others (wet?) or are much lighter (not dense enough?). Look for dark stains in the straw and be alert for a strong, musty smell that could indicate that bale is wet or moldy and unsuitable for use. Set it aside.

Storing Bales

Stack and store the bales properly, even if they're going into the wall the next day. Set the first layer on pallets or scrap wood to keep bales off the ground. Place plastic or a tarp under the pallets or scrap wood if ground moisture is a concern. Alternate bale layers are stacked in a crisscross pattern for stability. If stacking bales in the open, be sure to have tarps, weights, and ropes handy to cover the stack if rain threatens. Shape the top of the stack like a pyramid so the tarp drains. See Figure 5-4. Raise the tarp off the bales with 2× lumber to keep condensation on the tarp from wetting the bales, and secure it with ropes tied to trees or weights like 5-gallon buckets filled with sand. Direct the tarp edges away from the stack so runoff doesn't wet the bales.

Because bales do not come in precisely uniform lengths, some builders measure, mark, and sort the bales into shorts, mediums, and longs. This takes time initially, but later it will save time otherwise spent searching for the "right size" bale or making custom lengths.

5-4. Bales properly stacked for storage.

Tools and Materials for the Bale Raising

Chainsaws

Several brands of electric chainsaw are popular for building straw bale structures. Most builders use electric chainsaws because they are quiet, don't produce fumes, or get hot enough to ignite loose straw. Look for saws with bar lengths at least as tall as the bale, so between 14" and 16". The chain's cutting teeth must be periodically sharpened, the chain kept tight on the bar, and the bar oil reservoir filled to lubricate the metal-on-metal friction caused as the chain spins. Loose straw can collect under the chain sprocket cover and jam the chain, which requires unplugging the chainsaw, removing the cover and clog, then replacing the cover. Some models have easily removed covers, others are designed to minimize clogging.

Cutting straw creates a lot of dust that gets sucked into the chainsaw's motor vents. At the end of each day, clear the vents with a vacuum or blast from an air hose. Experienced builders prefer more powerful saws with longer bars (13 amps with 16" bars), although many straw bale buildings have been built with less powerful saws with shorter bars. Whether used by contractors, workshop participants, or volunteers at work parties, respect the damage that chainsaw teeth can inflict. Operators should be familiar with chainsaw use and safe practices.

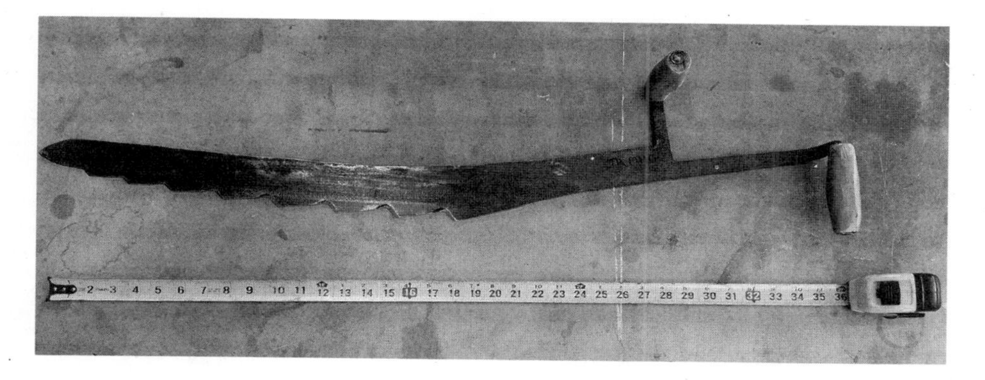

5-5. Hay knife.

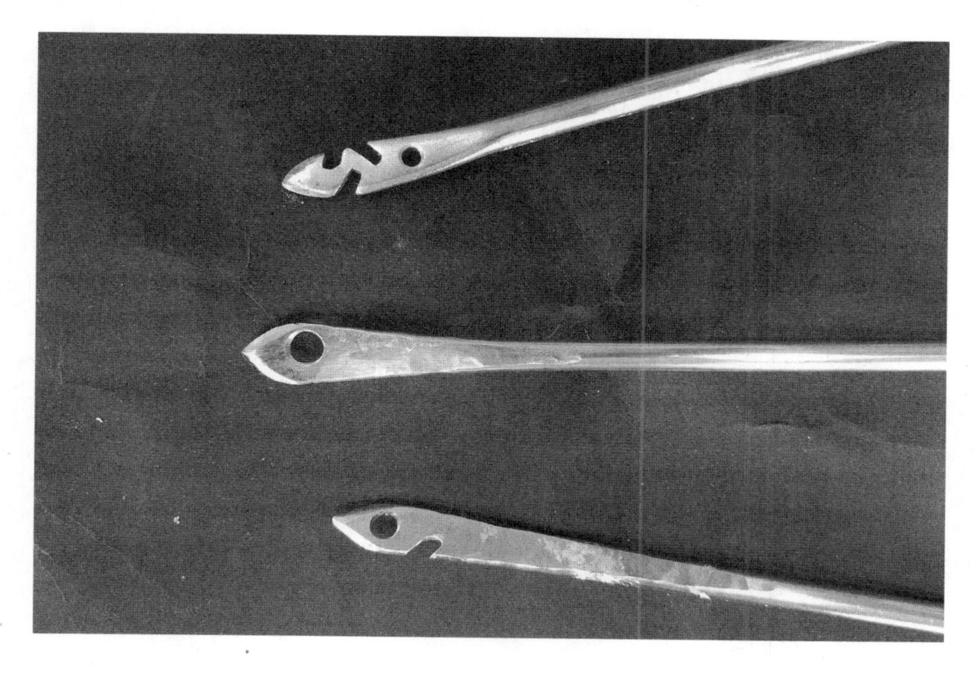

5-6. Various bale needle tips.

Bale Needle

Resizing a bale requires a specialized tool unique to straw bale construction: the baling needle. You may be able to purchase a bale needle on the internet or from a local bale builder, but bale needles can be fabricated from simple materials. While bale needles can have fancy handles and a variety of point shapes and configurations, a serviceable needle can be fashioned from a 3' length of ¼" steel rod. Most bale needle shafts are between 28" and 32" long. Use a torch to heat the rod where you'll bend it to create a 4"–6" "handle" at right angles to the needle shaft. Heat the rod's tip until it glows red, then hammer the tip to a tapered flat form. Think "arrow head." Let it cool. Shape the head with an angle grinder or benchtop grinder, then drill a hole in it—3/16" to ¼" will do—for the needle's eye. Use a file to sharpen the needle edges so it more easily penetrates straw bales. In

> **Hay Knives**
>
> These tools were once in common use for cutting wedges of hay from hay piled in barn lofts. Today they can be found in antique stores. Look for tools that aren't badly rusted, have intact wooden handles and straight blades. They require periodic sharpening but are very effective. They require more effort than a chainsaw, but are useful for making cuts in awkward places where operating a chainsaw would be difficult or dangerous. The safer and quieter alternative—a hay knife is often the only bale-cutting tool on a straw bale building site.

addition to the needle's eye, some designs feature two notches, one slanting forward, and the other back, for pushing and pulling twine through the bale without needing to thread the needle.

Baling Twine

You'll need baling twine for making half-bales, for through-tying mesh or external pins, and for many other uses. Farm and ranch outlets and feed stores carry the polypropylene twine or "string" used to bind straw bales. Sold in spools, baling twine comes in several strengths. Use the twine most similar to the twine binding the bales you'll be using, probably over 200 lbs. minimum knot strength. To keep the twine spool accessible and clear of loose straw, place it in a lidded 5-gallon bucket. Drill a ½" hole in the center of the lid and feed the twine out through the hole.

Bale Hammers

Bale hammers are heavy wood mallets, often fashioned onsite from scrap lumber. These hammers, often with whimsical names like the "Persuader," "Commander," and "Boss," have a 3' long wood or metal handle and a 10–20 pound wood head.

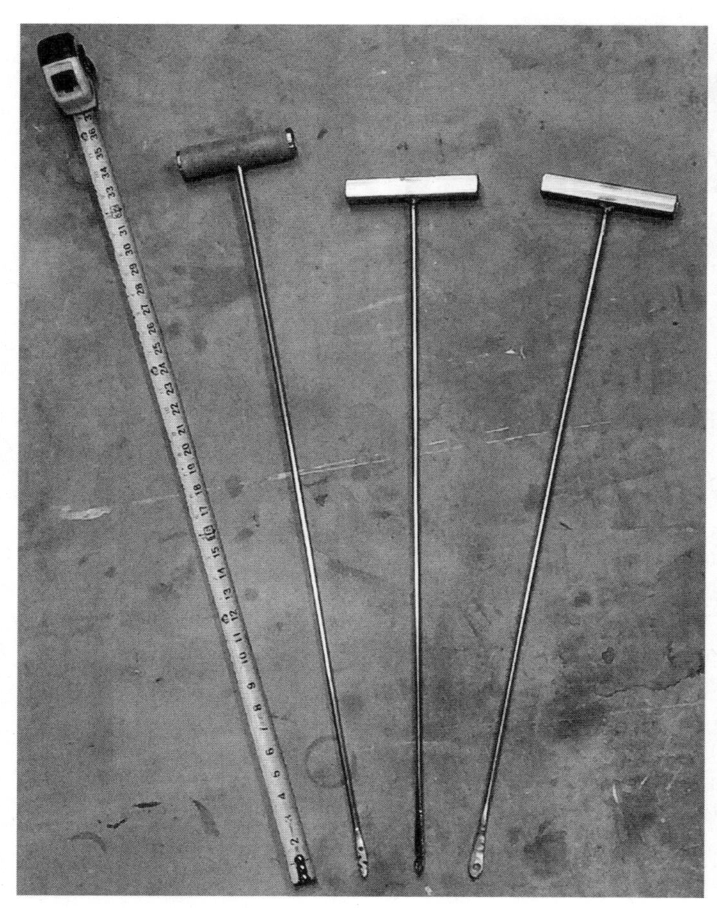

5-7. Bale needles.

5-8. Bale hammers.

Use them to realign bales in a wall, compress bale courses, and drive bales into tight spaces. Some builders find hand-held tampers useful.

Utility Knife

Each person on the crew needs a way to cut bale strings; the very practical utility knife with retractable blade can be kept in a pocket when not in use. It can also be used for cutting straw and building paper, and for countless other purposes.

Spray Paint

Use paint that contrasts with the bale color for marking cuts. Some builders will use spray paint cans left over from other projects; many builders use ground-marking paint—often called "upside down paint" because it sprays when the can is upside down. Use spray paint to mark sill plates where electrical wires and plumbing lines will be installed, and later paint the bales to show where plumbing and electrical are recessed into the wall.

Marking Peg

Some builders prefer to use wood dowels with sharpened ends to mark cut lines—drive the dowels into the bale just off the cut mark, then eyeball the cut.

Assorted Other Tools come in handy during the bale raising—bale hooks for handling bales, leaf rakes, brooms, pitchforks, and shredder-vacs effectively clean up loose straw, while claw hammers, drills, levels, etc. can be used for any number of tasks, from installing electrical boxes and plumbing chases to nailing sill plates.

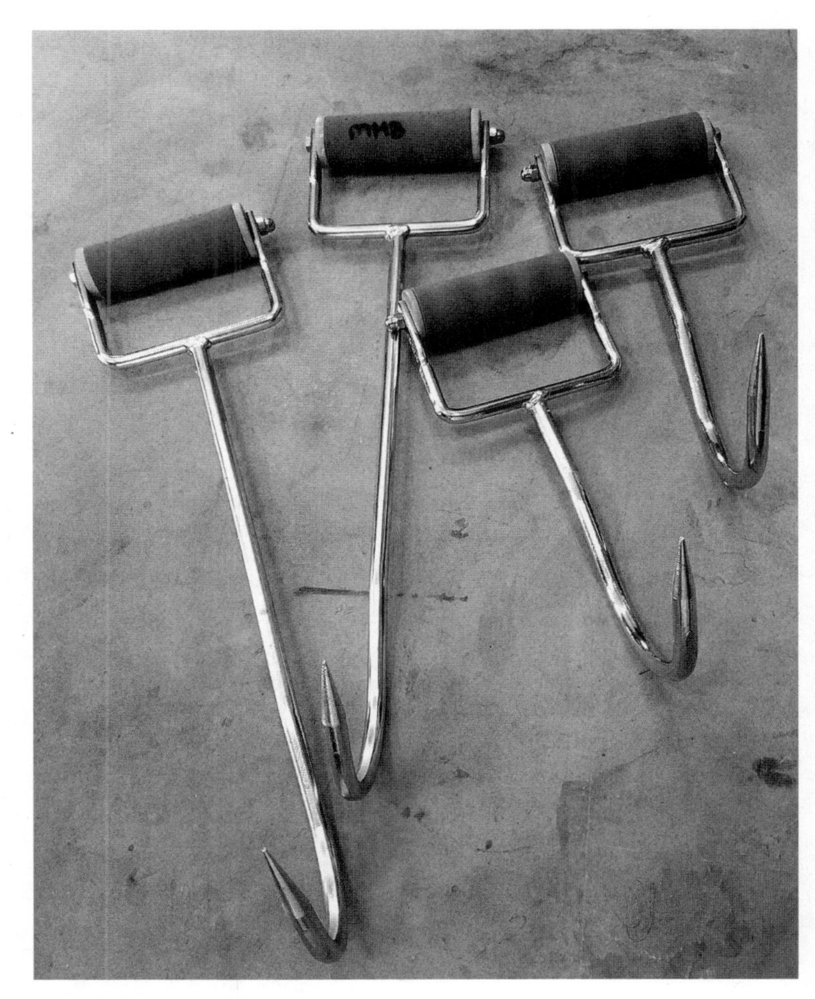

5-9. Hay hooks.

Preparing the Site for Bale Stacking

Site

If the ground around the building is level, and both the inside and the outside of the building are clear of construction debris and tools, the bale stack proceeds more quickly than if the crew needs to move tools and framing materials, clean up construction waste, or watch their footing on uneven ground. Ramps up steps to doorways reduce a trip hazard and assist in wheeling bales into the building with hand trucks.

For all straw bale walls, the stacking begins after a waterproof membrane has been installed under the sill plates, the plates have been fastened to the foundation with anchor bolts, and insulation and/or drain rock has been placed between and level with the top of the sills.

For load-bearing walls, builders often erect temporary corner boards fabricated from 2× or plywood, secured to the sill plates with diagonal braces. These guide the bale stackers in keeping the corners plumb as walls rise, and they are removed after the roof-bearing assembly has been secured at the top of the wall. Occasionally check the corner boards with a level during the bale raising because they can be bumped out-of-plumb. In post-and-beam bearing walls, posts holding up the roof help guide the effort to keep the bale walls plumb. In post-and-beam construction, most builders install bales after the roof has been waterproofed because seams between the roof sheathing can concentrate rain runoff onto the top of a bale wall.

5-10. Corner brace.

The Bale-Cutting Station

Set up designated cutting stations. Create safe zones where no one but the chainsaw operator may enter. This may be a separate room in the house, or a "corral" of bales outside. Since operators should be wearing a dust mask, safety glasses, and possibly ear protection, communicating with them may be difficult. Before beginning the stacking, establish a method that you all agree to, e.g., approaching from the front to get their attention, or waiting until the saw has stopped before tapping them on the shoulder, etc. Working at bench height reduces unnecessary bending and lifting, so work proceeds faster with less fatigue. Many builders use a "bale bench": a bale set on-edge so the strings aren't exposed to a sharp blade should you accidentally cut through a bale into the bench. Place bales that will be notched or resized on top of the bale bench.

Electrical and Plumbing in the Bale Walls

Discussed in Chapters 2 and 4, good planning for electrical, plumbing, and vents in straw bale walls makes for trouble-free bale stacking. If electrical and plumbing are not already installed, mark electrical box, plumbing chase, and air vent locations on the floor prior to stacking bales. Use the bale elevations as a guide and note anything that needs to be done during the bale raising on these elevations.

The Bale Raising

Goals

A properly stacked and prepared straw bale wall insulates better and is more easily plastered, which results in a finish that protects the walls and supports the building's design aesthetic and performance goals of energy efficiency and durability. A poorly stacked and prepared wall can have cold spots and require more materials and labor to lath and plaster, and it can lead to maintenance problems.

Unlike conventional building materials used as a substrate for plaster, bales are uneven and fuzzy. A straw bale wall may undulate from top to bottom and side to side. If a straight, flat finished wall is desired or required for a shear wall, then take care to stack and trim a plumb bale wall. How well the bales are stacked may be seen in how flat, straight, or plumb the wall appears when the plaster is complete. If you think of the bales as the muscle of the building, the plaster is the skin that will follow their form. Decide what the standard for the project will be—flat, straight, and plumb, or more informal and irregular—and communicate this to everyone involved with the bale stack.

The Stacking Crew

Straw bale walls can be stacked by individual owner-builders, work party volunteers, workshop participants, or by a professional crew. A couple of owner-builders may be able to stack the walls of a small building in a few days or a week, depending on their abilities and the type of wall system. A workshop with a dozen participants run by experienced instructors can usually stack the walls of a small structure over a long weekend, and work parties managed by experienced straw bale builders can accomplish similar results. An experienced crew by itself will usually be more efficient.

Regardless of who stacks the bales, all straw bale raisings include the following tasks:

- staging bales along the wall where they'll be stacked
- shaping the bales by "trimming the ears" (see below), resizing where necessary, and notching them for posts
- setting bales in the wall
- stomping each course of bales into the course below to minimize later settling
- stuffing gaps between the bales with loose straw
- sweeping up loose straw to minimize trip and fire hazards

Who does which tasks depends on skill and ability. Owner-builders working alone do everything, performing each of the above tasks for every bale. During workshops and work parties, participants may learn all of the tasks but gravitate to the work they feel most comfortable with. Professional crews may divide tasks. If a crew has members with different levels of experience, some may be assigned to mark and shape bales, and others will stage, stack, stomp, stuff, and sweep up. If the crew members have comparable experience, each may start in their own corner or room of a building, away from anyone operating a chainsaw, and do all of the work for that portion of the building. So long as everyone understands the bale-stacking goals and follows the same installation methods, the building won't show the "signature" of different stackers—some walls being loose and out-of-plumb, while others are tight and straight. In addition, many crews, workshops, and work parties assign one or more people to be *wall bosses* in charge of quality control; they make sure notches are sized properly, that courses are stomped, gaps stuffed, that the wall goes up plumb and straight, and generally facilitate the work by identifying and resolving problems.

Nailing the Plate

Some builders pound 20d (4") nails into the sills every 6 to 12 inches. Setting the first course of bales onto the protruding nail heads keeps them from shifting as the walls go up. Drive the nails in just an inch or so. Locate the nails toward the center or inside edge of the sill plates—away from the exterior or interior wall surfaces where a sharp-bladed tool used to trim walls could accidentally contact the nail head, both dulling the blade and causing a spark! Also, instead of nailing all

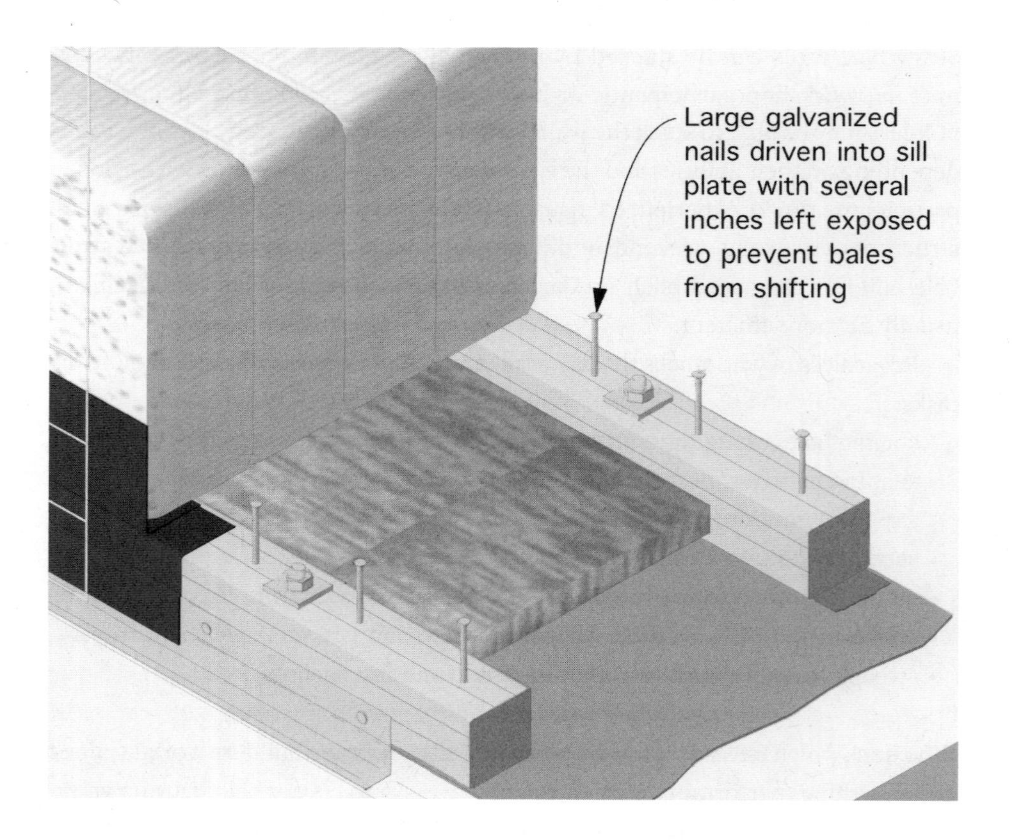

5-11. Sill plate nailing.

5-12. Bale ears.

sill plates at once, nail them just ahead of the stacking crew to reduce the chance that someone might step or fall on the protruding nail heads. Some builders prefer not to use nails in the sill plate, and the building's structural design will likely not require it. Builders in climates with extremely cold winters avoid placing metal in the bale walls because water vapor can condense on the metal surface, and potentially create moisture problems in the wall.

Shaping and Placing Bales to Fit Framing

Trimming Ears

Some builders first cut the "ears" off of bales in order to square them before stacking. The straw bale's ears are the "tufts" that stick out at the corners—a result of the baling process. Though not strictly necessary, this step leaves fewer voids to fill, and it's easier to accurately mark notch locations when you're pulling a tape measure from a firm edge instead of a fluffy one. Place the bale on the workbench and use a chainsaw or bale knife to square the end, being careful to avoid cutting the strings. Some builders wrap expanded metal lath around a wood block to make a "bale scrubber" and rasp the tufts off.

The First Course

Most builders begin stacking at the corners, setting the first course in one direction, then the other. Stacking can also begin at doorways. On this bottom course, use full bales at corners and door openings. When the wall system uses I-joists or LVL "posts" to support roof loads, stacking can begin anywhere adjacent to an I-joist or LVL, though it's best to stack both sides of the post more or less simultaneously, so it's braced from both directions as the wall rises. This keeps the I-joist or LVL from "bowing" should stackers stuff the bales too tightly on one side. With all systems, make certain the interior and exterior bale surfaces are flush with the interior and exterior surfaces of the sill plates.

Notching Bales

Bales set around framing need to be notched so the bale surface is flush with the exterior and interior wall planes. Whether the building's "posts" are 2×4, 4×4, 4×6, or round posts, a channel must be cut in the bale that fits the post size.

Measure and Mark the Cut

Keep spray paint or marking dowels on hand near the cutting station. Some builders also keep a block of wood the same dimensions as the structure's posts to help mark the cut and to test fit the notch for width and depth. Since many post-and-beam buildings use 4×4 posts at corners, the first cuts will notch bales to fit these corners. Place a bale on the workbench and mark the cut either with spray paint or marking dowels. You can also lay the test block on the bale and spray paint a line to follow. Use a saw to cut down into the top of the bale to the required post depth, then cut in from the side to the same depth. A rectangular "column" of straw should be freed from the bale. Don't cut the notch on the butt end too deep—this results in the butt end of the bale sticking out at the corner, creating a bump that is difficult to trim. When making a notch mid-bale, use a

5-13. Notching a bale.

tape to measure the distance on the wall—usually from the edge of the adjacent bale to the nearby post—and transfer the cut measurement to the bale. Consider that the bale end may be somewhat rounded and measure from the fattest point. Make two vertical cuts demarking the post width, and then make several diagonal cuts to clear the waste. Chainsaws can also make a *plunge cut* to clear the waste; be careful to enter the cut with the bottom of the chainsaw tip, slowly straightening the blade for the plunge cut. Contacting the upper portion of the saw blade tip will cause it to kick back. The saw's safety brake is designed to stop the spinning chain before it strikes the operator, but it's best to avoid this situation. After the cut, test the fit with the wood block before expending the effort to carry the notched bale to the wall. If the wood is proud of the bale surface, the notch needs to be deeper. If the fuzzy surface of the bale is deceiving, use a straight edge placed across the bale to see if the test block is truly deep enough. If the bale is proud of the wood surface because the notch is too deep, expect to stuff straw between the bale and post after the bale is lifted to the wall.

Sometimes a cut goes deeper than intended, or the strings are closer to the edge than normal, and the saw cuts a string. Simply restring the bale as tightly as possible.

Resizing Bales

Invariably, you'll need to resize some bales to fit spaces smaller than your standard bales. Most straw bale wall systems require resizing some bales, and it's important to maintain the bale's original compression. You will also need to use half-bales in alternate courses to maintain a running bond adjacent to door and window openings. To make a custom-sized bale, you'll need a bale needle, a tape measure, a way to mark the split, a sharp utility knife, and a spool of baling twine.

Measure the space you need to fill. Many builders will slightly oversize a bale so it squeezes into the space, helping to compress the bales on either side. But, be careful that this doesn't cause framing to bend under the added pressure. On the other hand, some bales are so tightly compressed with machine-applied ties that they grow slightly when retied by hand, no matter how tightly you hand-tie them. If this is the case, measure your custom bales slightly smaller than needed so they will still fit once they expand.

Next, place a full bale on the workbench. Transfer the measurement to the bale, and either mark it with a small spot of spray paint or a dowel. Calculate the length of baling twine you'll need to make a loop around the new bale size, and then add a few feet. If you're building with two-string bales, you'll need two strings of this length. If you're building with three-string bales, you'll need three strings.

Thread a string through the bale needle's eye or into the "push slot" depending on the type on needle you have,

Know Your "Wingspan"

There's a shortcut for estimating the length of string you're pulling from the spool without using a tape measure. Hold a piece of tightly stretched string from one outstretched hand to the other. Measure it. That's your "wingspan." If it happens to be around 6', and you need 9' of string to wrap a bale with a few feet left over, then pull one and a half of your wingspan from the spool. If you need 12', pull twice your wingspan.

position it at the mark adjacent to the bale's original string and plunge the needle straight through the bale. This is usually done while the bale is on-edge, so you or someone you're working with can see if the needle emerged on the other side near the existing string. Unthread the string, hold on to the loose end, and withdraw the needle. A bale needle with a push slot allows you to retract the needle without unthreading the eye. Repeat for the remaining strings.

Tie an overhand loop knot in the newly placed strings—the loop can be an inch or two in size. Pull on the unknotted ends of each string and wrap them around the bale, following the original strings that are keeping the bale's compression. Don't cross the strings. Run the end of one string through its loop knot, which will now function as pulley. Lay the bail flat on the bench, strings facing up. Position yourself at the end of the bale, place your knee or hip up to the bale and use one hand to pull the strings toward you. To ease the stress on your hands, wrap the string around a dowel and use as a handle. Watch the original string directly below the one you're pulling on. When it goes slack pinch the string at the pulley with your free hand and tie a double half-hitch knot onto the loop to secure the new compression string. Do the same with the other string(s). Cut the old strings at the original knot and use them to loosely tie the remaining section of bale. This leaves you with a retied, compressed bale of the size you need, and an uncompressed section of that bale that you'll set aside for later use. If you are using three-string bales, measure, sew, and tie the two outer strings, then separate the sections. Tying the third string on after the sections are free is slightly quicker than sewing it. If you get good at keeping track of which strings are which, you can use the discarded middle string to tie off the leftover section for safekeeping.

5-14. Resizing a bale.

Some builders retie both sections of a bale in the same operation. Instead of threading the needle with one string, they'll thread it with two—one each to go around opposite ends of the bale. This results in sectioning a bale into two smaller bales that retain the original compression. Make certain the strings don't cross during the process, or you will have two bales that are tied together.

Most bale strings are located 5" or 6" from the outer surfaces of the bales, but sometimes they drift to within 4" of the surface... right where a post notch might be cut. While resizing a bale, some builders will tie an additional string an inch or two closer to the bale's center to avoid being cut when a notch is made, or they'll use a flat bar to shift the outer string out of harm's way.

Sweeping Loose Straw

All of this cutting, notching, and restringing generates a lot of loose straw! It is essential for jobsite safety to clean up loose straw and keep it from spreading on the site and beyond. This reduces fire hazards and prevents workers from slipping on or tripping over it or losing tools under it. During bale stacking, routinely clean up the loose straw. Have plenty of large heavy-duty trash bags on hand to store loose straw for later use or designate a place some distance from the structure to pile loose straw. Use leaf rakes, brooms, and pitchforks to gather and remove loose straw—sweep it into piles and exterior doorways where it can be scooped up and bagged or shoved onto a large tarp and dragged away from the building. If you plan to use the leftover straw in straw-clay or plaster, keep it dry and free of debris like string and nails, especially if you plan to shred it. Clean the floor before you start stacking, keep trash out of loose straw, and bag it in a way that it will not be mistaken for trash. Some builders use clear trash bags for this reason.

Stomping

After completing the bottom course, make sure the bale faces are in plane with the sill plates, and then stomp the bale. Stand on the bale and jump up and down or use a hand-held tamper or bale hammer to pound on the bales, flatten the surface fuzz and seat them "into" the bale surface below. Repeat this after completing each course.

Stuffing

Use your hand to probe the ends where bales meet and fill any voids with loose straw. Pay special attention at posts and plates. Try to push your hand into the joint; if you can push the length of your fingers into the joint, it needs stuffing. Some builders use 1× and 2× sticks to push handfuls of straw into narrow gaps. This can be done as the bale courses rise, or it can be done at end of the bale stacking. Many think it's best done as the courses rise because at that point there's access to three sides of the bale surface, but it's an easy step to overlook during the bale-raising excitement. Stuffing these gaps is critical to the wall's thermal, acoustic, and fire-resistance performance.

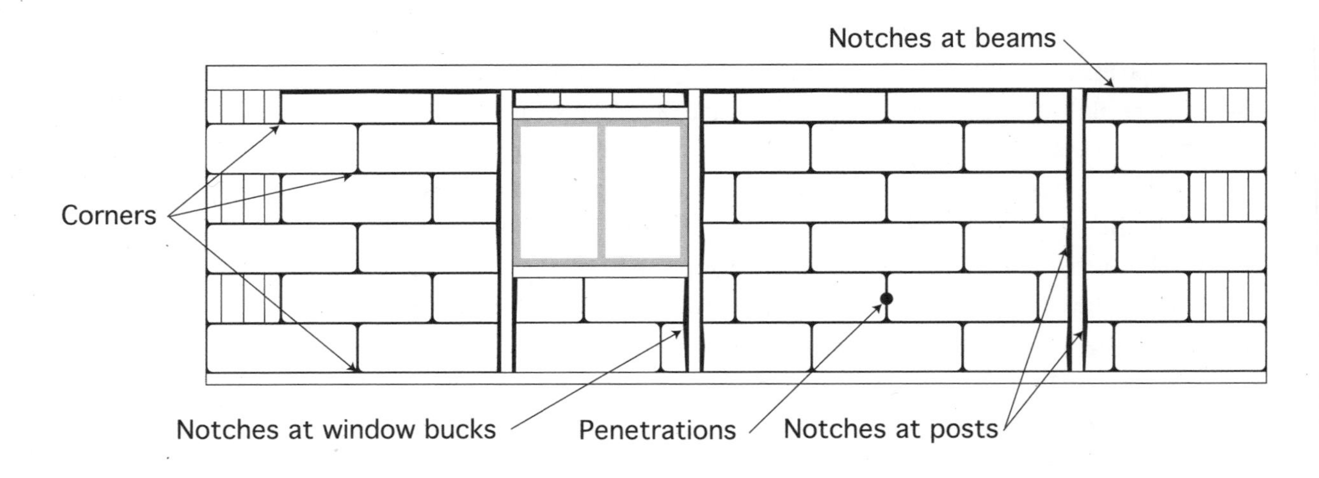

5-15. Areas that may need stuffing.

Second and Subsequent Bale Courses

A running bond for straw bale walls—similar to brick or concrete block walls—with vertical joints offset by at least a quarter-length of the bale is required in Appendix S only for structural walls using unreinforced plasters. However, staggering the bale joints makes a stronger wall, both during construction and when finished. At the corners of both load-bearing and post-and-beam wall systems where the bales are laid flat, this is easily done by lapping the corner bales. In an I-joist or LVL-post wall system, you may need to start the second course at the corner with a half-bale.

5-16. Running bond.

The bale stacking proceeds quickly with both load-bearing and I-joist/LVL wall systems—bales need only be resized, not notched. The bale stack for wall systems that require notching will go more slowly. As the courses rise, pay close attention to keeping the walls plumb, and to stomping and stuffing. Keeping the walls plumb as you stack alerts you to any notches that were not cut deep enough. Conversely, if a notch is too deep, the bale can be brought back into plane by stuffing loose straw between it and the post.

Electrical, plumbing, ducts, and flues in bale walls were addressed in Chapter 4; don't forget about them during the bale stacking.

Ladders and Scaffolding

As the bale courses rise, the bales need to be lifted overhead. Some builders use sturdy ladders, others use extra bales to create steps to facilitate lifting bales higher onto the wall. Both have advantages. Ladders are easily moved but can be unstable. Bales used as steps are quite stable, even with several people working from them, but you need sufficient extra bales, they're heavy to move, and they cover what's happening at the lower wall surface until they are removed. You can also use one lift of rolling scaffolding. Two people can load the bales up to the platform, while others safely lift the bale in place. Drywall or painting benches help in those in-between locations.

Windows and Doorways

Most straw bale wall systems use 4× or 2× framing for doorways and windows. Depending on the wall system, stacking bales around openings in the wall requires custom bales, and notching may be called for. When the bottom of a window buck doesn't sit directly on a bale course, most builders will size several partial bales to fit the gap. For bales on-edge, simply cut the bale lengthwise to fit beneath the window, or stuff narrower gaps with straw flakes. For bales laid-flat, bale strings remain hidden in the wall, but the bale flake orientation will be horizontal instead of vertical—stuff loose flakes or retied custom bales in this space. Be careful about over-stuffing bales and loose straw around window and door framing; avoid knocking the frame out-of-square. Periodically use a level to check for plumb around doors and windows, and remove any straw deflecting the frame.

Top of the Wall

Getting the tightest possible fit for bales at the wall top makes for a better building. Box beams transfer roof loads in load-bearing wall systems. After completing the wall stacking, builders lift site-fabricated box beam sections to the wall top and assemble them in place. If the building's sill plates were used as a box beam template and the walls are plumb, the box beam should fit perfectly. Most builders compress the wall using ¾" packing straps and buckles that were placed on 2' centers under the sill plate before the plates were installed; these are then wrapped over the top of the box beam. During the bale stack, these straps can

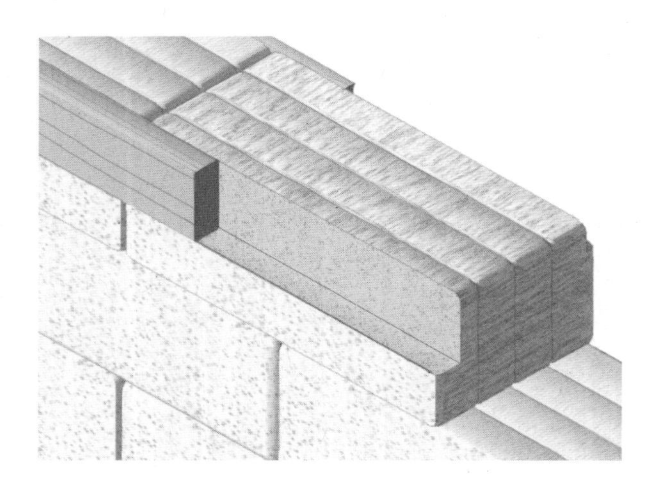

5-17. Top plate notched into bales.

be pinned or sewn to the lower part of the wall to keep them from becoming a trip hazard. These straps both compress the bale wall and level the box beam. All-thread rods installed on 2' or 3' centers and paired on both sides of the wall accomplish the same goal. The rods—which can be sectioned—must be anchored in the sill plate and pass through the box beam. After compressing the walls and leveling the box beam, builders proceed with the roof framing—trusses or rafters—followed by the roof sheathing (or SIPs, if used) and the finish roof.

Either box beams or a single 4× beam carry roof loads in a post-and-beam wall system. Box beams simplify bale stacking at the top the wall. Ideally, a full-size bale fits snugly into the space under the box beam. With on-edge systems, the top course bales can be trimmed to fit. In either on-edge or laid-flat walls, stuff smaller spaces with flakes. Buildings using bales laid flat and a 4× beam need the entire top bale course notched for the beam, and many bales will be notched for posts as well.

With post-and-beam structures, the roof weight doesn't bear on the bale walls, so any bale compression is due to bale weight or precompression measures taken by the builder. Dense straw bales won't settle much, but if they do, they can leave a small gap at the top of the wall. To avoid this, builders use a variety of methods to minimize this gap. One involves lag-bolting into the interior and exterior sill plates, looping heavy-duty compression straps up and over the bale wall (without the top course), compressing the bales an inch or two, measuring the remaining space, and placing a carefully sized bale into that space. When the compression straps are released, the courses "spring" back into position, making for a tighter wall. Other builders slightly undersize the space for the top bale course and use the straw bale equivalent of a shoehorn, made of thin sheet metal, plastic, or plywood, to squeeze the top bale course in, usually pounding it in with a bale hammer. Still others will over-compress a vertically oriented "custom bale" with heavy-duty compression straps, sizing it to the space required. When the compression straps are released, the bales, already snugly fit into the space, expand even more tightly.

Many building designs rely on external pins (or battens), paired on opposite sides of a straw bale wall and tightly tied to each other to stiffen the wall and resist out-of-plane forces. The plans should show pin spacing and through-tie requirements. Appendix S gives requirements for external as well as internal pinning when used as the wall's method of out-of-plane resistance (see Section AS105.4.2). Appendix S also allows plaster mesh for this purpose (see Table AS105.4 and Section AS106.9). Though not required, many builders through-tie the mesh to keep the mesh close to the straw and for better structural performance. (See lathing discussion in Chapter 6, Plastering Straw Bale Walls.) External pins can be located on the outside face of the sill plate and box beam, or pre-

5-18. Stuffing a modified bale into place.

drilled into them. If the sill plates and box beam have been predrilled with shallow holes to accept the pin ends—usually ½" diameter bamboo poles, ¾" diameter wood rods, or ½" diameter steel rebar—locate these pocket holes. Use a chainsaw to cut a 1"–2" deep vertical channel in the bales from sill plate to box beam at the hole locations. Be careful to avoid cutting electrical cables and plumbing lines (which should be marked) or bale strings. Cut the pins long enough so that when inserted in the channel they seat in the holes, then tightly tie the paired pins to each other through the bale wall. Fill the channel cavity with straw-clay flush to the bale surface.

Straightening the Wall

Bales that protrude from a vertical wall plane, or are sunken into it, may be visually appealing or appalling, depending on taste and design requirements. If the plastered straw bale wall plays a structural role, the bale wall surface needs to be quite flat, with no part protruding more than 2" from plumb over an 8' height. See Appendix S: AS106.7 for this and other requirements.

To ensure a flat plumb wall, use guides and regularly check for plumb as the wall goes up. Where bales fit over framing, carefully notch them to the correct depth—and gauge this with test blocks. An under-notched protruding bale cannot be beaten into submission (although you may be able to trim its face). As the wall

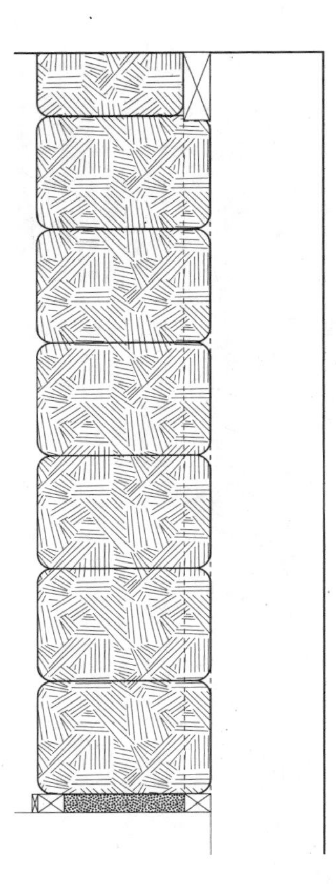

5-19. Bales in and out of plane.

Fire Safety

Note: Straw bale buildings are most vulnerable to fire during the period between stacking the bales and the first plaster coat. The most likely fire source—sparks from angle grinding, welding, plumbing soldering, halogen work lights, or careless smokers—can ignite loose straw on the ground. Flames can then climb the exposed bale walls and ignite exposed roof framing, resulting in a total loss. Take precautions. Sweep up and bag loose straw frequently while stacking and trimming; designate a fire-safe smoking area away from building; and keep fire extinguishers and pressurized water hoses with nozzles on the building site.

goes up, use the building's posts to gauge if the bales are flush with the wall plane. In areas without posts, use a "truth stick" to guide the effort. Find or cut a straight-edged board that reaches both the sill and beam or box beam. Make sure it's straight! Check by pulling a string tight from end to end. When this guide is touching both sill and beam, the bales are in plane. If the stick "crowns" on a bale or a gap shows under the stick's edge, bales need to be pounded in or out. The bales can also be trimmed, or the hollows filled in with straw-clay (discussed in a sidebar, below). Note that if you have a hollow on one side of the wall, it often means a bulge on the other. A straight edge held horizontally against successive posts also shows if bales are flush with the wall plane.

Builders use a number of techniques to keep the bales in plane during the stack:

- Screw a straight form board between posts, and stack bales flush with the board, moving the board as the courses go up.
- Secure bales to adjacent posts by sewing them in place with baling needle and twine.
- Pound a 12" wood stake into the bale near a post, and drive a long wood screw through the stake and into the post.
- If the wall's plaster will include mesh, paper the posts and install the mesh on one side of the wall before the bale stacking begins—this will keep the bales in plane as the walls rise, though this method creates challenges for trimming bales or stuffing gaps on that side.
- At recessed window and door bucks made of plywood or 2× lumber, and/or when using LVL posts, drill a ½" hole through the buck or LVL post approximately mid-wall depth, then pound sharpened wood dowels into the adjoining bales to keep them from shifting.

Trimming Bale Walls

Once the bales are snugly stacked into place and any protruding bales pounded back into plane, the surface of the wall may still be shaggy and not flat enough. Now is the time to trim, shave, and shape the bales,

5-20. Tools for trimming bale walls.

and to carve any corners, niches, recessed shelves, or to insert attachment points for cabinets, art, or other heavy elements that will hang on the bale wall. Hedge clippers, electric grass string trimmer, chainsaws, and angle grinders with Lancelot blades are the most helpful tools for shaving the fuzz off the surface and creating a tidy wall, as well as for carving, sculpting, and any significant reshaping of bales that may be lumpy or protrude from the wall plane. Trimming the bale walls isn't strictly necessary if the wall isn't structural or if an informal, irregular wall surface is desired. There's a limit to how much trimming can straighten a wall, and trimming shouldn't be attempted on a wall with bales stacked on-edge because of the exposed strings.

Trimming the wall surface generates a lot of straw dust! Wear protective clothing, including long sleeves, covered shoes, eye protection, and a particle respirator or dust mask. After the cloud of flying straw and dust settles, a less shaggy wall emerges.

Cabinet Blocking, Partition Connections, Niches

Set cabinet blocking at a height appropriate for attaching the cabinet, in a horizontal channel in the bale. Ideally, cut along the edge of a bale course so the 2× rests on the bale below. The channel should be deep enough for the 2× to rest flush with the bale wall surface. See Figure 2-31: Attaching/Hanging Objects from Straw Bale Walls Detail. Similarly, but vertically, notch the walls for partition connections. See Figure 2-30: Partition Wall Connection To Straw Detail. If carving a niche or shelf into the wall, or installing a 2× cabinet support, use spray paint to outline the shape to guide the effort. Be careful not to force power tools, which can choke with straw. Use caution when cutting around electrical wires and plumbing behind the bale wall surface; these should have been marked with spray paint. Also consider the bale strings that you may encounter within the wall—inadvertently cutting them can suddenly decompress the bale and foul the tool blade. Cut loose strings hanging from the wall surface before trimming with an angle grinder as they can snag the blade and jerk it into the wall.

Once the walls are trimmed, go over them once more, feel along edges between bales and posts, and make certain that gaps, cracks, and cavities are stuffed with loose straw, straw flakes, or a straw-clay mix.

Straw-Clay

Often used to fill gaps, straw-clay is a mixture of loose straw and clay slip, which is a sticky, soupy suspension of clay particles in water. See Chapter 6, Plastering Straw Bale Walls—Clay Slip. To make straw-clay for stuffing into gaps and filling hollows on a straw bale wall, fill a concrete mixing tub (usually about 30"×18"×6" deep) with loose straw. Pour about a half-gallon of clay slip over the straw and mix it —similar to the way you might toss a salad after pouring dressing on it. Mix for several minutes to completely coat the fibers with slip. If the mixture doesn't hold together, it may be too dry; add more slip. If you can wring slip from a handful of straw, it may be too wet; add more straw. The material needs to feel sticky and hold the shape of whatever gap or cavity it fills. The real test is whether it stays in place after you stuff it. Straw-clay sticks better to gaps and cavities when they, too, have been coated with slip, which can be hand or spray applied. Mixing and applying straw-clay is time consuming. The extra effort made to carefully notch bales for posts and beams, stomp bale courses, stuff gaps with flakes and loose straw, and maintain a plumb, flat wall during the bale stack more than offsets the day or two spent meticulously filling gaps, cracks, and hollows with straw-clay. If there's a lot of gap filling needed, or window and door reveals and soffits will be shaped with straw-clay, work comfortably at standing height at a mixing table fashioned from straw bales. Note that straw-clay is typically used in another wall system called "light straw-clay construction." A cousin of straw bale construction, it is sometimes used as a compatible secondary wall system in straw bale buildings. It has its own place in the IRC: Appendix R – Light Straw-Clay Construction—which is available for free download at www.strawbuilding.org.

Use dry loose straw or straw flakes where the void is deep enough to hold it. For small shallow gaps, use a sticky straw-clay mix. If the wall surface will not have lath or mesh applied, plaster keying into loose straw could pull it out, delaying the plaster application as you stop to refill the cavity. Conversely, wet straw-clay stuffed into a deep void will dry slowly and could cause moisture problems. Consider that the plaster is only as good as the substrate it is applied to.

Porches and overhangs protect exterior walls from the weather and help to regulate interior conditions. Careful detail design leads to a structurally strong, enduring and comfortable building.

Photography (*left to right, top to bottom*): Martin Hammer, Rebecca Tasker, B.J. Semmes, Claudine Cavet, Rebecca Tasker

Exterior style is only limited by site conditions, personal taste, and detail guidelines that have been developed based on decades of research and the combined experience of professionals and owner-builders.

Photography (*left to right, top to bottom*): (*above*) Paul Schraub, Eric Millette, John Swearingen; (*opposite*) Edward Caldwell, Martin Hammer, Dietmar Lorenz, Dietmar Lorenz, John Swearingen.

Photography (*left to right, top to bottom*): (*above*) David Arkin, Jim Reiland, Jim Reiland, Rebecca Tasker; (*opposite*) Rebecca Tasker, Rebecca Tasker, Erica Bush, John Swearingen, Claudine Cavet

Truth windows are unique to straw bale buildings, proudly revealing what's inside the wall.

Photography (*left to right, top to bottom*): Catherine Wanek, Rebecca Tasker (3), Martin Hammer (2)

Conclusion

5-21. Niche carved into bale.

There are many variations of the bale-stacking guidelines described in this chapter. Some builders "butter" the bottom of each bale with clay slip as it is set into the wall, helping secure it, seal the gap between the bales, and acting as a moisture and fire barrier. Where weather and building schedules permit, others let the walls settle for a few weeks, then install the top course. Builders are an innovative bunch, and where they see a need to streamline tasks, they'll fabricate job-site jigs for faster and more accurate bale cutting, sizing, or compressing.

Builders have also adapted materials and techniques that work best in their particular situations, leading to local and regional practices. Off-site production of straw bale wall panels that are craned into position is already available in parts of Canada and Europe. Prefabricated straw bale panel construction promises reduced construction time and cost, and more uniform straw bale wall assembly performance. However, it may not be suited for projects far from production facilities or for projects depending on owner-contributed or volunteer labor to stay within a building budget. There will always be a place for site-built straw bale buildings.

Important benefits flow from carefully stacking a straw bale wall: better insulation, faster plaster preparation, and faster and better plastering. Taking the time to complete each of these tasks results in better buildings. When the bale stacking is finished, the wall should be ready for plaster prep and application.

6

Plastering Straw Bale Walls

Part 1: Introduction

Straw bale walls and plaster are perfect partners!

Exterior plasters are gorgeous, and they protect the straw bale wall from rain, wind, fire, pest intrusion, and more. Interior plasters beautify the walls. They are the final layer of a vapor-permeable wall system. Bales supply a stable surface and strong key for plaster to hang onto, while plasters provide much of the thermal mass that stabilizes straw bale building temperatures, while at the same time making the walls more fire-resistant. Plasters can also lend structural support—if designed for this purpose. See Chapter 3, Structural Design Considerations. Finally, while bales determine the wall thickness, plasters are largely responsible for the finish: a straw bale building's beauty and *soul.* They are the skin covering the bale body; we relate to it whether it is unassuming and unadorned, or sculpted with complex patterns and decorative details. Plasters create the opportunity for us to experience spaces quite differently than we do with flat, featureless painted walls in conventional buildings—they are largely responsible for the beauty, delight, and comfort we feel in a straw bale building.

In Part 1 of this chapter, we'll discuss basic design and application concepts, along with plaster's performance and pleasing benefits.

> **Plaster vs Stucco**
>
> Historically, *stucco* referred to the exterior coating of a building, whereas *plaster* referred to the interior surfaces. It has become common in the United States to use the term "plaster" for both interior and exterior and all the materials used to coat a building: *stucco* is now understood to refer specifically to exterior cement plaster. We'll use *plaster* as the general term throughout this text when referring to clay, lime, and cement-lime plasters. Although cement stucco is the most widely applied exterior plaster coating in the United States, an all-cement stucco isn't vapor-permeable enough for straw bale structures.

Design Considerations

Ask the following questions when planning and designing plasters:

- What functional role will the plaster play in the wall assembly—will it be non-structural or part of a structural wall assembly—and how does that dictate material choice and installation method?
- What plaster material and application method will best support the design?

- Where the plaster meets other materials—wood, tile, metal, or other plasters—how will the transitions be addressed?
- Will the plaster be applied to other wall surfaces besides straw bales, such as drywall, papered and lathed plywood, light straw-clay, cob, or rammed earth?
- How will the plaster stop at the top and bottom of a wall? For interior plasters, will there be trim at the ceiling or baseboard at the floor?
- How will the plaster return around corners or into doors and windows? Will it stop flush with another material or die into an edge?
- What is the shape or radius of inside and outside corners? Will they undulate from top to bottom or be linear? Will they form large rounded bullnoses or crisp 90-degree angles?
- How should the bales be shaped to best carry the plaster and express the design? Will they be linear, curved, notched, or carved?

6-1. Shape and radius of inside and outside corners, window reveals.

6-2. Plastered returns around doors and windows.

6-3. Plaster stops at the top and bottom of wall.

- What will the plaster's shape be in the field of the wall—the uninterrupted area of plaster—and how will that shape express and support the design? Will the field be flat, undulating, deeply textured, or imprinted?
- Will the finish plaster be tinted or pigmented? Smooth or textured? Will there be aesthetic additives?

Pay special attention to edges and corners—especially where well-lit surfaces meet darker ones. Our eyes are drawn to these places on a building; we perceive the wall textures especially at these transitions. Our eyes tend to glance over the uninterrupted expanses of large wall surfaces. Conventional wall materials excel at creating straight, flat surfaces and crisp transitions at corners. These fit well with straight and flat wood trim and molding used at doors, floors, and windows. But applying these linear finish details indiscriminately in straw bale structures can create a disconnect with the straw bale wall's inherently irregular surface. With forethought, planning, and skillful execution, the finish details and plasters sublimely reinforce the design and amplify the bale wall contours whether they are straight and linear or curvaceous.

As explained in Chapter 1, Why Build with Bales?, a straw bale building needs *good boots, a good hat, and a coat that breathes.* Exterior plasters are the coat. The plastered bale wall system should transpire moisture but repel rain, protect the straw bale core, and create the desired aesthetic. With so much important work to do, the plasters must be durable. The best way to ensure a plaster's long life is to adequately protect it from excessive water on its surface. Chapter 2, Designing with Straw Bales, and other sections of this book discuss many ways to ensure the walls are well protected—follow them.

Aesthetics

When stacked in a wall, each bale's informal rectilinear shape creates overall linear surfaces (undulating or flat) and soft corners. Bales can be shaped and plastered to create conventionally plumb and flat walls, but they also create soft and curved surfaces far more easily than conventional materials can. People visiting a straw bale building for the first time talk about the play of light across the wall surfaces, the security of being surrounded by thick walls, and a sense of warmth and comfort. These pleasurable experiences come from the shape of the walls, the finishes, and the plastered details. The informal shapes and curves natural to straw bale buildings echo human bodies, which are not straight and flat but soft and round. Plaster imparts character to the walls that can define the space; and, as plasters age and wear, they patina and acquire personality.

Cracking

Wet plasters are elastic and pliable when applied, but they shrink as they dry or cure, creating tension within the plaster. Too much shrinkage in the material can result in cracks—within the field, radiating from corners, or as slight separations where the plaster meets other materials. Separations commonly form at the top and bottom of walls, and around windows and doors.

Different kinds of cracks result from different tension sources within the building and plaster skins. Common causes include site and structural settling, wood framing expanding and contracting, joint movement in the substrate, insufficient lath overlaps, plaster applied too wet, improper mix proportions, variable plaster thickness, and other installer errors. Cracks due to settling often travel down inside corners, radiate off window and door corners, and appear in plaster surfaces over plywood edges or lath layers. Vertical or horizontal cracks over a wood-framing member suggest the wood wasn't separated from the plaster by building paper to allow for differential movement. Cracks caused by variable plaster thickness occur from uneven drying or curing and can run the length of the thick and thin interface. Prepare substrates well, plan joint details, perfect plaster recipes, and apply plasters properly to prevent or control plaster cracks instead of allowing the plaster to crack out of control.

Not all cracks indicate failure. In many parts of the world, they signify something entirely different—a structure with the desired patina of age. Sometimes cracks are a part of the desired aesthetic. Think of "Tuscan Villa- style" buildings replete with faux cracks that create visual interest and the illusion of age. One of the most valued plaster finishes that is waterproof and used both on exteriors and interiors, as well as showers, basins, and pools, has a crazed finish. Plaster is a versatile, dynamic material, and cracks are inherent in them—on a microscopic level, but also in their personality.

The plaster on straw buildings tends to crack less than plasters on conventional stud-framed buildings because the bales provide a more resilient, stable

6-4a. Telegraphing crack.

6-4b. Depth cracks.

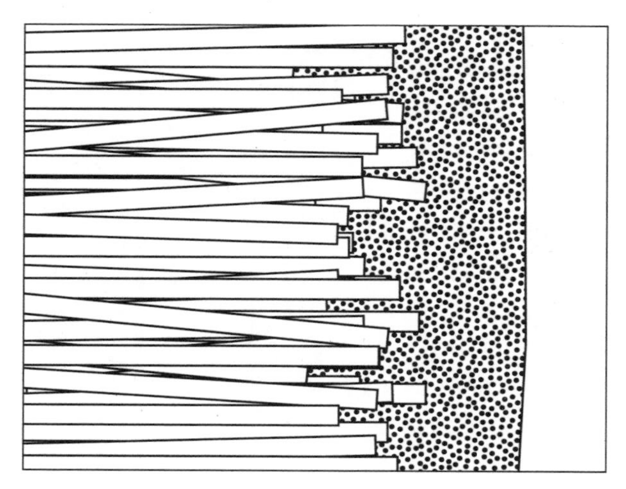

6-5. Mechanical connection of plaster to bale.

substrate for plasters. Conventional buildings separate the structure from the exterior plaster with a drainage plane, usually a double layer of building paper that is installed against the wall sheathing followed by a metal wire lath. Plaster attaches to and encases the lath. Plastered straw bale walls are considerably different. The shaggy bale surface provides an ideal natural lath and amazing physical key that plaster embeds deeply into, with no need for a drainage plane except where required over wood framing and sheathing. See Appendix S: AS104.4.1 Plaster and membranes and its commentary. Plasters bond more continuously and deeply with a properly prepared bale wall than a typical papered and wire-lathed wall. When embedded into the straw bale wall, plasters coat individual straw fibers, key into voids large and small, and become inextricably connected to the wall. In conventional construction, you remove the plaster by removing the lath. Try to tear plaster from a bale wall and you reveal the inner plaster layers and plaster-coated fiber. Keep going, and you are inside the bale wall. It is very difficult to cleanly separate a well-adhered plaster from a straw bale wall.

Planning for shrinkage and settling cracks

To control cracking, correctly design the joints where plasters meet other materials, reinforce corners, and install appropriate supports. Where plaster will meet a different material, choose from the following methods:

1. Plaster to the edge of the adjoining material and hand tool the plaster edge as it sets up. This closes the seam and controls the joint where plasters meet dissimilar materials such as wood and metal. With this approach, a slight separation appears where the plaster shrinks away from the edge. Sometimes this reads as a hairline crack, but it can be wider. More open cracks may need filling or caulking.

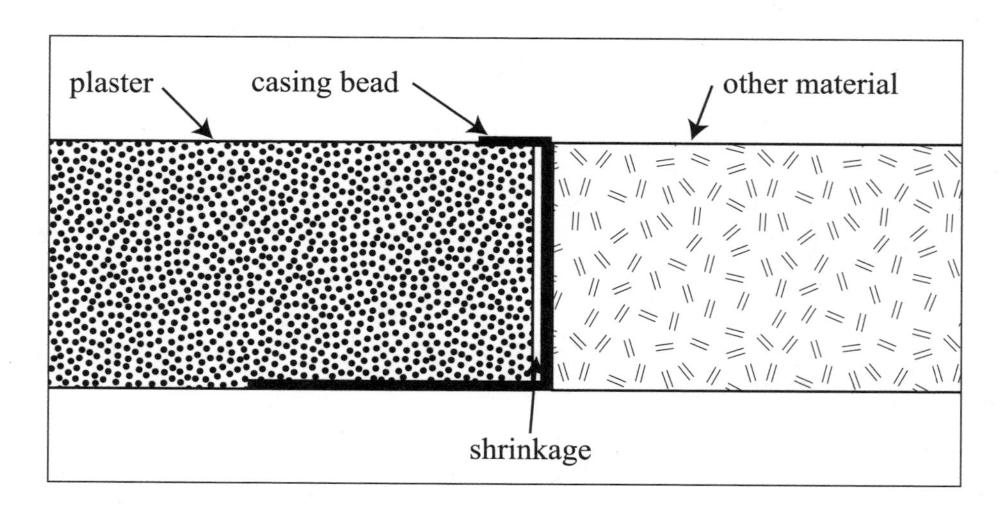

6-6. Cross-section of plaster joint covered by casing bead.

2. Install a metal screed—often called *casing bead*—on the adjoining material to act as a ground and to meet the plaster (many profiles are available or customizable). Metal screed and casing bead is most often used with lime plasters. Be sure to use stops suitable for the plaster depth. Most cement stucco casing bead is made for ⅞" thick stuccos, yet many straw bale structures have plaster skins 1"–1½" thick and even up to 2" thick. Custom metal channel can be ordered from metal fabricators and metal roofing suppliers.
3. If the adjoining material is wood, a recess milled into the wood edge will cover and protect the joint and effectively hide the separation. Similarly, milling the wood at a recessed angle mostly conceals the crack, but offers less protection.
4. Use trim to cover the joint. This requires the finish plaster to be level with the sub-trim. The trim piece attaches to the sub-trim and extends onto the plaster, covering the joint between them.

Next, carefully prepare the bale surface (covered in Part 2), design effective mixes, and use proven installation practices (covered in Part 3).

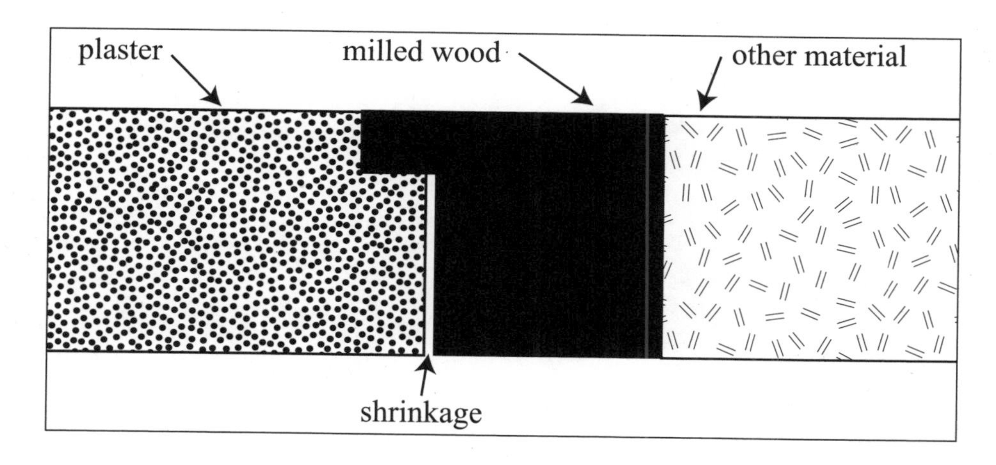

6-7. Cross-section of plaster joint covered by milled wood.

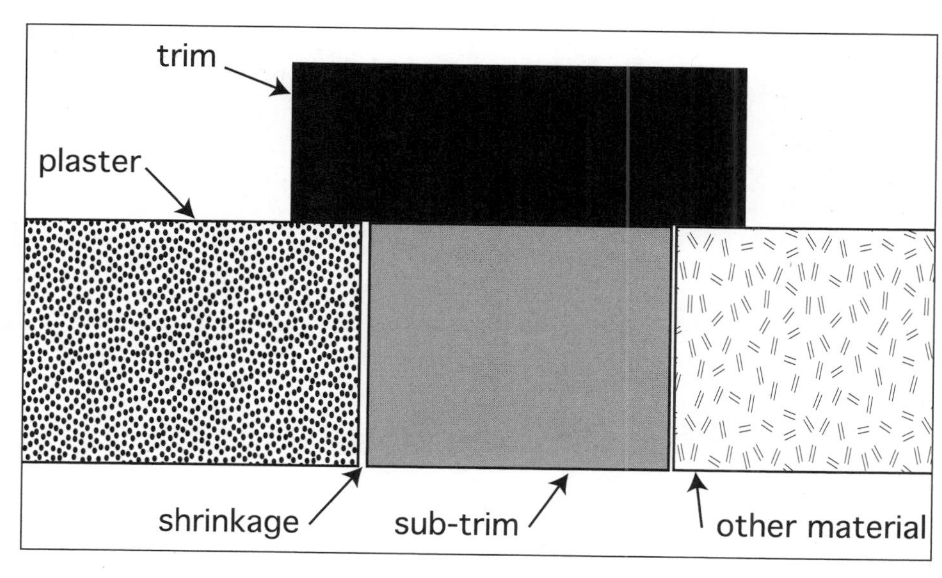

6-8. Cross-section of plaster joint covered by trim.

6-9. Visual impact of window reveal shapes.

Deep window and door openings create opportunities for many options. When viewed from the inside, reveals can be squared, flared, or rounded.

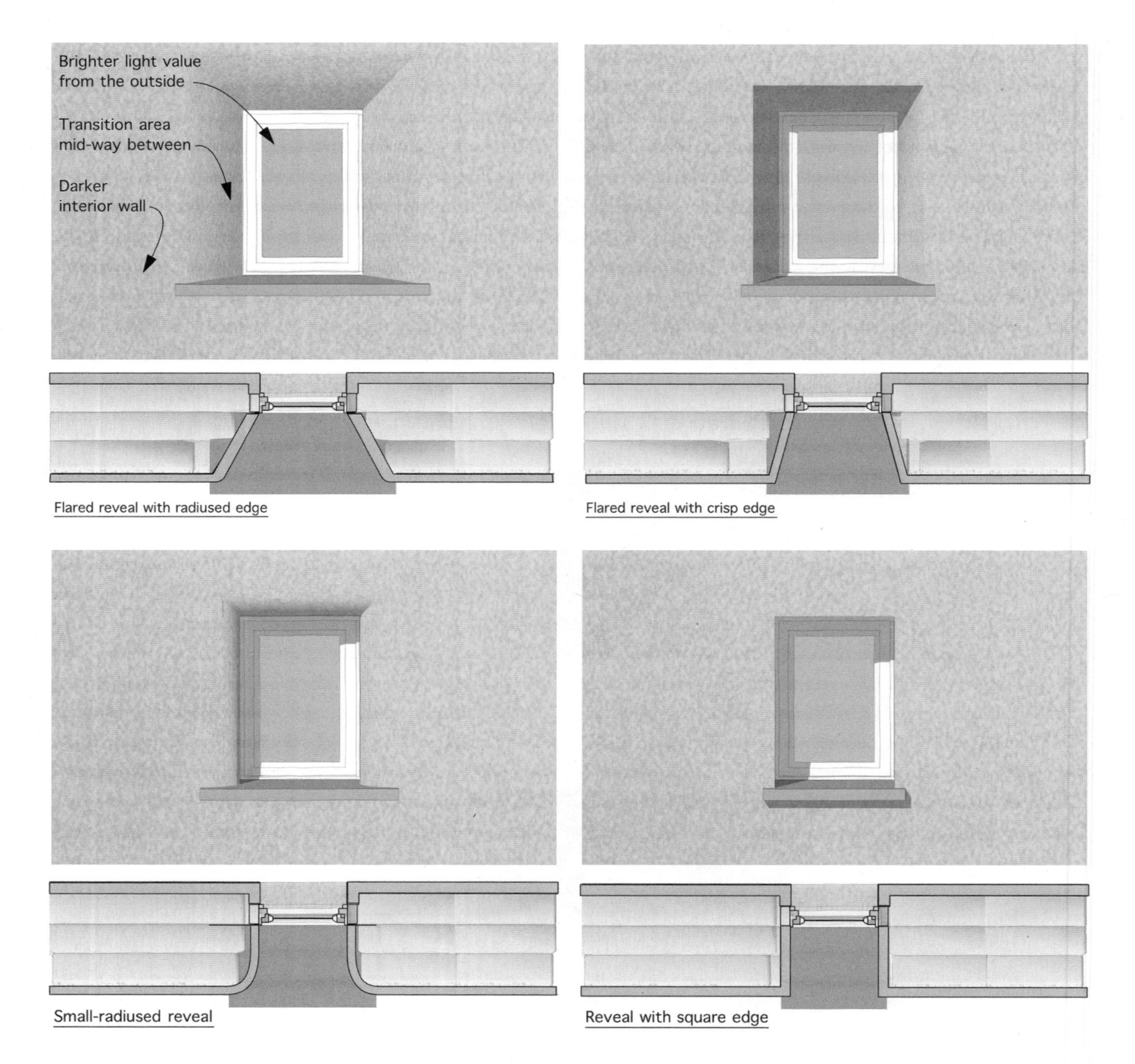

NOTES:

- See color photo section for built examples of various reveal shapes and styles.

- Recessed windows shown. See discussion of flush mount and recessed windows in Chapter 2.

Reveals

Reveals are the sides of a window or door opening that "reveal" the wall's thickness. Much of the charm and comfort we experience with straw bale buildings is derived from the deep window reveals and the opportunity these spaces provide within walls—for window seats, deep shelves, desk nooks, and niches. Unlike thinner conventional wall systems, straw bale window and door reveals become places unto themselves that bring greater visual meaning to these transitions between inside and outside. See Chapter 2 discussion on shaping window reveals and window seats. Plasters refine the shape and finish the surface for the desired light reflectance and appearance.

Material Considerations

Typically, two or three plaster layers cover straw bale walls. The type of plaster and the desired finish determine the number of layers, called *coats.* Lime and cement-lime plasters are usually applied in a three-coat system with a scratch coat, a brown coat, and a finish coat. Clay plasters are much more flexible and can be applied with one, two, and three coats.

It's important to avoid through-cracks in exterior plaster that might present rainwater a direct path to the bales. Multiple coats address this concern. The first coat will often exhibit shrinkage cracks, even with good preparation, mixes, and application. The next coat fills and covers those cracks. For all plasters, the finish can be troweled smooth or textured, and it further insures against through-cracks.

Plasters both follow and refine the shape of a straw bale wall. Any shape that can be carved into bales can be plastered—limited only by imagination and skill. Although not covered in this book, plasters offer endless possibilities beyond simply finishing a wall. They can be used to produce decorative reliefs, window and door surrounds, trim elements, countertops, tubs, sinks, escutcheons, baseboards, and more. Plasters can be textured, incised, or carved in place. They can be stained, oiled, waxed, adorned with fresco, or layered in multi-colored coats. The decorative and ornamentation possibilities are endless and are born of ancient craft; we have only just begun to explore plaster's expressive capabilities in straw bale buildings.

Plaster Composition

All plasters consist of at least two main components: a binder and an aggregate. Many plasters also contain fiber. Binders act as the "glue" surrounding and "binding" the other ingredients together. Binders also give wet plaster adhesive and pliable qualities. Clay, lime, cement-lime, and gypsum are the most common binders used in plasters for straw bale buildings. Binders define the application and finish qualities of the plaster, even though they are not the largest part of the plaster mix. Use both aggregate (usually sand) and fiber (often chopped straw) to control cracking. Aggregate, in addition, provides most of the plaster's compressive strength.

Exterior plaster used on conventional buildings in North America is almost exclusively cement plaster, sometimes with acrylic or other additives. Where interiors are plastered, gypsum plaster is typically used, though plastered interior walls are rare, given gypsum board's predominance. Conventional cement plasters are generally not suitable on straw bale walls because of their very low vapor permeability, close to 1.0 perms, which is the building code definition of a vapor *retarder.* Appendix S allows cement plaster if it contains at least 1 part lime to 6 parts cement. This will achieve the 3 perm minimum vapor permeability required in Appendix S (see AS104.3). However, a higher vapor permeability is better, and more lime increases permeability. When the project requires a plaster containing cement—for example, for a stronger structural wall, or great durability, or decreased setting time—use a cement-lime plaster with at least 50% lime. Appendix S contains a category for cement-lime plaster where the cement-to-lime proportions range from 1:3/4 to 1:2 (see AS104.4.7). Soil-cement plasters combine a small amount of cement (between 10% and 20% by volume) with a clay soil or clay and sand (see AS104.4.4). Gypsum is a good binder with high vapor permeability, but it doesn't resist weathering and is only used on the interior of straw bale walls (see AS104.4.5).

Refer to other sources for more details about working with cement, soil-cement, and gypsum plasters. This text will focus on clay and lime plasters—the two plasters used most often on straw bale walls in the United States and in Canada.

Clay or Earth?

Some builders make a distinction between the word "clay" as describing a more processed material and "earth" as less-processed material, such as site soil containing clay. Although the differences between them can be significant, the binder is still clay, so we'll be using the term "clay" to describe both.

Clay Plasters

Thirty percent of the world's homes are covered with clay-based plasters. They are common in places with abundant clay soils and long traditions of using earth for building. In North America, clay plasters and earthen building traditions are strongest in the Southwest.

Clay plasters were the first kind of plaster known to humankind, and to this day they can be found covering buildings throughout the world, from the humble to the majestic and from ancient to modern. Clay plasters differ from other plasters because they do not undergo a chemical change as they harden, so they can be returned to their wet, plastic state. This quality makes them less durable, but easy to repair. Their base nature stays intact from the day they are mixed and applied to the day they return to the earth. They are minimally processed, inexpensive, abundantly available, and have low embodied energy.

Clay plasters can be categorized by their production. "Wild" clay plasters are made from local soils; the "dirt" under our feet. Wild soils with minimal silt and a high percentage of clay can be crafted into great plasters. Projects on rural acreage might find suitable site soil from a foundation excavation, road cut, pond, or other landscaping activities. Urban projects may find suitable soils onsite or from nearby excavations or local quarries. These soils are *alive*; they are not crushed like processed clays, and they have a wide range of physical and chemical compo-

Types of Site Soil Tests

A few simple tests can help assess a site soil's suitability for plaster; all of them have been described extensively in other New Society Publishing books, and elsewhere. Here we present just a summary:

Ball Test: Wet a handful of soil until it's the consistency of putty, then shape it into a ball. If cracks appear in the ball, or the ball shatters when you drop it on a hard surface, it does not contain enough clay.

Worm Test: Roll the ball into a "worm" shape and form a ring around your finger. If the worm easily segments, crumbles, or cracks, it does not contain enough clay.

Jar Test: Place a cup of soil in a 1-quart glass jar with a lid. Fill the jar about ⅔ full with water and shake vigorously. Allow it to settle. The soil should separate into bands. Aggregates will settle first, followed by the clay, then silt. Organic matter will float on the surface. Clay soils that require little amending will contain about 70% aggregate and 30% clay, with very little silt and organic matter. If tests show that the soil has more or less aggregate or clay than these percentages, the soil can be amended with clay or sand.

sitions. This also means that each clay plaster made from local subsoil (not topsoil, which contains significant organic matter) is unique and will have different strengths and weaknesses. Evaluate each site soil to determine its performance and aesthetic characteristics when used in a plaster.

Clay-rich soils tend to clump and require screening or soaking before use in plasters. In the first method, screen dry soil through ¼" wire mesh, or finer screen. The finer the screen, the smaller the particles, the thinner the plaster that can be applied. With this method, all materials in the soil (rocks, sand, silt, clay, organic matter) that are the size of the screen or smaller will pass through and become part of the mix. This process can be small-scale—shovels of soil screened through wire mesh stretched over a wooden frame, or large-scale—a backhoe pouring bucket loads of soil through an industrial-sized quarry grading screen.

Soaking the soil in water is another approach to processing. Once soaked and agitated, the soil separates, with the heaviest particles—stones, pebbles, and sand—sinking to the bottom and the finest particles—the clay and silt—rising to the top. Drain the water and scoop out the clay for use in the mix.

If site soils are not suitable, consider other sources. Visit nearby stone quarries. Clay soil is often a quarry waste product, and the rejected material—called tailings—often has enough clay for plasters. These already-screened materials can be hauled away for free or a small fee. (But be sure the clay soils are not toxic and are free from other contaminants like trash.) Talk with local builders doing excavations nearby—clay soils can often be had for the price of hauling. Landscape and building material supply yards often sell high-clay soils.

6-10. Screening soil.

Always test these raw materials for clay content. Not every clay is suitable for making plaster; some clays have high expansion and contraction rates that will cause the plaster to shrink and crack.

Consider refined bagged clay. There are many bagged clays of varying qualities, characteristics, and colors available from landscape, ceramic, and building material suppliers. *Mortar clay*, or *fire clay*, is the most common and least expensive. It is often combined with cement to create heat-resistant mortars and to produce bricks, flues, and tiles, but this gray-colored expansive clay can be used to make extremely durable clay plasters. Kaolin, a white-to-pink-colored clay typically used in ceramics, is also a commonly used bagged clay for plasters and clay paints. Test bagged clay for suitability, or consult an experienced plasterer before committing to its use in your plasters.

Lime Plasters

Lime plasters have been used on buildings for over 6,000 years throughout the world. They are found in historic and modern buildings in exterior and interior applications over brick or stone masonry, wattle-and-daub, timber frame with wood lath, cob, adobe, and straw-clay. Because the source materials for lime—limestone deposits and seashores with shell deposits—were widespread, and because they are more durable than earthen plasters, lime plasters were quite common before faster-setting and even more durable cement plaster displaced its exterior use, and highly economical gypsum board displaced both gypsum and lime plaster for interior use. The straw bale building revival of the late 1980s generated renewed interest in lime plasters because they are more vapor-permeable than cement plaster and more durable than clay plaster.

Lime is made from heating limestone at high temperature to change its chemical composition. Limestone is calcium carbonate ($CaCO_3$) with varying amounts of other compounds, such as magnesium carbonate and clays. Limestone is quarried, crushed, and fired in lime kilns at 600 to 2,000 degrees Fahrenheit (315 to 1,100 degrees Celsius). This drives off carbon dioxide (CO_2) in the molecular structure to create calcium oxide (CaO), or *quicklime*. Quicklime must be hydrated—combined with water—(also known as *slaking*) to create calcium hydroxide (CaOH) to be used as a binder. It is either *dry slaked* with just enough water to remain dry and create powdered hydrated lime, or *wet slaked* with a slight excess of water to create lime putty. If the material is to remain quicklime, it is usually crushed, ground, and bagged for commercial distribution. If the material is to become hydrated lime, the dry slaking occurs in a factory-controlled process and is bagged as either Type S (special) or Type N (normal) hydrated lime. In the United States, Type S is the most commonly used builder's lime. Many natural builders prefer lime putty, but quicklime is more challenging to obtain and to process into lime putty.

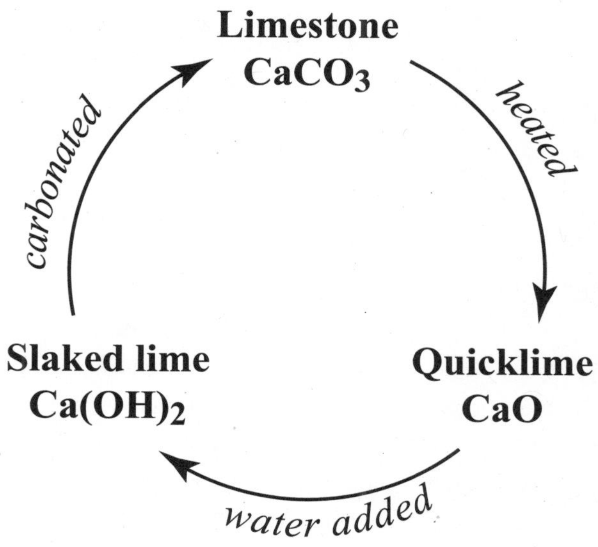

6-11. Lime cycle.

On the jobsite, the plasterer mixes lime (putty, hydrated, or hydraulic) with aggregate and water to make a lime plaster. The lime sets, or cures, on the wall. It loses the hydrogen in its molecular structure, reabsorbs CO_2 from the air, and reverts to its original chemical state of calcium carbonate ($CaCO_3$), hardening around the aggregate. All lime in lime plasters undergoes this cycle.

All limes are highly alkaline, with a pH of 12 or higher. They are caustic and should be handled only while wearing rubber gloves, long sleeves and pants, dust mask, and eye protection. Exposure to both dry and wet limes and lime plasters can cause serious chemical burns. Immediately wash off any lime splashes and soak the exposed area with vinegar to neutralize it. Make a point of having some vinegar on the jobsite.

Three kinds of lime are suitable for plastering: quick lime and its lime putty, hydrated lime, and hydraulic lime.

Quicklime is available as a high calcium, magnesium, or dolomitic lime. It is highly reactive—adding water results in an exothermic reaction that generates significant heat. It is also the most caustic of the three forms of lime. Quick limes do not set under water—they require air to cure and may not fully cure for many months after plaster application. Look for finely powdered materials. Most often, builders wet slake quick lime to create lime putty. This also takes great care, and is typically done at least two weeks, or even months, prior to making lime plaster.

Hydrated limes are available as a powder, and due to the addition of water in their manufacture, they are not as highly reactive as quicklime. They require air exposure to set and also may not fully cure for many months after application. Four sub-types of hydrated limes are available: S, SA, N, NA, but Type S is used most often for plasters due to its widespread availability.

Natural hydraulic limes have essentially the same chemical composition as the other limes, with one significant difference. They are mined where naturally occurring clays, ash, and other impurities found in the source limestone become part of the lime's makeup. This difference gives hydraulic lime the ability to partially set in the presence of, and under, water (without access to air). For the plasterer's purposes, these limes have a faster setting and initial curing time. Different grades of natural hydraulic limes denote their strength and hydraulic set. NHL2 is mildly hydraulic, NHL3.5 is more hydraulic, and NHL5 is very hydraulic, meaning it has the fastest set-time of the three. NHL5 also produces the strongest plaster, but it is the least vapor permeable (though still quite high). Natural hydraulic limes yield predictable results more quickly but can be less forgiving to install because of their shorter set-time. Natural hydraulic limes are not as available as hydrated lime and are more expensive because they come from only a few places in the world. The carbon footprint of natural hydraulic limes can be higher if shipped a long distance.

For lime plasters on structural straw bale walls, IRC Appendix S requires the use of hydraulic lime (see AS104.4.6.3) because it more reliably produces the minimum compressive strength of 600 psi required in Table AS106.6.1. However, the commentary to the same section states that the building official may allow plasters using quicklime or hydrated lime if the plaster demonstrates the minimum required compressive strength.

Apply lime plasters in three coats. Because they need contact with air to set, they should be applied with an average depth of ⅜" and not more than ⅝"—which is a typical scratch coat depth given the straw bale's surface irregularities. Coats applied too thickly may sag and be weak and powdery when dry. Most lime plasters can be recoated in 10 to 14 days. Once the scratch coat has achieved the wall shape and set up

Lime and Clay Plasters

Early in the straw bale building revival, builders and plasterers thought to combine clay and lime plasters to take advantage of both of their desirable properties: clay plaster's lower cost and easy workability and lime plaster's better weather resistance. Many straw bale buildings received exterior clay plaster scratch and brown coats, and a lime plaster finish coat. While some of these walls have stood the test of time, encouraging some natural builders to recommend this regime, others have been costly failures. Lime plasters are more rigid than clay plasters, and clay tends to swell when wet. The bond between coats is largely mechanical—the thinner lime finish plaster gripping the roughened or scratched clay plaster brown coat. Although lime plaster does not erode, it does wick water. If rain soaks through the lime plaster and wets the clay plaster underneath, the clay plaster can swell and break the bond with the more rigid lime plaster, causing it to delaminate. Some builders have experimented with "stabilizing" the clay base coats with lime so they swell less when wetted, with mixed results. There are too many variables—different thermal properties between clay and lime, clay and aggregate characteristics, application, and site weather conditions—for us to recommend or discourage this approach. We suggest testing large samples before committing to this plaster application method.

sufficiently, apply the brown coat at a more consistent ⅜" thickness. Apply the finish coat ⅛" to ¼" thick.

Lime plasters cure under moist, mild weather conditions. The plaster surface must be kept damp; ideal curing temperatures are between 45 and 75 degrees Fahrenheit. The curing rate slows and even stops at warmer or cooler temperatures. Freezing conditions will ruin a fresh plaster. The available water will expand as it freezes, causing the plaster to spall and delaminate. In regions with hot summers and cold winters, apply lime plasters in the spring and fall. Shading walls exposed to direct sun and misting them with a garden hose can extend the application season. If frost threatens a fresh lime plaster, protect the wall with tarps and heaters.

Aggregates

Aggregates are often the largest mix ingredient by volume. The inert "filler" in a plaster, aggregates give the plaster compressive strength and help prevent cracking. Sand is the most common aggregate, but not the only one. Some site soils contain enough naturally occurring sand or small gravel and do not require additional aggregate. We call this lucky circumstance "ready mix."

Sand for scratch and brown coats should be coarse, sharp, well-graded, and clean. *Well-graded* sands have a mix of particle sizes which will minimize voids. Visualize a baking pan filled with ping-pong balls—there will be noticeable voids between the balls. Adding marbles to the pan fills some of those voids without displacing the ping-pong balls. Adding BBs fills still smaller gaps. A liquid poured into the pan fills the remaining spaces. In a plaster, the binder fills the remaining

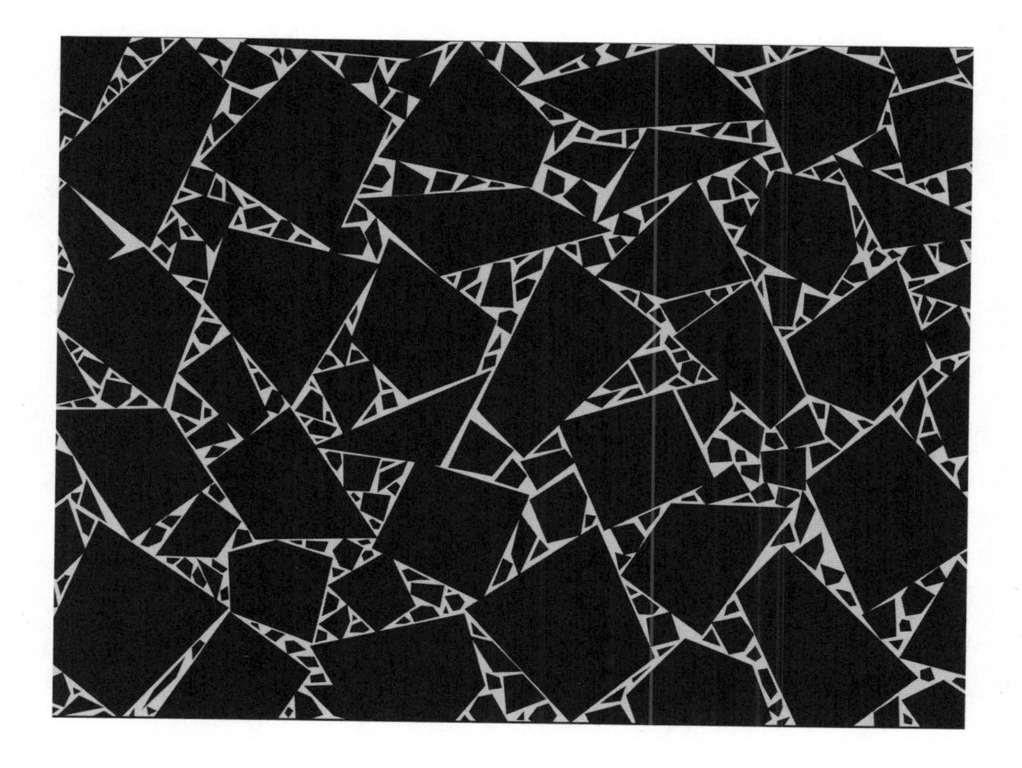

6-12. Particle size distribution in plaster matrix (enlarged).

spaces between sand particles (and fiber when included). Poorly graded, uniform sands produce a weaker plaster that is more prone to cracking. Evaluate sand size distribution with a sieve test or ask for grading tests from bulk sand suppliers. The ideal plaster for thicker base coats has a graded sand distribution that resembles a "bell" curve: fewer large and fine particles blended with more medium sizes. Finish coats are much thinner and use several sizes of finer sand. Sand particle sizes are described by the mesh size they pass through, ranging from No. 10 (large) to No. 100 (fine) to 300 (extra fine) and beyond.

The larger the aggregate, the thicker the plaster can and must be. The smaller the aggregate, the thinner the plaster can and should be. A plaster thickness must be at least twice the depth of its largest aggregate particles, and preferably three or four times deeper. What most material suppliers call "plaster sand" is suitable for all plaster coats. A coarser mix, often sold as "concrete sand," can be used for scratch and brown coats. Both include a variety of particle sizes, up to ⅛" (No. 6 mesh) for concrete sand and up to 1⁄16" (No. 18. mesh) for plaster sand.

Suitable plaster sands are found in bulk at masonry and concrete suppliers or bagged (often) at hardware or landscape supply stores. Finish coat aggregates can include plaster sands, silica sands, marble dust, and mica. Their sizes range from relatively large to very fine—60 mesh and up. Finer aggregates can be found at some masonry and plaster supply companies or at landscape, hardware, and ceramic supply outlets.

Less-common aggregates include crushed limestone, perlite, crushed terracotta, crushed glass, and glass beads. Rice hulls (the removed coverings of grains

Sand Particle Void Analysis

The often-referenced 1:3 binder:aggregate ratio is a useful general guide, but since the space between aggregate particles varies with different sands, it's wise to evaluate how much binder you'll need for a given volume of sand to avoid plaster shrinkage and cracking. A sand particle void analysis refines the ratio. Measure one cup of the sand and dry it—either in the sun or in a microwave or conventional oven, stirring to evaporate all the moisture. Measure and pour one cup of the dry sand into a jar. Then measure one cup of water and slowly pour it into the dry sand. When the liquid begins to pool on the sand's surface, it has filled all of the voids—the sand is saturated. Note how much water you poured. It's probably about 1/3 cup, but could be more or less, depending on the sand. That's the amount of binder needed for 1 cup of sand. Depending on the sand, the ratio could be anywhere between 1:2 to 1:4. A sand particle void analysis is most useful for evaluating mix ratios for thicker plaster coats. The surface area of the aggregate (and fiber when included) per volume is an important factor in plaster mixes. The binder must "coat" the surface of all particles, and so the same volume with finer particles will require more binder. Finish plasters sometimes use more binder to achieve aesthetic and performance goals.

of rice) can be considered aggregate as well as fiber, but they are comprised primarily of silica and bear closer resemblance to sand. Rice hulls are strong, both sharp and round, lightweight, and insulative; they do not absorb water, and they resist decay.

Fiber

Fiber adds tensile strength, which reduces cracking. Like fiber in cloth, it weaves a plaster together, helping it flex. Many fibers are suitable for use in plasters. Select fibers both for tensile qualities as well as aesthetic appeal.

Synthetic fibers like fiberglass, polyester, nylon, and polypropylene are sometimes used in cement plasters and can add tensile strength to lime mixes as well.

Chopped straw added to clay plasters provide tensile strength and aesthetic interest, and some builders add it to lime plasters, too. Two-inch lengths work well for scratch and brown coats. Use more finely chopped and screened straw for finish plasters. If the first coat must be applied over and into reinforcing mesh, smaller straw lengths pass through more easily and encase the mesh. Other plant fibers, including coir, hemp, sisal, soap root, yucca fiber, lavender, and cattail, can be used as a tensile component and for visual effect in clay plasters.

Hair—human or animal—has many more strands per square foot than larger fibers, but it can be hard to distribute evenly in a mix. Cellulose insulation or

Ordering Materials

Premixed bagged plasters (containing lime and sand) have coverage information printed on them, e.g., 96 sq. ft. at ⅛" thickness. To figure the coverage for bulk materials, visualize a cubic foot sliced into the plaster thickness you need. A cubic foot of 1" slices covers 12 square feet, and ½" slices cover 24 square feet, etc. Use the elevation drawings to determine the project's plastered wall surface square footage (don't forget to include the reveals!). Divide square footage by the cubic foot coverage of a plaster thickness to arrive at the cubic feet of material you'll need. For example, if you have 1,000 square feet of wall area, and the total plaster thickness will be 1½", a cubic foot of plaster will cover 8 square feet, so you'll need 1,000 ÷ 8, or 125 cubic feet of plaster.

How much sand and binder? Since binder fills the space between sand particles, it mostly disappears into a volume of loose sand. So, in the example just given, you would order 125 cubic feet of sand. In the United States, bulk sand is delivered by the cubic yard (27 cubic feet). So here, 125/27= 4.62 yards (round up to 5). As for binder, if you have determined that the optimal mix ratio for your project is 1:3 binder:sand, you have a 4-part mix that needs 125 cubic feet sand. 125/4 = 31.25 cubic feet of binder. If you're planning to use bagged clay or lime, confirm how many gallons are in a 50 lb. bag. If it's 9 gallons, divide by 7.48 gallons (number of gallons per cubic foot). The 50 lb. bag holds 1.2 cubic feet of binder. 31.25 cubic feet of binder/1.2 cubic feet per bag = 26 bags.

paper pulp fibers have a high shrinkage rate, but when mixed with straw fiber, they contribute a finer strand, filling in between the larger straw pieces.

Aged manures from grazing animals—horses, cows, elk, and deer—can be dry screened or shredded and used in clay plaster mixes. The enzymes in digested plant fibers may also add pliability.

The amount of fiber in a mix depends on its purpose: in general, the thicker the plaster, the more fiber it will need. A "high-fiber" clay plaster base coat might be almost equal parts binder and fiber. Lime plaster base coats may have as much as a quart of synthetic fiber per 15 gallons of mix. Many finish plasters have little or no fiber, or fiber added only for visual effect.

Additives

Additives can improve a plaster's function, workability, durability, water resistance, and aesthetic value. Adding Portland cement to Type S lime plasters causes them to set faster, allowing the next plaster coat to be applied sooner. This may be necessary to meet schedules or if freezing weather threatens. The amount of cement varies depending on the need—from 5% to 84% of the binder—the code limit that ensures minimum vapor permeability. Flour paste is a common additive to clay plasters, particularly interior plasters and clay paints. It prevents finer clay formulas from dusting and functions like a glue to increase adhesion. Inexpensive and easy to make from white flour, it can also be purchased in powder form as gluten, a food additive. See Resources at the CASBA website, www.strawbuilding.org, for recipes.

Casein—a milk protein—makes simple paints and aids in the pliability of some clay plasters. The mucilage from prickly pear pads, or nopal juice, is a natural polymer that increases pliability and resistance to erosion. Borax added in small portions inhibits mold growth and decomposition as clay plasters dry, though adding too much causes surface crystallization. Essential oils, clove in particular, adds antimicrobial and antibacterial properties to clay plasters as well as scenting the air and wall for a short period of time. Linseed oil in small portions adds water resistance to exterior clay plasters. As little as four cups of linseed oil added to a full 8-cubic-foot mortar mixer has been shown to increase a plaster's resistance to erosion.

Specialty ingredients added to finish plasters for aesthetic or sensory qualities introduce a wide variety of visual effects, colors, and textures. Finely chopped and screened straw provides a lovely visual texture. When burnished, it shines golden against any color plaster. Mica—available in four forms: muscovite, phlogopite, graphite, and bronze—can be used as an embedded fine aggregate for smooth plasters or in larger sizes that appear at the surface to achieve a particular aesthetic. Each type of mica is reflective and has a different color, size, and shape, from black to iridescent and from flour fine to pebble or fish-scale thin. Flowers, leaves, glass, hair, eggshells, and spices can also be used to increase the expressive capacity of plasters. Of course, consideration should be given to the toxicity, stability, and durability of any additive.

Pigments and Finish Plasters

Pigments are created from many different materials—plants, animals, minerals, metals, and many chemical compounds. Use pigments that are non-toxic and stable. Generally, use mineral pigments in plaster. Pigment nomenclature can be confusing; if you are sourcing your own pigments, look for the above-mentioned criteria. Mineral pigments commonly used in cement work can be found at local pottery, hardware, and landscape supply stores. For more selection, seek pigment suppliers online. Prices for pigments vary widely: 10 grams of lapis lazuli ultramarine costs nearly ten times more than 5 pounds of iron oxide. A number of premixed pigmented finish plasters can be purchased online.

If making your own finish plaster, create a workable binder and aggregate mix. It typically takes about 1⅓ cups of dry material to make about a cup of wet plaster because adding water actually causes the material to lose volume. 1 cup of wet plaster will cover a square foot at about 3⁄32" thickness. To begin color tests, cut or mark scraps of drywall or plywood into 12" squares. Stir a small amount of pigment, e.g., 1 tsp. iron oxide, into 1⅓ cup of dry material (binder and aggregate), then add a small amount of water and mix to a paste-like consistency. Trowel it onto a one-foot square. Add slightly more (or less) pigment to the next 1⅓ cup of dry mix and repeat. Create as many samples as you like. Make careful notes of the amount of pigment in each sample. When they have dried or cured, you'll have graduated color samples. Some plasterers also use this exercise to scale up from the one-square-foot sample to the square footage of an entire wall or project. Simply multiply the chosen sample mix ingredients (binder, aggregate, and pigment) by the number of square feet of wall to be plastered. If a thicker plaster is called for, add appropriately. For example, a ⅛" thick plaster requires about 25% more material than a 3⁄32" thick plaster.

Pigments are easily added to alter the color of a finish mix. They also can be applied topically using fresco techniques or mixed into other mediums such as glazes, oils, or other transparent media to stain, seal, and patina plasters.

Your imagination is the only limit to using materials that increase your creative use and pleasure of plasters!

Finding Plasterers

Lime and gypsum plasters were once very common but faded from use as modern finish materials became available in the middle of the 20th century. These newer materials are inexpensive, yield good results, are widely available, and require less time and skill to install. Today, most professional plasterers are trained to apply predominantly conventional materials: cement-based and synthetic plasters. Few are skilled with gypsum plasters, and fewer still have experience with lime and clay, let alone plastering a straw bale building. Finding a professional plasterer familiar with these materials on straw bale buildings can be challenging. This may explain the recent popularity of books and workshops about clay and lime plasters, and the many owner-builders willing to train themselves to plaster their own homes. We applaud this DIY approach and recommend attending a plastering workshop (or two) or at least consulting with a professional plaster contractor who specializes in these materials.

Part 2: Plaster Preparation for Straw Bale Walls

A plaster is only as good as the substrate it's applied to, and plaster preparation begins during bale stacking. See Chapter 5, Stacking Straw Bale Walls. A properly prepared wall is more easily plastered and results in a finish that protects the walls and supports the building's energy efficiency, moisture management, durability, and aesthetic goals. A poorly prepared wall requires more materials and labor and is more likely to compromise energy performance and lead to maintenance problems.

Install Lathing Accessories

Plaster stops, casing bead, corner bead, grounds, screeds, trim, and expansion joints are all lathing accessories that define the plastered area, provide a gauge for thickness and finish, and protect the plaster edges. Sized for conventional ⅞" stucco thickness, galvanized weep screed and casing bead can be used for clay plastered walls, though they are more commonly paired with lime plastered walls. When using a ⅞" lath accessory for a project that calls for 1" or 1½" thick plasters, plan to taper the plaster thickness toward the edge and use another method to gauge coat thickness as plasters are applied. If the plaster is part of a structural straw bale wall, its bottom edge must be supported to transfer structural loads. See AS106.10. Many plasterers order custom-thickness weep screed and casing bead from local metal roofing suppliers if they want a thicker screed for a thicker plaster or a colored screed edge to accent the finish.

6-13. Casing bead profiles.

Weep screed installed at the wall base should cover the vulnerable joint between the sill plate and foundation. See AS105.3.1. Casing bead usually defines the top of the wall and doorways. If the plaster will run up to and die on the trim around the windows, then casing bead defines the window trim as well. Plaster that runs to windows with mounting fins doesn't require casing bead.

Both screed and bead are cut with tinsnips. Some lathers prefer to attach screed and bead with galvanized nails driven about ½" into the wood framing, then bent over; this facilitates removal and repositioning without damage. Others use a pneumatic stapler or roofing nailer with galvanized fasteners. Weep screed and casing bead

Flashing at Doors and Windows

This step is often handled by whoever installs the doors and windows. Work from the bottom up. Start by flashing the wood framing at the bottom of the rough window opening. Apply a strip of peel-and-stick membrane along the exterior face of the sill framing, extending several inches at each end beyond the rough opening. Form a partial sill pan with another strip of membrane applied to the top of the sill framing that laps down onto the first strip, and that runs several inches up the jambs. Some peel-and-stick membranes are flexible, allowing you to stretch the membrane over these two surfaces; others require a diagonal cut at each corner with a sharp utility knife. Next, install the window, setting the top and side window fins into a bead of caulk. Once the window is fastened plumb and level, place a strip of flashing over the window fin on each jamb, overlapping the horizontal sill flashing installed earlier. Finally, apply a strip of membrane horizontally at the head of the window.

Doors are handled similarly but have thresholds at floor level instead of sills above a wall. Most builders set the door in a full pan flashing with an upturned back lip, similar to the window partial sill pan flashing described above, but usually made of metal or plastic. After installing the pan, proceed as with a window. This flashing system protects the vulnerable areas immediately around the window or door from water intrusion and extends far enough into the field of the wall—at least an inch beyond the wood framing members—to integrate with the plaster. It is the same as on a conventional wood-framed building except that the flashing will not be integrated with a weather-barrier membrane "house wrap" or building paper used over wood-framed walls because the plaster alone functions as the weather barrier on a straw bale wall.

are usually installed after the bales are trimmed and all the gaps stuffed with either dry straw or straw-clay. This step usually follows window and door installation because the plaster screed and bead attaches over flashing. Attach the screed level and the casing bead level, plumb, and tight to any trim, or it will stand out.

Paper

Papering bridges dissimilar materials. When plaster covers a joint where straw meets a different material—wood, metal, or masonry—prepare the joint to prevent cracking. Straw, wood, metal, and plaster expand and contract differently with humidity and temperature changes. If plaster adheres to joints between these elements, any movement results in a telltale crack, showing the joint's location beneath the wall finish where plates, posts, beams, window and door support framing, or interior wall framing are embedded in the straw bale wall surface. Building paper stapled over wood framing separates wood movement from plaster movement and doubles to provide code-required protection for the wood from potential moisture damage. The commentary for AS105.6.8 suggests that the paper lap at least 1" over the bale surface; many plasterers lap this joint as much as 6". Although this code section contains exceptions to papering wood not wider than 1½", it's good practice to paper any wood that will receive a plaster finish. This is a greater concern on the exterior, where thermal and moisture variations are greater, and a crack could lead to water intrusion; some builders don't paper wood on the building's interior.

One layer of building paper is usually sufficient for clay plaster. Because lime and cement-lime plasters can bond to the paper and wick moisture from the exterior wall surface into the wood framing, many builders use a double layer of Grade D building paper. One layer may bond with the plaster, while the other serves as a drainage plane, directing water that may wick into the wall down and away. Appendix S also allows a single layer of No. 15 asphalt felt; more recently introduced synthetic vapor-permeable products have been used.

Building papers and asphalt felts are sold in 36" wide rolls—ideal for covering large expanses of sheathing typical in conventional construction. However, in

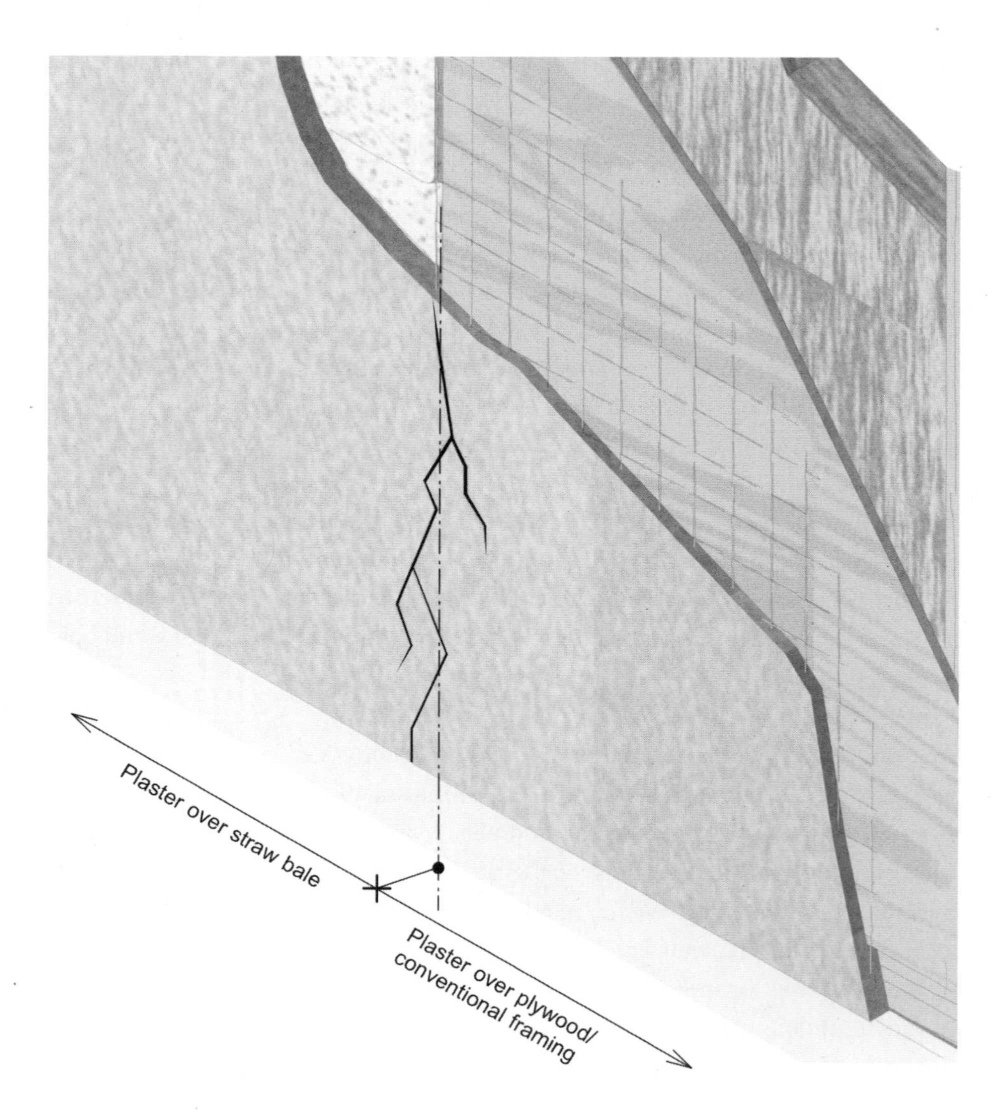

6-14. Crack over poorly prepared joint.

straw bale building, the paper is only needed to cover wood framing members, so the paper must be cut into long strips that cover wood framing and lap onto the bale surface. Unroll the paper onto a long jobsite work table or plywood subfloor and cut it with a utility knife and straight edge. Or use a reciprocating or chop saw to cut sections off the roll. Note, though, that asphalt-impregnated paper will gum up a saw blade. An alternative to so much cutting is to use synthetic products, some of which come in 6", 8", and 12" rolls.

If laps are necessary, install building paper shingle fashion to direct water to run *off* the wall, not into it. The horizontal paper edges should overlap paper beneath it a minimum of 3"; vertical edges should lap 6". Always imagine how water would run if it flowed down the paper surface. Start installing paper at the sill plate and work up the wall, with the paper covering posts lapping over the paper on sill plates, etc. Paper installed over framing under windows should lap over the paper covering window bucks and continue on up to the top plate. Fasten building paper with a hammer tacker or staple gun, but if it's going to be left on the building for some time prior to lathing, many builders use cap nails where washers secure the paper better against wind tearing it loose. See flashing examples throughout detail illustrations, especially exterior window details, 2-23, Wood Walls Above Straw Bale Walls, and 2-26, Plaster Finish at Platform-Framed Floor.

Building paper lapped over framing hides the framing edges, making it difficult to accurately attach lath or mesh—staples driven into the straw bales have

6-15. Typical window flashing detail.

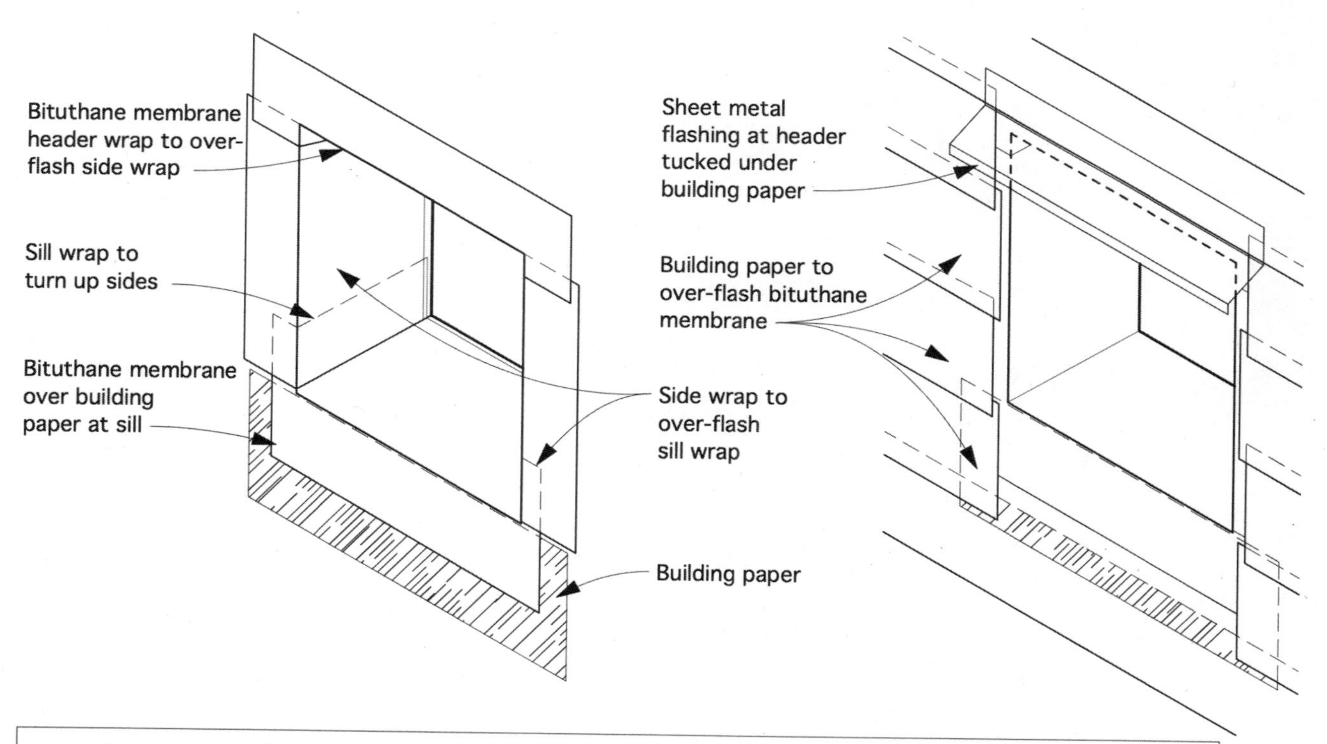

no holding power. Mark the paper so you can still "see" the framing. Feel through the paper and trace the edge of framing with a colored crayon to outline post-and-beam locations.

Bridging over steel members presents a challenge because staples don't secure paper to the steel (or the lath or mesh that might follow). During the bale raising, many builders attach a wood 2× alongside steel members to facilitate paper and lath or mesh attachment. Fastening building paper and mesh to the straw on each side of the steel member may suffice.

Lath and Mesh

Both lime and clay plasters will key into the rough surface of a straw bale without lath. But plasters don't readily adhere to the building paper covering posts, beams, and other joints; lath attached over the building paper provides the key. Lath covers the building paper and continues an inch or more onto the bale surface where landscape pins or tied-through twine secures it.

6-16. Expanded metal lath reinforcing window corner.

Lime Plaster Lath and Mesh

Several types of galvanized steel lath and mesh attach and/or reinforce lime and cement-lime plasters: expanded metal lath, ribbed lath, 17-gauge woven-wire mesh, and 14- or 16-gauge 2×2 welded wire mesh. Fiberglass lath is also available. Expanded metal, ribbed, or fiberglass lath is often used beneath other mesh to hold the plaster and resist cracks at vulnerable locations around windows and doors, vent openings and electrical boxes, corners, and other joints between straw and dissimilar materials.

Lath should be able to hold the plaster weight. A common mistake made early in the straw bale building revival was to use thin-gauge "chicken wire." When not galvanized or very thinly galvanized, it often sags under the plaster's weight, or rusts through the finish. Use the right lath for the job.

Handling Metal Lath and Mesh

Expanded metal or rib lath sheets measuring 27"×8' can be cut into narrower strips with a tinsnips. Wear thick gloves when handling expanded metal lath and other metal lath and mesh to protect against cuts from sharp edges. Stucco suppliers may also sell lath strips that are a suitable width for your project. Working from a jobsite workbench fashioned from sawhorses or straw bales and 4×8 sheet of plywood is more efficient and comfortable than working on the floor. Cut fiberglass lath from the roll with a scissors. Fasten lath through the building paper to the wood framing using a pneumatic staple gun or with hand-driven galvanized staples or galvanized roofing nails. A pneumatic stapler is much faster. Mesh may attach over this lath. Cracks often form over butt joints in lath, so overlap edges a minimum of 1". Fasten the lath flat to the wood members using as few fasteners as possible because every hole in the building paper or peel-and-stick flashing is a potential path for water to enter the wall.

Galvanized mesh comes in rolls that vary in height and length. 17-gauge woven-wire lath is available in 3'×150' rolls. Welded wire mesh come in 4' and 5'×100' rolls. Mesh needs to be cut to size from the roll, then attached to the wall. The building's unfinished floor can be a convenient work area for unrolling, measuring, cutting, and folding mesh, but you should protect finished floors with plywood. To speed the process, mount the roll on a rack fabricated from 2×4 "saddles" and a length of pipe. Then mark the floor for the desired mesh length(s), pull mesh to the mark, and cut at the roll. Tinsnips cut 17-gauge lath; many builders use a cable cutter for thicker gauges. If working near exposed straw bale walls, don't use cutting tools that throw off sparks. It is all too easy to start a fire that way.

Whether run horizontally or vertically, mesh should be secured to any framing, and it should overlap any lath edges. Use stainless steel staples to secure lath to pressure-treated lumber and galvanized staples in other framing. Overlap mesh edges by at least 3" to prevent cracks telegraphing through at the joint.

When not part of the structural design, mesh attachment is less prescribed. If the plaster plays a structural role (see Chapter 3, Structural Design Considerations), the plans should specify which mesh is to be used, where it attaches to framing, minimum laps, areas at the sill and top plate that require a second layer of mesh, staple gauges (lengths and type), and attachment frequency. Follow *all* of these specifications to ensure that the wall performs as designed.

Installing mesh in structural applications calls for particular application methods. For example, if plans call for a 2×2 welded wire mesh attached via two rows of staples to a 4×4 sill plate and a 4×10 beam, it's best to staple the mesh at the sill plate first, because it usually has only 3.5 inches of fastening height compared with 9.25 inches for the beam. Metal lath coming off a roll has a "memory" and wants to "spring," making it difficult to pull flat. If you attach the mesh to the beam first, you might find that the 2×2 wire intersections don't align well at the sill plate—allowing only one row of staples where two are required. Conversely, if you have a wide rim joist at the bottom of the wall and a narrower box beam at the top, you might reverse the order of attachment: staple the mesh at the top first, then pull it down to the bottom where you have more attachment surface.

6-17. Straw bale mesh.

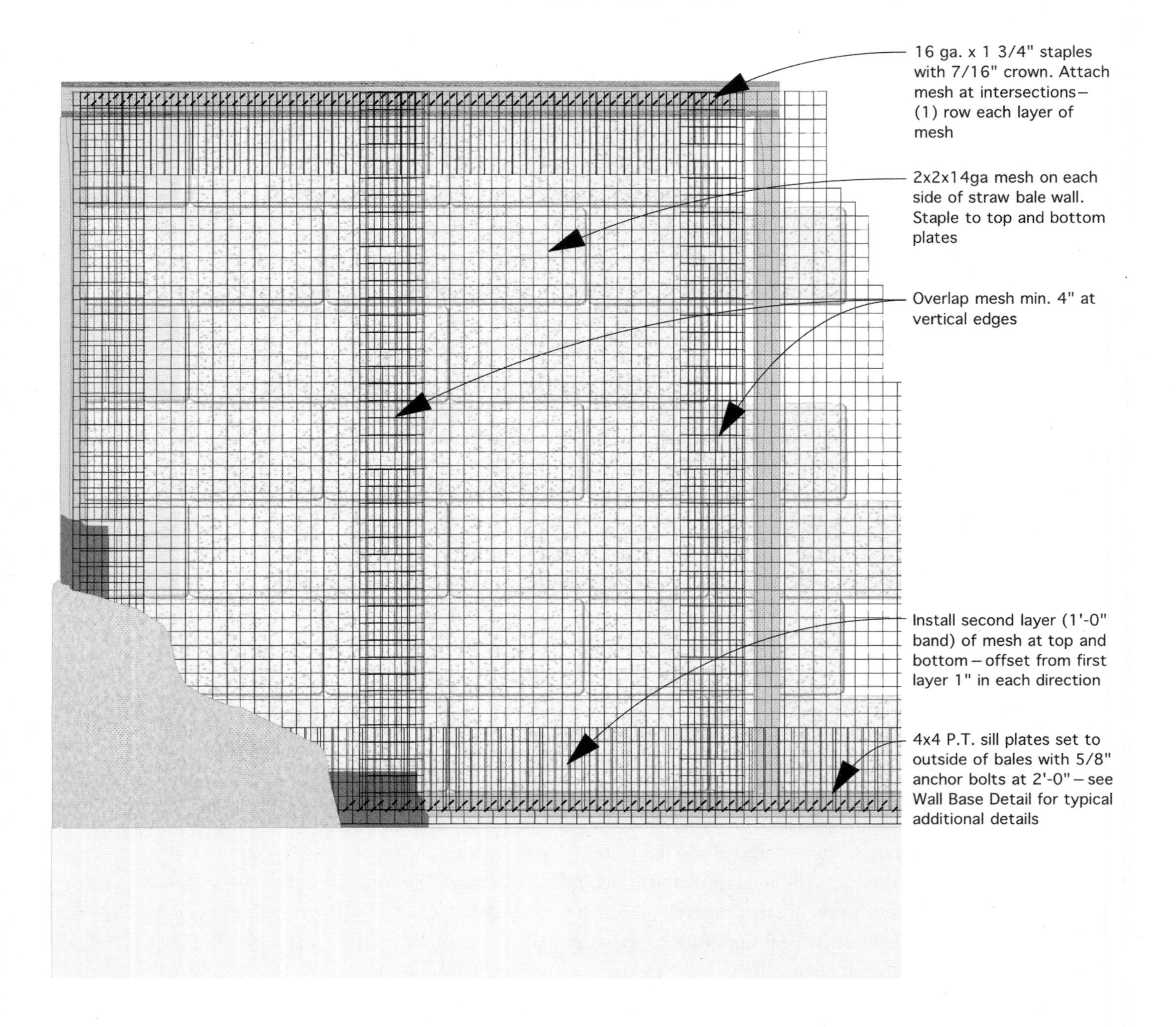

NOTES:
- This detail illustrates the mesh required to create a very high-performance shear wall, with typical lapping and fastening.

- Mesh and fastening should follow the requirements in Appendix S or be as designed by an architect or engineer specific to the needs of your project.

Over conventional plywood substrates, mesh is usually run horizontally with 3"–6" overlaps. On straw bale walls, structural mesh runs continuous from top plate to sill plate. Appendix S requires a minimum overlap for structural mesh of 4" at vertical edges and 8" at horizontal edges (Section AS106.9.1). Pull the mesh tightly against the bale wall surface before stapling through the building paper to the framing.

After all of the lath and mesh has been attached, look over the walls again to make sure all mesh lies reasonably tight against the bale surfaces. Builders use several methods to keep mesh tight to the bales, and one or more may be required:

- Through-tie the exterior and interior mesh together with a baling needle and twine. This is more easily done with two people, one on each side of the wall. One uses a baling needle to pass two ends of a length of baling twine through the wall about 12" apart. The other ties the ends together using the same technique employed to resize bales—make a loop in one end, pass the other end through it, then use the loop as a pully to draw the twine and mesh panels tight. The plans may prescribe the through-tie spacing (e.g., every bale course at 2' centers). Follow the plans, but also tie through wherever mesh bulges away from the wall.
- Pin the mesh to the bale wall with galvanized landscape staples or other manually bent wire, called *Robert pins* (much larger than similarly shaped "bobby pins") pushed by hand or driven by a hammer. Many builders barb, or "hook" the prongs of the Robert pins with pliers to increase their holding power.
- "Fold" wires at areas that bubble away from the bale wall using a lineman's pliers. This takes up slack in the mesh, drawing it closer to the wall. Take care

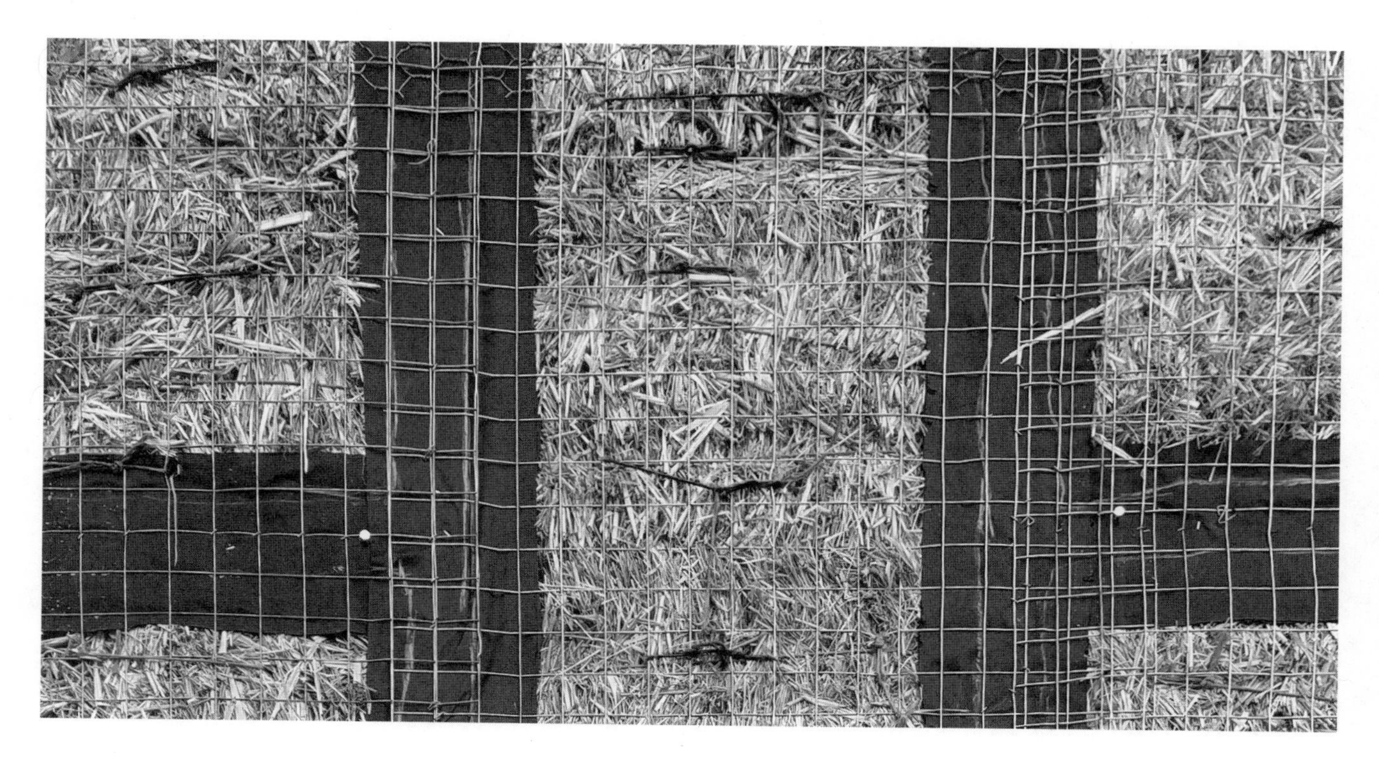

6-18. Mesh tightly pinned and tied through to bales.

at edges and overlaps so that the mesh lays flat against the bale surface—use pliers to "twist" adjacent wire strands together.

Most structures carefully prepared for plastering will use two if not all three of these methods. Keeping lath tight to the straw bale wall surface will take time, but the reward comes in a faster plaster application.

Clay Plaster Lath and Mesh

Clay plasters can be applied over metal mesh or lath; however, because clay is hydrophilic, galvanized metal could eventually rust. Many builders instead use fiberglass lath or *bio-lath*: coir netting, hemp cloth, burlap, or reed mats. Clay-plastered shear walls must use 2×2 polypropylene *deer fencing* or welded wire mesh. Bio-lath is available in a wide variety of forms and sizes, from rolls to squares. As with metal lath, it's useful to have a large work area for unrolling, measuring, and cutting polypropylene mesh and fiberglass or bio-lath, which can be done with a utility knife or scissors.

Burlap or hemp cloth are often attached directly to wood framing and plywood that forms window and door reveals and soffits. It can be pinned to bales or stapled to wood when it is dry, or it can be applied after being dipped in clay slip. If installed dry, spray slip over the cloth and work it in by hand to improve plaster adhesion. Wet clay-slipped cloth will stick to wood and bales alike. Allow it to dry before stapling it to the wood substrate; stapling is necessary to prevent it from pulling free when applying the clay plaster. Lap bio-lath several inches onto the bale surface to minimize cracking.

Attaching bio-lath or plastic mesh over the entire wall follows the same principles as attaching metal mesh for lime plasters. Stretch the mesh tightly across the bale surface and staple to framing. As with metal mesh, plastic mesh and bio-lath must be tight to the wall, but still allow plaster to be worked behind to encase it; use the sewing and pinning techniques previously described.

6-19. Installing bio-lath dry and wet.

Appendix S requires an inspection when mesh is required for structural or other reasons. After completing the building's mesh installation, contact your building department to call for an inspection. See AS105.7.

Clay Slip

Clay plasters usually benefit from an *adhesion coat* of clay slip applied to the bales. Make clay slip by mixing water with bagged clay or a clay-rich soil that has been screened to ⅛" size to facilitate mixing; all rocks should be removed because they can clog a sprayer. Ratios vary; if using bagged clay, start with a 1:1 ratio of water to clay. Although making slip can be done in a mortar mixer, many builders use a heavy-duty electric drill and paddle mixer and a five-gallon bucket or a larger barrel. Slip should feel sticky and have the consistency of a milkshake or thick soup. Apply slip by hand or spray it on the wall using a drywall texture hopper and a large air compressor. It's messy! Mask windows, doors, and floors adjacent to the straw bale wall. If clay plasters will not be applied for a few weeks or months, some builders spray slip shortly after the bale stack so the walls are more fire-resistant.

Masking

Protecting adjacent trim, windows, doors, floors, soffit, or ceiling surfaces from plaster drops, splashes, and spills is time well spent when compared to a larger detailing and cleanup effort. Masking tape lines guide the plaster crew in applying the correct depth, and paper or plastic prevents plaster from staining adjacent materials. Masking tape has a *day rating,* that indicates the number of days it can be left on a surface without leaving residue when removed. Use at least a "7-day" tape. Some adjacent finishes may need a special tape. Be sure the tape you use will not pull the finish off the material it is adhered to.

There are two masking techniques. One installs new masking for each plaster coat. Set the inside edge of your masking to the finished depth of the coat you are working on. If the scratch coat needs to come out ½" on the casing bead, trim, floor and ceiling, measure to that point and tape to a consistent depth. Gently remove the tape when the plaster has set up but is still pliable, and tool the edges as needed. When the plaster dries or cures, apply new masking at the proper depth for subsequent coats.

The second technique applies masking for all plaster coats at once. This takes greater skill but saves time and materials in the long run. First, locate the line of the finish coat on walls, floor, and ceilings, and apply the finish coat masking. Next, apply the masking tape for the brown coat. Finally, apply the tape for the scratch coat. The masking tape for the finish coat would also hold down any plastic or cardboard that protects the floor, ceiling, windows, and doors. Masking tape for the brown coat would cover the finish masking, and tape for the scratch would be the final layer. After applying each plaster coat, remove the top-most tape. The masking tape underneath reveals the depth of the next coat. This technique works well with 14-day or longer rated masking tapes and is most effective for interior plasters. When used for exterior finishes, weather conditions could compromise the masking materials, requiring reapplication for each plaster coat.

When plastering above a finished floor, protect the floor by first covering it with plastic or a tarp (to stop moisture), then cardboard (for impact), then drop cloths (to absorb moisture). In addition to masking a depth line, stuff electrical boxes and vent sleeves with paper to keep plaster from filling them.

Part 3: Six Steps for Mixing and Applying Plasters

Now that the straw bale wall has been prepared, it's almost time for plastering.

Step 1: Safety!

Make sure the exterior ground surface is level and free of trip-and-fall hazards. Many builders address this soon after the foundation drains are backfilled, as all phases of construction benefit from a level work area. Plasterers need to focus on the wall, not their feet, and level ground makes for faster, safer work.

Make sure scaffolding and ladders are properly set. Railings and weight-rated planks are required on scaffolding. Be familiar with your jurisdiction's regulations and safe practices for working above the ground.

Crew members should wear clothing and protective equipment appropriate for the job. Lime and cement-lime plaster is caustic—long sleeves and pants, safety glasses or goggles, and rubber gloves prevent plaster burns. Keep a bottle of vinegar on the jobsite to neutralize lime or cement-lime that splashes on exposed skin. Clay plasters aren't caustic but can dehydrate hands, so gloves are recommended. Caps with visors fend off plaster splatters and drops from above. Mixers should wear N95 rated dust masks or respirators when dry-mixing materials to prevent inhaling airborne lime, cement, and other plaster ingredients. Safety glasses protect eyes from most plaster splashes; keep eyewash handy for plaster that gets into eyes.

Gasoline-powered mixers are loud and the exhaust fumes unhealthy to breathe. Mixers should also wear hearing protection when the mixer is running. If possible, mix all plaster needed for a session before the plastering crew arrives, or locate the mixer close enough to the walls to facilitate the work, but not so close that noise and fumes impact the plastering crew. A barrier made with straw bales erected around the mixer may block some sound and fumes. Electric mixers are quieter and produce no exhaust, but they can be difficult to find in the larger sizes often needed.

Step 2: Stage the Jobsite Well

Effective staging ensures a more efficient, safer installation. Locate the mixing station on level ground near the building. Consider how the site drains—or doesn't—when it rains. Set up the mixer and stage the bulk materials (sand, binder, fiber, pigment, water, etc.) nearby. Keep all materials dry and protected from rain and

6-20. Efficient plaster mixing station.

ground moisture. Binders and additives must often be protected from ambient moisture as well. Clear a path in front of the mixer to facilitate pouring the mix into buckets, tubs, or wheelbarrows for transporting it to the wall.

It's convenient if the tool cleaning area is adjacent to the mixing station and has good drainage away from the mixing site. One garden hose can support both the mixing and cleaning stations. Wooden pallets on the ground or sawhorses with planks between them make good workstations for scrubbing and drying tools. When working with a clay plaster, use a large plastic trash container to wash tools—if the water remains free from contaminants, use it in the next plaster mix. Water from a garden hose has sufficient pressure to clean equipment and tools, but a pressure washer more quickly cleans the mixer and ladders. Anyone operating a pressure washer must wear goggles and expect to get wet (wear a raincoat)! If a stream of water alone doesn't dislodge plaster from tools and equipment, scrub them with moderately abrasive pads, brushes, or a scraper, being careful not to scar a tool's working surface. Dry carbon steel tools with a towel to prevent rust, then rub with camellia or olive oil to prevent corrosion.

Step 3: Prepare to Catch Falling Plaster

Even the most skilled plasterers have drops. Place scrap plywood drop boards, cardboard, or plastic at the wall base to catch falling plaster and facilitate cleanup. Plaster that falls in dirt, stones, or straw becomes contaminated and can't easily be scooped up and reapplied.

Step 4: Mix Consistently!

Post the mix recipe on a sheet or board and record each batch. Mixing plaster may be as simple as pouring it from a bag and adding water, or as complex as following a multi-ingredient recipe. Whenever possible, measure the ingredients using containers of known and confirmed volume, i.e., two-gallon or five-gallon buckets. Confirm the volume of the containers being used and mark the line with a permanent marker on the *inside* of the bucket for consistent measuring. If full five-gallon buckets are too heavy for a single person to lift into a mixer, assign two people to that task or use partially filled or smaller buckets. Avoid using shovel scoops because these can vary greatly and throw off the mix from batch to batch. If the mix calls for one unit of binder (e.g., lime or clay) and three units of aggregate, arrange the piles or sacks of ingredients in the order they'll be added to the mixer.

Fill buckets with mix materials before starting to mix, and fill them the same way every time. For example; if you shake a bucket, the materials will settle and you will be able to add more than if you don't shake the bucket. This difference can throw off the mix.

If you have enough time and buckets, premeasure materials for each batch to avoid mistakes. Line up filled buckets in the order they will be put in the mixer. If the mixer can hold two five-gallon buckets of clay, six five-gallon buckets of sand, and 1.5 five-gallon buckets of chopped straw, all of these ingredients should be premeasured and lined up to go in the mixer. As soon as they have been added and while the batch is mixing, fill the empty buckets for the next batch.

Ideally, all of the dry ingredients have a low—or at least, consistent—moisture content, so water added to the mix can be accurately measured as well. However, two factors can affect the amount of water needed in the mix: weather and sand moisture. Weather, including direct sun exposure, exerts a day-to-day (and sometimes batch-to-batch!) influence on mix consistency. More water may be needed on dry days than on humid days. Sand is often delivered wet, and during warm weather, the sand pile's outer layers dry out while the inside remains moist. Add or decrease water as needed to create a plaster that is not too wet or too dry, but just right. Get to know the ideal consistency of the plaster. The mix master and plasterers should check with each other regularly for quality control.

Some plasterers judge consistency by how the plaster mounds up on a hawk. If it is too wet, it may puddle and run off the hawk. If too dry, it may not hold together or be hard to work with a trowel. Observe the plaster; touch it with your hands, move it around the hawk with a trowel and spread it on the wall to see how it moves. Use this observation to check consistency one batch to the next.

It may take 30 minutes to mix enough plaster to support a plastering crew. The mix master can arrive early and begin mixing before the plasterers arrive, or the plasterers can attend to preparation while the first batch is mixing. This may include setting up the site, masking, staging scaffold or ladders, and planning the day.

If using a bagged dry mix, follow the manufacturer's instructions. Otherwise, mix masters have different styles. Some mix all of the ingredients dry before adding water to achieve optimal material dispersion. Others prefer to hydrate the binder before adding the aggregate. Still others add the aggregate first, then the binder, and then fiber; in this sequence, the aggregate agitates the binder as it is added. For fine finish plasters, some mix all ingredients dry for optimal dispersion and then add the mixed dry material to the mix water in a separate container. If there are no guidelines and you don't have a preference, we recommend starting with the water, then adding the aggregate, then the binder, and finally the fiber and any other additives.

Mix each batch for the same amount of time. A clock located near the mixer helps the mix master monitor each batch. Mixing times will vary with the plastering material, but a common time is ten minutes after all the ingredients have been added. This results in a fresh batch every 15 or 20 minutes.

Keep track of the mix materials. Inventory the material before you begin, e.g., 5 cubic yards of plastering sand, 38 bags of lime, etc. At the end of each day, measure the square footage completed and divide it by the number of batches mixed. This gives you coverage rates, helps track materials, and alerts you to the need for additional materials.

6-21. Plaster materials lined up.

Step 5: Wet the Walls and Apply Plaster

Finally, it's time to put plaster on the wall! For clay plasters, moisten the wall surface before applying a plaster. This helps the plaster adhere to the straw surface or to the previously applied clay slip or plaster coat, and it helps prevent the plaster from drying too quickly. When applying the scratch coat, spray the wall with water immediately before troweling on plaster. This step is repeated through the brown and finish coat applications. Always apply plaster to a moistened wall surface.

For lime plasters, lightly spray the bale wall immediately before applying the scratch coat to improve adhesion and prevent the straw from drawing water needed for workability and curing from the lime plaster. For subsequent coats, spray the wall well in advance of plaster application—hours, or even the evening before. This is crucial for properly curing lime plasters.

Be realistic about how much can be done in a day, given the project, the crew size, and their skills. Take into consideration site and building features that facilitate plastering. These include ground conditions, staging, the number of doors and windows, and the wall's square footage. Project features that slow a crew down include uneven ground, the need to transport plaster a long distance to the building, working on scaffolding or ladders, detailing around framing or around recessed doors and windows, and soffit details requiring a lot of hand tooling.

Work as a team. Different materials have different drying and setting times after which they become unworkable, so crew members must work together to accomplish a quality finish. Different crew members play distinct roles that support each other and help the installation flow. In addition to mix master, other critical roles include *hod carrier*, *plaster lead*, *support plasterers*, and *finish plasterers*.

The hod carrier stages plaster tubs or buckets where they are most needed, and communicates information about the batches between the mixer and the plasterers. In addition, the hod carrier can be responsible for many supporting tasks critical to an efficient project:

- moistening the wall surface immediately ahead of the plaster crew
- scooping up dropped plaster, and if clean, returning it to a plaster bucket
- repositioning ladders, scaffolding, hoses, drop boards, and other staging items
- filling cleaning buckets to wash trowels and tools as they are being used

The plaster lead starts on the wall and sets the pace. They may be the most experienced plasterer on the job and will often make decisions about the timing of the application, check quality of both the mix and the other plasterers' work, and determine how much can be accomplished each day.

The support plasterers follow the pace set by the plaster lead and focus on getting a consistent and high-quality layer of plaster on the wall.

Finish plasterers are responsible for accomplishing the final aesthetic of the finish plaster. In addition to the all-important finished surface, they clean up and detail edges and put finishing touches on bullnoses and shaping details.

These roles are often combined or shared by crew members, e.g., the plaster lead may also be the finish plasterer, and the hod carrier may step in as a support plasterer or lend a hand with mixing. Regardless of how roles are divided, everyone must keep in mind that there is one predominant aesthetic to achieve. To accomplish this, the team must look at, go over, and try to mimic each other's work so they leave no "signatures" on the wall. The telltale signs of an incohesive crew can be seen on the wall. Instead of one aesthetic throughout, the top of a wall might be rough while the bottom is smooth and the middle undulating.

Where to start? Typically, plaster installations begin at the top of the wall above a doorway, an inside corner, or on the exterior of a building on an outside corner. The crew moves in one direction around the building, all working the same wet, or *leading*, edge. This is particularly important for the finish coat because a cold/dry joint will stand out. If a cold/dry joint is necessary, try to leave it as an irregular beveled line in the least conspicuous place: an inside corner, behind a downspout, or at a small run of wall above a door, for example.

6-22. Plaster crew working a diagonal.

For most single-story walls, one person—usually the plaster lead—starts applying plaster to the upper portion of a wall, a second person standing below them plasters the middle portion of the wall, and a third person handles the lowest portion of the wall. Since most plasterers begin at the top of a wall and work down so as not to damage newly applied plaster below, a crew of three works a diagonal line across the wall, the top person in the lead, followed by someone in the middle, trailed by someone at the bottom. A finisher may follow behind going over the entire wall for plaster depth and application consistency. It helps if the crew members are equally paced. Windows that interrupt the wall surface may require the middle person or the finisher to skip ahead and work from the edge of a window toward the plaster's leading edge so the crew stays together—none getting too far ahead of the others, but also not getting slowed down by edges and curves around windows and doors. Coordinating these details keeps the working plaster edge wet and decreases cold/dry joints.

Some buildings are designed with obvious plaster stops like sharp inside corners, exposed posts that run the full wall height, weep screeds along the bottom of a wall or soffit trim at the top. Dividing walls into discrete panels makes it easier to manage the installation by providing stopping points. A small crew may be able to complete a single coat over the entire house over several days, rather than one larger crew trying to finish in one day. Cold joints in the scratch and brown coats on the exterior are common, but they, too, should be irregular and beveled to minimize risk of cracking that could extend through the finish. The finish coat should wrap the building in a single day, unless the surface is divided into panels. If the crew is big enough and the building can be completed in a day, it helps to start two crews in the same spot, working away from each other and moving around the building until they meet. This eliminates cold joints.

Weather can exert a huge influence. As mentioned earlier, humid, wet conditions can change the mix consistency and slow the drying or setting of the plaster. Conversely, windy or hot days or direct exposure to sun can cause plasters to dry too quickly and make finish work more difficult. In addition to daily or weekly weather patterns, seasonal factors influence when to plaster. Ideally, apply exterior clay and lime plasters during seasons with moderate temperatures. In most climates, the plastering season starts in the spring with the last frost and ends when the first frost can be expected in the fall. In mild climates, the plastering season can be year-round.

Use shade to your advantage. If the building's south and west sides will be in full sun by late morning or early afternoon, start on those sides in the early morning and work on the east and north sides as the afternoon warms up. Tarps stretched from the roof overhang to the ground protect a wall (and a crew) from the sun and from drying winds. All three lime and cement-lime plaster coats must be moist-cured—kept damp for a minimum of 7 days and, ideally, 28 days when temperatures range between 45 and 75 degrees Fahrenheit. Freshly plastered wall surfaces should be kept moist with a light spray; just enough so water doesn't run down the wall. Depending on the weather and how well the walls have been shaded and protected from drying winds, this may require spraying a few times each day, or even several times each hour. Exterior clay plasters are best applied in weather that is conducive to drying—warm with low or moderate humidity, and breezy.

Wet plasters should never be allowed to freeze. If freezing temperatures are forecast, consider postponing the plaster work until the weather becomes more favorable. If that's not possible, be prepared to "tent" the wall. Create an envelope around the wall surface with tarps or heavy plastic sheeting, and run propane or electric heaters in the tent if needed. If using heaters, also supply adequate moisture to prevent lime or cement-lime plasters from drying too rapidly.

Interior plasters can be applied year-round if the interior climate can be controlled and the materials are stored in a warm, dry place. During the spring, summer, or fall, opening and closing windows is all that may be necessary to control drying or curing. During winter months, be prepared to heat the space and run fans and a dehumidifier if drying clay plaster.

How to Plaster 101

The essence of plastering is quite simple: move plaster material from a hawk to a trowel, then spread it on the wall. Though skilled plasterers can quickly cover a large area with eye-hand coordination and techniques that take many hours or years to develop, people new to plastering can still do the job. Begin by loading a scoop of plaster onto the hawk (a small hand-held platform that holds plaster). Face the wall and use a trowel to cut a smaller portion of plaster from the hawk and press it onto the wall in an upward motion. Repeat (many times). Many straw bale walls are plastered with not much more technique than this. Importantly, during scratch coat application be sure to use smaller trowels (or hands) to work plaster into the mesh or bales. Brown coats, which are laid over the scratch coat, are usually thinner and must be applied more evenly. Experienced plasterers don't overload the trowel. They pull just enough off the hawk to spread evenly without overworking the material, thus moving quickly and efficiently.

Scratch, Brown, and Finish Coats

The first plaster layer—the *scratch* coat—levels the uneven bale wall surfaces. The underlying shape of the wall was established during the bale stack and further refined by trimming during plaster preparation. The scratch coat is the last step in achieving the desired wall shape. Its thickness will vary depending on the straw bale wall: thicker in hollow areas and thinner at bulges if a flat wall plane is desired.

Scratch coats on straw bale walls differ considerably from scratch coats on conventional walls; they are thicker, and they need to be worked into the bales. Use a small rigid steel trowel, a wood float, or a gloved hand to work the plaster into the straw and/or mesh. You are trying to create a strong bond between the straw wall and the plaster skins. Consider applying the scratch coat in a double pass. First work a thin layer well into the straw and/or mesh, and, while wet, follow it with a second layer, building the coat to the desired thickness. Pressing plaster into the bales/mesh with open hands and fingers on the first pass can be more effective than using a trowel, but you should always flatten it with a trowel on the second pass. Conventional walls typically have a ⅜" scratch coat. Depending on the straw bale wall surface, this layer can be from ¼" to 1" thick at different points on the wall. As they dry or cure, scratch coats normally show cracks, which will vary in thickness. These will be filled by the next coat.

6-23. Truth stick.

When the wall must be flat and plumb, the same "truth stick" technique used to straighten a straw bale wall ensures a flat plaster application (See Chapter 5, Stacking Straw Bale Walls—

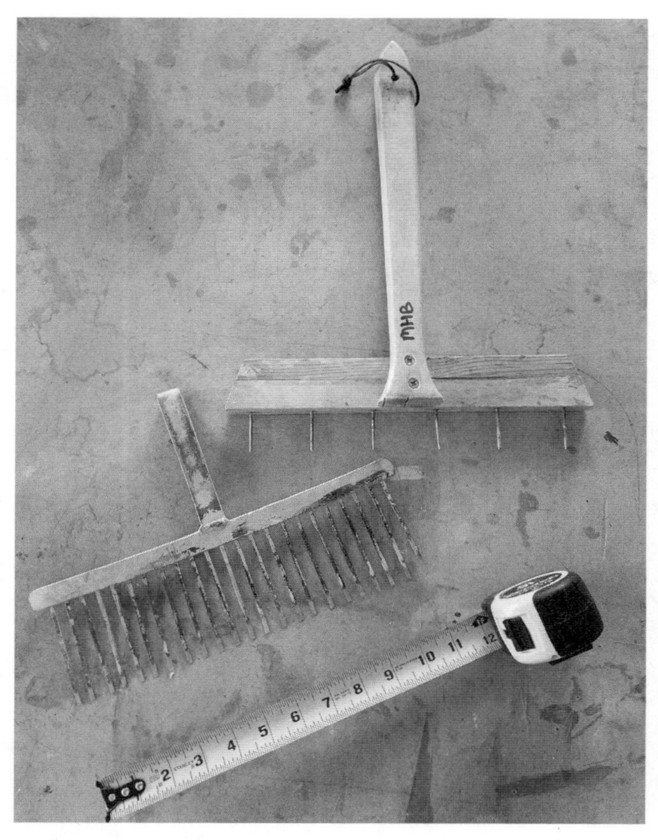

6-24. Scratching tools.

6-25. Brown coat floats and sponge.

Straightening the Wall). Check the plaster plane regularly while the plaster is workable. The finish plasterer may use a long wood float at this stage, adding and removing plaster to achieve the desired surface shape.

After the plaster sets up a bit but before it starts to dry, "scratch" the surface. This is where the layer gets its name. The scratch creates a key and more surface area for the next layer to bond to. The scratch can be done with a scarifier (a comb-like tool made for this), the edge of a trowel, or with a homemade tool (wood with nails or screws protruding from it). Drag the tool of choice over the wall to create shallow scratches, about $\frac{1}{8}$" deep.

For the second coat—the *brown coat*—continue using the same or larger metal trowels or wood floats. Keep in mind that the longer and stiffer the trowel, the flatter the surface will be. Some plasterers call long stiff trowels and floats "truth tellers" because they show the low and high points. Use flexible and pool trowels for curves and free-form surfaces. The brown coat is typically $\frac{3}{8}$" to $\frac{1}{2}$" thick and follows the shape of the scratch coat. It may need to be as much as $\frac{5}{8}$" thick when plans call for a $1\frac{1}{2}$" total plaster thickness. Applying this coat with steel trowels leaves a smooth surface; the finish plasterer floats the still workable surface with a magnesium or wood trowel to texture it, providing tooth for the final coat. This is called a *sanded* surface. Some plasterers lightly scratch the brown coat to provide additional mechanical bond for the finish coat.

6-26. A variety of floats, trowels, and tools for finishing plaster.

If the finish plaster color differs greatly from the natural color of the plaster mix, consider pigmenting the brown coat sufficiently to decrease the contrast between the two layers. This makes any scrapes in the finish plaster that expose the brown coat less noticeable.

Finish plaster coats are usually quite thin and often applied with rigid or flexible trowels of many sizes, depending on the desired finish. Carbon steel and stainless steel trowels are the most common. Achieve a smooth, burnished finish with flexible stainless steel trowels that glide over the wall after the plaster is applied.

Generally, use the same size trowels that were used in the previous coat to help follow the existing wall shape. Use smaller trowels to plaster in constrained areas like corners, windows, doorways, niches, behind columns, and in light wells. Many specialty trowels are available for this purpose.

Step 6: End-of-Day Tasks

Most plastering crews keep tools clean during the day by periodically rinsing them in water. With lime or cement-lime plasters, in particular, there may be a midday tool cleaning. At the end of each plastering day—whether scratch, brown, or finish—be sure to set aside enough time to do the following:

- Clean all tools: mixers, buckets, scaffold, ladders, hawks, and trowels. Hydraulic limes and cement set under water, so don't simply put tools with those plasters in a bucket of water!
- Dry and oil tools that may rust.
- Pull masking. Do this while the plaster is still workable so any plaster edges disrupted by pulling the tape can be pressed back into place with a trowel.

- Bag used masking materials and clean up the jobsite.
- Scratch the scratch coat, float the brown, or smooth over and finish the final coat.
- Press out ridges, trowel marks, gaps, and refine corners, bullnoses and other shaping details.
- Mist lime and cement-lime plasters to assure proper curing.
- Protect lime and cement-lime plasters (all coats) from sun and drying winds, and protect all plasters from freezing temperatures.
- Review how many batches were mixed, how much material was used, etc. Measure the square footage completed, divide by the number of batches.
- Confirm there are sufficient materials on hand to complete the project; if not, arrange for acquiring more materials.
- If appropriate, dry mix any type plaster for the next day or wet mix clay plasters, taking care to store them so dry mixes stay dry or wet clay mixes stay wet.

Maintenance

Cleaning and caring for plastered surfaces depends on the specific plaster, but in general, plastered walls are easy to clean if you keep a few things in mind. Unsealed plasters are porous; oil and grease will stain them. Water splashed on unsealed walls will darken the color temporarily, but they will dry back to normal without permanent effect. Sealed plasters have a higher tolerance for stains and are easier to clean. Lime plasters must not be cleaned with acid-based cleansers; like stone finishes, acid will erode them. Clay and earth plasters must be carefully cleaned with a damp sponge or cloth, but must not be rehydrated to the point of making the material soft enough to be rubbed away.

Plasters are also repairable, often seamlessly. Some plasters can be spot-repaired without having to refinish an entire area. Don't forget to set aside small pieces of clay plaster from the original mix (commonly called *cookies*). When repairs are necessary, rehydrate a cookie and carefully cover chips, scratches, dents, or fill cracks in a wall surface. Many homeowners prefer not to repair anything less than a structural injury because age adds a personality. In the same way that oil from hands on metal door hardware can beautifully patina the surface over time, so can the minor wear-and-tear marks of daily living beatify natural plasters.

Conclusion

When most people think of or remember a straw bale building, what they often envision or recall most are the plastered wall surfaces. We hope the information we have presented on designing for and selecting plasters, preparing the wall surfaces, and mixing and applying plasters will help you create sound, long-lasting, and beautiful straw bale structures.

7

Straw Bale Construction and Building Codes

Since the rebirth of straw bale construction in the 1980s, designers, builders, and owners have often struggled with building permits, inspections, and codes. The 2013 approval of a new straw bale construction appendix for the 2015 International Residential Code (IRC) greatly reduced the likelihood of these struggles.

The most recent version of that effort, 2018 IRC Appendix S: Strawbale Construction with Commentary, follows this brief introduction.

Brief History of Straw Bale Codes in the United States

For almost 20 years before the publication of Appendix S, a handful of US states and local jurisdictions had some form of a straw bale code, beginning with the Tucson/Pima County, Arizona and New Mexico codes. The Arizona code allowed load-bearing straw bale buildings, whereas the New Mexico code did not. A few other states and several local jurisdictions followed with codes that mostly mirrored the Arizona code. Those codes served reasonably well, but they were also flawed, based on limited testing and experience from the straw bale building revival's early years. Increased demand for straw bale buildings, new testing, and more in-depth experience clarified the need for a unified, comprehensive code. A small group of practitioners galvanized to develop and propose what ultimately became Appendix S in the 2015 IRC.

Why Is the IRC Important?

As a model code, the IRC has no legal standing of its own. However, it is the basis for the residential code in virtually every US jurisdiction. The International Code Council (ICC) develops codes through a public process and publishes the IRC and an array of other "I-Codes." Though the organization's name and the names of their codes suggest widespread international use (and that is ICC's intention), in practice these codes have been used primarily in the United States.

Appendix S and What It Covers (and What It Doesn't)

As part of the IRC, Appendix S applies to one- and two-family dwellings, their accessory structures, and townhouses. The large majority of straw bale buildings constructed over the years fit within the scope of the IRC and Appendix S.

However, many non-residential straw bale buildings have been and will continue to be constructed that fall instead under the International Building Code (IBC). The IBC covers all *non-residential* occupancies, as well as larger *multiple residential* occupancies, such as hotels and apartments, but it lacks a chapter or appendix on straw bale construction. To fill that gap, Appendix S authors plan to propose one for the IBC in a future code cycle.

In the meantime, use one of two options to submit a non-residential straw bale building design for permit:

1. Use the building code's Alternative Materials, Design and Methods of Construction and Equipment," section (104.11).
2. Propose using the IRC Appendix S for the project.

The former can be onerous; it requires proof of equivalency with relevant building code requirements, which can mean starting from scratch for every aspect of straw bale construction. The latter uses a well-developed ICC approved code—a defensible position as long as the project meets the technical provisions of Appendix S. Non-residential projects will require design by a registered design professional.

As part of the larger IRC, Appendix S references sections and uses the IRC's language and format. At the outset, Appendix S states: "Buildings using strawbale walls shall comply with this code [the IRC] except as otherwise stated in this appendix." Fully understanding and using Appendix S depends on accessing the entire IRC.

Appendix S offers a mix of prescriptive and performance requirements. It's easier to show compliance by following the prescriptive requirements, but they are limiting. The performance requirements give more flexibility, which is important for the many variations of straw bale construction, but they can make demonstrating compliance more difficult. Any straw bale construction method not explicitly allowed in Appendix S must go through the IRC's Section R104.11 Alternative Materials, Design and Methods of Construction and Equipment.

There are no electrical or plumbing requirements in Appendix S. However, the commentary briefly covers both subjects in its General section.

Importance of the Commentary

The building code itself contains requirements, whereas the code's commentary explains the requirements, describes intent, and gives code-compliance examples. Many design and building professionals don't know that building code commentary exists, and those that do don't use it very often. However most building officials both know about and commonly use code commentary.

The Appendix S commentary was written by 17 leading US practitioners with oversight by ICC staff. It is twice as long as Appendix S itself. The commentary offers a highly useful means to educate local building officials as well as design or building professionals, and it can help give latitude to responsibly construct straw bale wall systems that vary from what Appendix S explicitly allows.

The Building Code Is a Living Document

The ICC and most US states and local jurisdictions update their building codes on a three-year cycle. Every three years, the ICC invites code change proposals. Select committees and ICC voting members (mostly code officials) evaluate proposals in public hearings and vote on them. Anyone can submit a code proposal and testify at the hearings. However, in practice, mostly design and building professionals, code officials, trade organizations, and industry representatives participate in this process.

ICC's model codes are a year or more ahead of most jurisdictions' adoption cycle. For example, the 2018 IRC is the basis for the 2019 California Residential Code (CRC).

Status of US States' Adoption of Appendix S

At this book's publication, California, Oregon, New Mexico, New Jersey, Maryland, and five local jurisdictions in Colorado (the city of Denver and counties of Boulder, Garfield, Gunnison, and La Plata) have adopted Appendix S. It's possible that other state or local jurisdictions will have adopted it by the time you are reading this, and many have used it informally or on a project-by-project basis.

Where a state or local jurisdiction has not adopted Appendix S, project applicants can propose using it on a project basis. This was successfully done for several California projects before Appendix S became effective statewide.

Like all IRC appendices, the jurisdiction that is adopting the IRC must explicitly adopt Appendix S. Also, jurisdictions that adopt Appendix S may amend it. Consult your local building department about which IRC edition forms the basis for its current residential building code and whether it includes Appendix S.

Appendix R: Light Straw-Clay Construction

A related appendix in the IRC on light straw-clay construction was adopted and published by ICC at the same time as Appendix S. Light straw-clay is made of loose straw mixed with clay slip that is tamped into forms in between wood wall framing, and then plastered. It can be installed in varying densities with thermal performance characteristics that depend on the balance of mass and insulation. Light straw-clay construction is sometimes used in concert with straw bale construction, as they have much in common; but light straw-clay provides different advantages and disadvantages.

The 2015 Appendix R with Commentary is published in the book by Lydia Doleman, *Essential Light Straw Clay Construction*, New Society Publishers, 2017. The 2018 version of the Appendix (with and without commentary) is available with the entire IRC for purchase at www.iccsafe.org.

A Note About Insurance and Financing for Straw Bale Buildings

In addition to building permit challenges, straw bale building owners have sometimes had difficulty securing insurance or financing. Both industries are risk-averse, and when faced with an unfamiliar building system, some companies have declined to insure or finance a straw bale building. However, each year more insurers and mortgage lenders are recognizing the advantages and value of straw bale structures. In some cases, providing Appendix S to an agent can convincingly demonstrate that straw bale construction is a safe and durable building system.

Questions or Comments regarding IRC Appendix S can be directed to its lead author, Martin Hammer, at mfhammer@pacbell.net or 510-684-4488.

Appendix S: Strawbale Construction and Commentary, from the 2018 International Residential Code and Commentary (IRC)

Note: In addition to its appearance here, Appendix S and its commentary are available for free download at: www.strawbuilding.org/. This download is made available by agreement between CASBA and the ICC. As future editions of the IRC are published (every three years), the download will be updated.

The complete IRC and the complete IRC with commentary are available for purchase at www.iccsafe.org.

Appendix S: Strawbale Construction

The provisions contained in this appendix are not mandatory unless specifically referenced in the adopting ordinance.

General Comments

Strawbale construction, a wall system using stacked bales of straw that are covered with plaster or other finish, originated in Nebraska in the late 1800s, made possible by the invention of the baling machine. It was practiced regionally into the 1940s, and buildings from that first era, some more than 100 years old, are still in service. After decades of nonuse, strawbale construction was rediscovered in the 1980s and utilized again in the American southwest. Since then it has been further developed and explored, subjected to considerable testing and researched regarding structural performance (under vertical loads, and lateral wind and seismic loads), moisture issues, fire resistance and thermal and acoustic properties.

Since the 1980s, the use of strawbale construction has steadily increased and there are now strawbale buildings in all 50 U.S. states, as well as in more than 50 countries throughout the world. It is estimated that there are over 1,000 strawbale buildings in California alone. Strawbale construction has been used primarily in the construction of residences, but it has also been used for schools, office buildings, wineries, retail buildings, a municipal police station and a federal post office. Most strawbale buildings have been one story, though many two- or three-story buildings have also been constructed and an eight-story, nonload-bearing strawbale apartment building was constructed in France. Over both its early and modern history, strawbale construction has proven, through testing and in practice, to be a safe, durable, resource- and energy-efficient and fully viable wall system (see Figure AS101.2 for typical load-bearing and typical post-and-beam strawbale wall systems with their basic components).

Purpose

As of 2014 only New Mexico, Oregon and North Carolina had adopted statewide strawbale building codes. California has legislated strawbale construction guidelines for voluntary adoption by local jurisdictions. In addition, nine U.S. cities or counties have adopted strawbale building codes. Three countries outside of the United States—Germany, France, and Belarus—have limited strawbale building codes. New Zealand has official guidelines for strawbale construction. In jurisdictions in the United States without a strawbale building code, strawbale buildings have been permitted on a case-by-case basis, often with little reliable guidance for building officials, builders and owners. This has been an impediment to strawbale construction's proper and broader use.

Most strawbale building codes that do exist in the U.S. are derived from the first load-bearing strawbale code created for and adopted in 1996 by Tucson and Pima County, Arizona. One is derived from the State of New Mexico's nonload-bearing code, also adopted in 1996. Field experience, testing and research since 1996 have shown these codes to be deficient. They are often either too restrictive or not restrictive enough, and in some cases do not address important issues at all. The purpose of this appendix is to bring the practice of strawbale construction into alignment with current understanding and to unify disparate strawbale building codes, while providing flexibility to allow time-tested local and regional variations, as well as viable new variations as strawbale construction continues to evolve.

SECTION AS101 GENERAL

AS101.1 Scope. This appendix provides prescriptive and performance-based requirements for the use of baled *straw* as a building material. Other methods of *strawbale* construction shall be subject to approval in accordance with Section R104.11 of this code. *Buildings* using *strawbale* walls shall comply with this code except as otherwise stated in this appendix.

❖ Historically, many variations of strawbale construction have been practiced, influenced by climate, level of high wind or seismic risk, available materials, local building practices and regional architecture. Still more variations are possible. This appendix, through both prescriptive and performance-based requirements, is intended to be inclusive of as many safe and durable methods of strawbale construction as possible. (See Figure AS101.2 for typical load-bearing and post-and-beam strawbale walls with their basic components.)

Variations of strawbale construction that are not explicitly allowed by this appendix may or may not be viable and acceptable. Such methods should be evaluated by the local building official in accordance with Section R104.11 of the code. Also see the commentary to Section AS105.1.

All components and aspects of strawbale buildings other than their strawbale walls, including foundations, nonstrawbale walls, roof structure, energy efficiency and mechanical, plumbing and electrical systems, must comply with the code, unless this appendix states otherwise. See the commentary to Section AS105.6 for recommendations regarding plumbing in strawbale walls as related to moisture control.

Although there are no provisions in this appendix related to electrical wiring systems in strawbale walls, these installations must address the same code criteria as for other wall systems in the code, such as protection of wiring from damage after construction (e.g., separation of wiring from the surface of finishes to protect from inadvertent nailing), secure attachment of wiring or conduit systems and boxes, and air sealing of electrical boxes and penetrations.

Decades of electrical installations in strawbale walls have yielded common practices, some of which have been codified in local or state strawbale building codes. Nonmetallic sheathed cable has been allowed with and without conduit, with frequency of attachment complying with the electrical code, to either wood framing members or to the bales using long wire "staples." Electrical boxes have been secured to either wood framing members or to 12-inch (305 mm) wooden stakes driven into the bales. Plaster mesh has also been used as a means of securing electrical boxes in strawbale walls. All wiring unprotected by conduit and armored cable has generally been required to be set back from the face of the strawbales at least $1^1/_2$ inches (38 mm), except where entering an electrical box.

AS101.2 Strawbale wall systems. *Strawbale* wall systems include those shown in Figure AS101.2 and *approved* variations.

❖ "Strawbale wall systems" is general terminology encompassing the specific systems shown in Figure AS101.2 as well as approved alternative systems.

SECTION AS102 DEFINITIONS

AS102.1 Definitions. The following words and terms shall, for the purposes of this appendix, have the meanings shown herein. Refer to Chapter 2 of the *International Residential Code* for general definitions.

BALE. Equivalent to *straw bale.*

❖ Many agricultural and nonagricultural materials are baled. However, in this appendix, the term bale is used to mean straw bale.

CLAY. Inorganic soil with particle sizes less than 0.00008 inch (0.002 mm) having the characteristics of high to very high dry strength and medium to high plasticity.

❖ Clay has been used for thousands of years as a building material. This includes unfired clay in adobe bricks, rammed earth walls, cob walls, earthen plasters and earthen floors, and fired clay in bricks, roofing tiles and floor and wall tiles. In all of these materials, clay is the binder, sometimes along with another binder such as lime or cement, that holds together other materials such as sand or straw. In this appendix, clay is a component of clay slip, straw-clay, clay plaster, and soil-cement plaster.

Clay for these purposes is obtained from inorganic subsoil with sufficient clay content, or can be obtained as a commercially quarried and bagged material. See the commentary to the definitions of "Clay slip" and "Clay subsoil" and to Section AS104.4.3.

CLAY SLIP. A suspension of *clay subsoil* in water.

❖ Clay slip is made by mixing water with clay subsoil or refined clay. Clay slip is used to make straw-clay. It is also sometimes used to coat the bottom of the first course of bales to aid in protecting the straw against potential moisture intrusion. For walls that will receive a clay plaster finish, clay slip is sometimes applied to the face of a bale wall during construction to provide temporary protection against light rain, or immediately before application of the first coat of clay plaster for improved bonding to the straw. A simple way to determine if clay slip contains sufficient clay and is of sufficient viscosity for these purposes is to dip a finger in the slip. If upon withdrawal the finger remains covered with an opaque coating, the slip contains sufficient clay.

CLAY SUBSOIL. Subsoil sourced directly from the earth or refined, containing *clay* and free of organic matter.

❖ The word "soil" is commonly associated with topsoil, which contains organic matter. Clay subsoil (below the layer of topsoil) used to make clay plaster is an inorganic mineral soil containing clay, sand and silt in varying proportions. It is important that the subsoil does not contain organic matter because it can decompose over time, though trace amounts are acceptable.

FINISH. Completed compilation of materials on the interior or exterior faces of stacked *bales.*

❖ See Section AS104 for acceptable finishes on strawbale walls.

FLAKE. An intact section of compressed *straw* removed from an untied *bale.*

❖ Industrialized baling machines push clumps of straw taken from the straw collection chamber tight toward the end of the bale chamber until the set bale length is achieved. The bale is then tied by the machine and the process is repeated.

In this process, the clumps of straw are flattened into 3- to 4-inch-thick (76 to 102 mm) mats known as flakes. When a bale is untied and pulled apart, the separation tends to occur between flakes, and the flakes tend to remain compressed and intact. Flakes of straw are often used in strawbale construction to fill voids between bales, or between bales and framing members in strawbale walls.

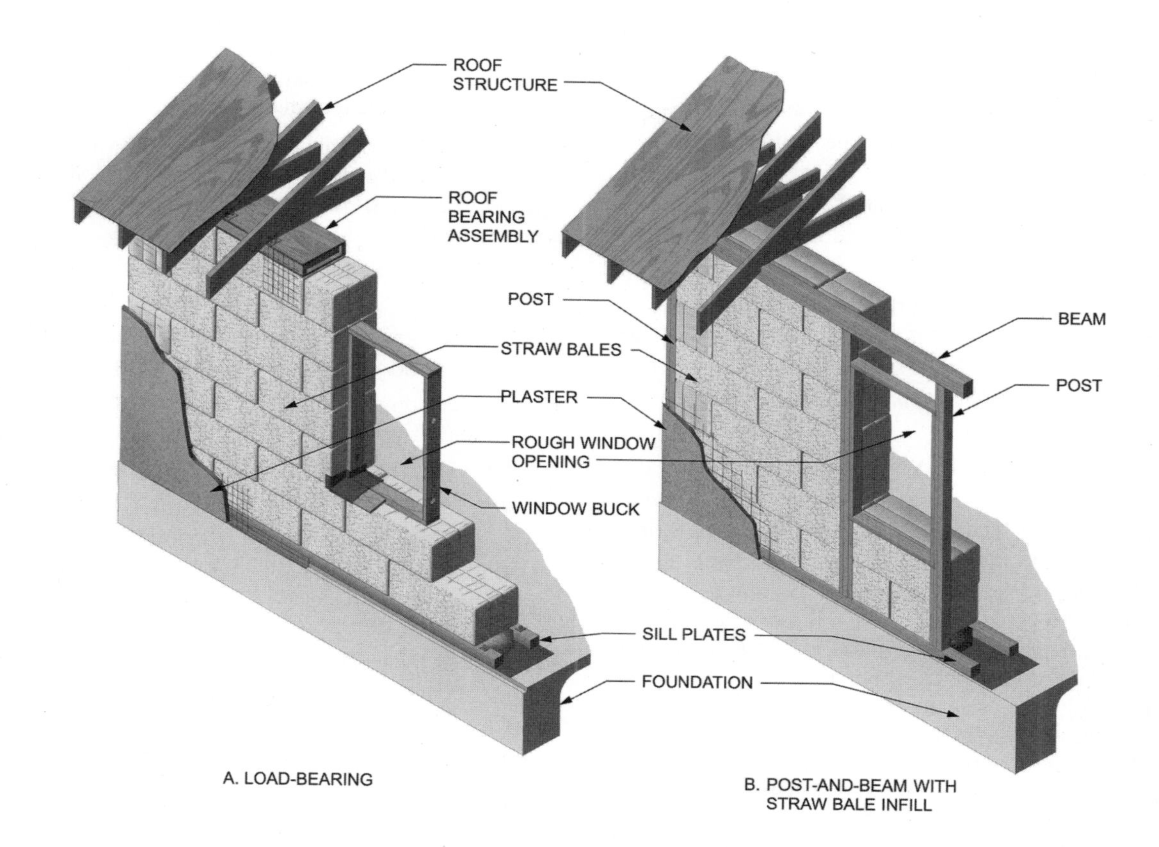

NOTE: SEE FIGURES AS105.1(1) THROUGH AS105.1(4) FOR DETAILED VIEWS AND SECTION REFERENCES. OTHER STRAWBALE WALL SYSTEMS OR VARIATIONS ARE PERMITTED AS *APPROVED*.

FIGURE AS101.2
TYPICAL STRAWBALE WALL SYSTEMS

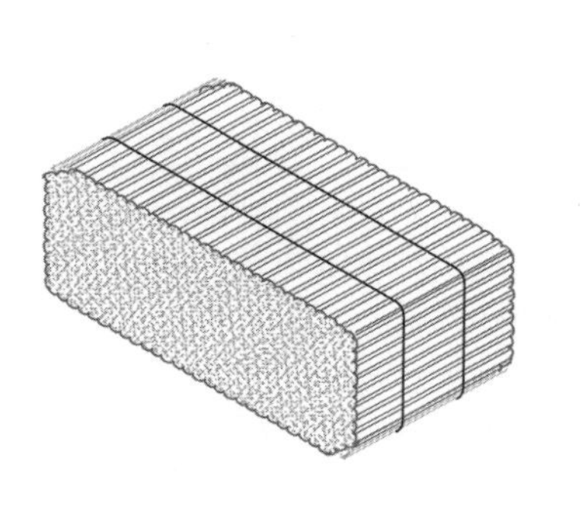

LAID FLAT

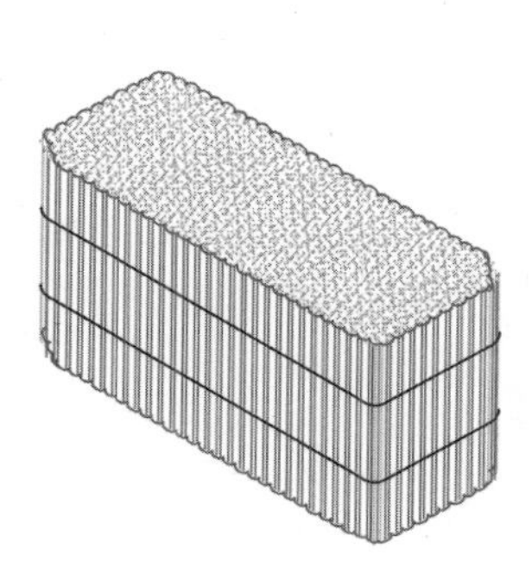

ON-EDGE

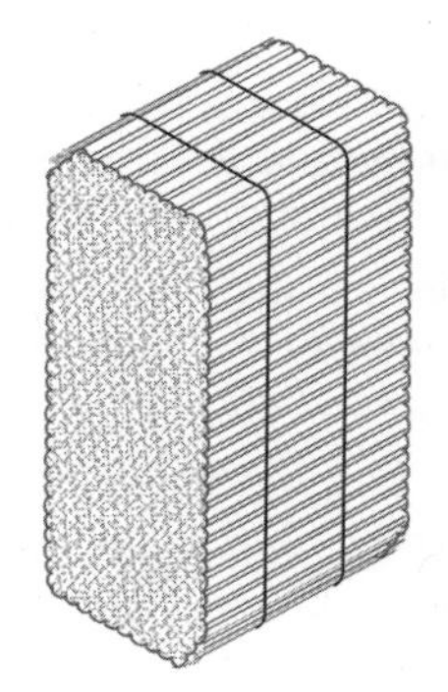

ON-END
(FOR USE IN NONSTRUCTURAL STRAWBALE WALLS ONLY.)

NOTE: ILLUSTRATIONS ALSO SHOW THE PREDOMINANT DIRECTION OF THE LENGTHS OF STRAW IN A TYPICAL STRAW BALE. HOWEVER, SOME RANDOMNESS OF DIRECTION IS NORMAL.

For SI: 1 inch - 25.4 mm.

FIGURE AS102.1
BALE ORIENTATIONS

LAID FLAT. The orientation of a *bale* with its largest faces horizontal, its longest dimension parallel with the wall plane, its *ties* concealed in the unfinished wall and its *straw* lengths oriented predominantly across the thickness of the wall. See Figure AS102.1.

LOAD-BEARING WALL. A *strawbale* wall that supports more than 100 pounds per linear foot (1459 N/m) of vertical load in addition its own weight.

❖ The definition of this term is consistent with the definition of "Wall, load-bearing" for stud walls in the *International Building Code*® (IBC®).

MESH. An openwork fabric of linked strands of metal, plastic, or natural or synthetic fiber.

NONSTRUCTURAL WALL. Walls other than *load-bearing* walls or shear walls.

❖ The definition of this term is consistent with the definition in ASCE 7. ASCE 7 is one of the standards used as the basis for structural loads on buildings in the code and in the IBC.

In nonstructural strawbale walls, the straw bales are infill only and serve as enclosure, insulation and as a substrate for plaster. The straw bales and their finish carry no superimposed vertical or in-plane lateral loads, and the bales can be in any orientation including laid flat, on-edge, or on-end (see Figure AS102.1). However, other elements in the same wall, such as wood or steel framing, or structural panels, may also carry such structural loads.

ON-EDGE. The orientation of a *bale* with its largest faces vertical, its longest dimension parallel with the wall plane, its *ties* on the face of the wall and its *straw* lengths oriented predominantly vertically. See Figure AS102.1.

ON-END. The orientation of a *bale* with its longest dimension vertical. For use in *nonstructural strawbale* walls only. See Figure AS102.1.

❖ Bales can be oriented on-end only in strawbale wall systems where the bales are nonstructural and serve only as insulation and/or the substrate for plaster.

PIN. A vertical metal rod, wood dowel or bamboo, driven into the center of stacked *bales*, or placed on opposite surfaces of stacked *bales* and through-tied.

❖ See the commentary to Section AS105.4.2.

PLASTER. Gypsum plaster, cement plaster, *clay* plaster, soil-cement plaster, lime plaster or cement-lime plaster as described in Section AS104.

❖ In the body of the code, the word "plaster" is used only for cement and gypsum plasters. The word "plaster" in this appendix is used with any of the types described in Section AS104.

PRECOMPRESSION. Vertical compression of stacked *bales* before the application of finish.

❖ See the commentary to Section AS106.12.1.

REINFORCED PLASTER. A *plaster* containing mesh reinforcement.

RUNNING BOND. The placement of *straw bales* such that the head joints in successive courses are offset not less than one-quarter the bale length.

❖ The definition of this term is consistent with the definition in the body of the code, except that straw bales are used in place of masonry units.

SHEAR WALL. A *strawbale* wall designed and constructed to resist lateral seismic and wind forces parallel to the plane of the wall in accordance with Section AS106.13.

❖ The term "shear wall" is used interchangeably with the term "braced wall panel" in this appendix.

SKIN. The compilation of *plaster* and reinforcing, if any, applied to the surface of stacked *bales*.

STRUCTURAL WALL. A wall that meets the definition for a *load-bearing* wall or shear wall.

❖ The definition of this term is consistent with the definition in ASCE 7. ASCE 7 is one of the standards used as the basis for structural loads on buildings in the code and in the IBC.

STACK BOND. The placement of *straw bales* such that head joints in successive courses are vertically aligned.

❖ The definition of this term is consistent with its definition in the code, except that straw bales are used in place of masonry units.

STRAW. The dry stems of cereal grains after the seed heads have been removed.

❖ See the commentary to Section AS103.3.7.

STRAW BALE. A rectangular compressed block of *straw*, bound by *ties*.

❖ See the commentary to Section AS103.

STRAWBALE. The adjective form of *straw bale*.

❖ Strawbale construction has historically been referred to as "straw bale," "straw-bale," or "strawbale" construction. Applying accepted English grammar, when used as an adjective, "straw-bale" is correct. However, the English language contains many compound adjectives that originated as two words, but became one word because of frequency and simplicity use. For these reasons the compound adjective "strawbale" is used in this appendix.

STRAW-CLAY. Loose *straw* mixed and coated with *clay* slip.

❖ Straw-clay is a material in the context of this appendix and should not be confused with the term light straw-clay construction, which is a nonload-bearing wall system described in Appendix R.

TIE. A synthetic fiber, natural fiber or metal wire used to confine a *straw bale*.

❖ See the commentary to Section AS103.3.

TRUTH WINDOW. An area of a *strawbale* wall left without its finish, to allow view of the *straw* otherwise concealed by its finish.

❖ "Truth windows" of modest size are commonly included in strawbale buildings to show the straw bale core of the walls. Also see the commentary to Section AS104.2.

SECTION AS103 BALES

AS103.1 Shape. *Bales* shall be rectangular in shape.

❖ Straw bales as referred to in this appendix are rectangular in shape. Rectangular bales created by common industrialized baling machines fall into one of three size categories: "two-string," "three-string," and "jumbo." See the commentary to Sections AS103.2 and AS103.3 for information regarding bale dimensions in these categories. Round (cylindrical) bales have also been used for strawbale building, but are not covered in this appendix.

AS103.2 Size. *Bales* shall have a height and thickness of not less than 12 inches (305 mm), except as otherwise permitted or required in this appendix. *Bales* used within a continuous wall shall be of consistent height and thickness to ensure even distribution of loads within the wall system. See Figure AS103.2 for approximate dimensions of common *straw bales*.

❖ This section states the minimum height and thickness (width) for straw bales used in strawbale construction, except that a larger minimum thickness is stated in Section AS106.13.1 for strawbale braced panels. It also requires bales in a continuous wall to be of consistent height and thickness. There is no length requirement. The dimensions shown in Figure AS103.2 are approximate because exact dimensions of bales are difficult to measure due to their irregular, "fuzzy" surfaces.

The most commonly used straw bales for building are known as "two-string" and "three-string" bales, named for the number of ties that hold each bale together (see Figure AS103.2). Their width and height dimensions are determined by the compression chamber of the baler. These two types of bales typically have dimensions as shown in Figure AS103.2. Two-string and three-string bales are small and light enough to be handled by most able-bodied persons.

Of the three dimensions of a bale, the length can vary the most. However, bale widths are typically about twice their length so they stack well in storage and transit for farmers. The bales are laid flat with successive layers changing direction for stability. As with conventional unit masonry, this ratio is also useful at corners of strawbale walls with a running bond, where successive courses interlock and maintain the running-bond offset for both intersecting walls.

"Jumbo" bales have also been used for building. They are typically 3 or 4 feet × 4 feet × 8 feet (910 or 1220 mm × 1220 mm × 2440 mm), and can only be moved by mechanized equipment. Bales as small as 1 foot × 1 foot × 2 feet (305 mm × 305 mm × 610 mm) have been successfully used in strawbale construction, including for structural walls in one-story buildings.

Although straw bales in industrialized countries are almost universally made with industrialized baling machines, manually compressed bales have been used in strawbale construction and are acceptable if the bales meet the requirements of Section AS103.

AS103.3 Ties. *Bales* shall be confined by synthetic fiber, natural fiber or metal *ties* sufficient to maintain required *bale* density. *Ties* shall be not less than 3 inches (76 mm) and not more than 6 inches (152 mm) from the two faces without *ties* and shall be spaced not more than 12 inches (305 mm) apart. *Bales* with broken *ties* shall be retied with sufficient tension to maintain required bale density.

❖ In strawbale construction, the ties that confine a bale serve two major functions: to enable the bale to be easily moved and handled during construction, and to keep the bale at its required density. If ties are removed (such as where making half-bales) or broken before installation, the bales must be retied to the density required by Section AS103.5.

Straw bales in industrialized agriculture are now tied almost exclusively with polypropylene ties, though steel wire ties are sometimes found. Any synthetic fiber, natural fiber (e.g., sisal or hemp) or metal wire ties that can maintain the required bale density is acceptable. The most commonly used bales for building are "two-string" and "three-string" bales, named for the number of ties used (see Figure AS103.2).

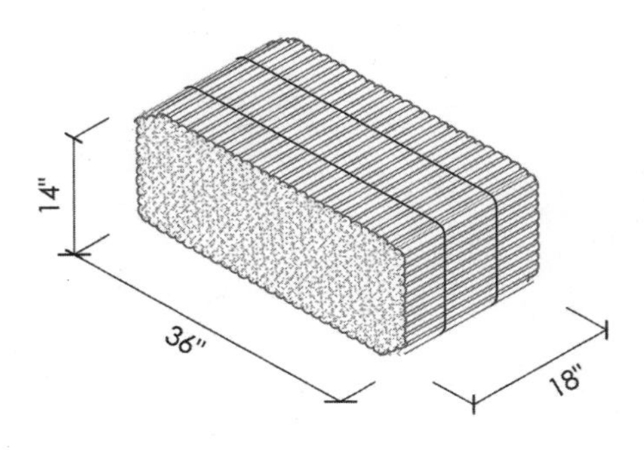

TWO-STRING BALE

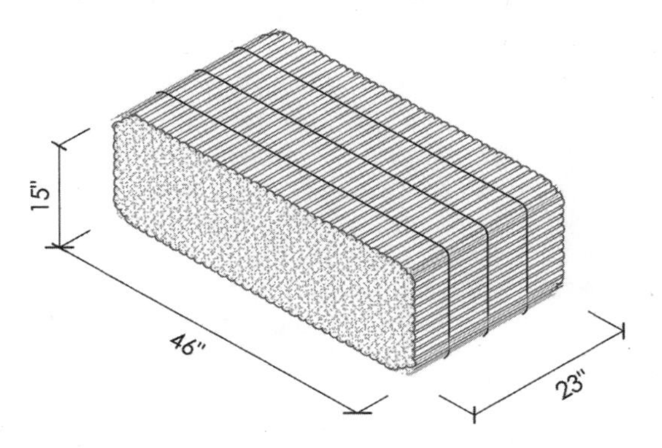

THREE-STRING BALE

FIGURE AS103.2
APPROXIMATE DIMENSIONS OF COMMON STRAW BALES

Some practitioners of strawbale construction intentionally cut ties on bales laid flat in nonstructural walls before plastering after courses of bales are sufficiently confined at their ends. The purpose is to allow the bales to expand along their length to fill gaps between bales or between bales and vertical framing, creating a more consistent insulating core and lathing surface. When ties are cut, a bale tends to expand only lengthwise and change in the other two dimensions is negligible, due to the nature of how baling machines create bales (see the commentary to the definition of "Flake").

AS103.4 Moisture content. The moisture content of *bales* at the time of application of the first coat of *plaster* or the installation of another finish shall not exceed 20 percent of the weight of the *bale*. The moisture content of *bales* shall be determined with a moisture meter designed for use with *baled straw* or hay, equipped with a probe of sufficient length to reach the center of the *bale*. Not less than 5 percent and not fewer than 10 bales shall be randomly selected and tested.

❖ Field experience and laboratory testing have shown that if the moisture content of straw does not exceed 20 percent for prolonged periods of time, the straw will not degrade. Some studies have shown that degradation does not occur until 25 percent moisture content is reached. Twenty-percent moisture content is widely considered a safe maximum for straw in strawbale construction. However, practitioners of strawbale construction prefer a working range of 7- to 15-percent moisture content, which allows a margin for modest future wetting without degradation.

It is required that the bales have a moisture content of less than 20 percent at the time the walls are plastered or receive another finish, but should also be below 20 percent at all stages. These stages include: at time of baling, in storage, in transport, at the job site and in service. Knowing or controlling the moisture content of bales before procurement or arrival on site can be difficult or impossible. Therefore, at procurement, bales that measure above 20 percent moisture content or that show visible signs of previous wetting, such as dark staining from microbial growth, should not be used for construction (see commentary, Section AS105.6).

The required method for determining whether bales are acceptable for construction is to test the quantity stated in this section with a moisture meter designed for use with baled straw or hay (see Commentary Figure AS103.4 for examples of moisture meters). The building official may specify how this is demonstrated. The bales tested must not exceed 20 percent moisture content, and their centers should be checked. If any of the bales tested show a moisture content higher than 20 percent, such bales should not be used for construction, and the building official may then specify how many more bales must be tested before accepting the entire lot of bales for construction.

Once the lot of bales is demonstrated to be acceptable, if bales with moisture content above 20 percent are subsequently discovered, such bales should not be used for construction.

AS103.5 Density. *Bales* shall have a dry density of not less than 6.5 pounds per cubic foot (104 kg/cubic meter). The dry density shall be calculated by subtracting the weight of the moisture in pounds (kg) from the actual *bale* weight and dividing by the volume of the *bale* in cubic feet (cubic meters). Not less than 2 percent and not fewer than five *bales* shall be randomly selected and tested on site.

❖ The required method for determining whether bales are of sufficient density for construction is to check the quantity stated in this section. The building official may specify how this is demonstrated.

The dry density is determined as follows:

1. Weigh the bale.
2. Find its percent moisture with a moisture meter designed for use with baled straw or hay.
3. Calculate the volume of the bale in cubic feet by multiplying its length x width x height in feet.

Enter those numbers in the following equation, where A = total weight (lbs), B = percent moisture content, C = Volume (cu.ft.) and D = Dry Density (lbs/cu.ft):

$$[A - (B \times A)] \div C = D$$

For example: A two-string bale measuring 14 inches × 18 inches × 36 inches is determined to have a 12 percent moisture content and weighs 47 lbs. Its volume is 1.17 feet × 1.5 feet x 3 feet = 5.26 cu.ft. Therefore A = 47 lbs, B = 12, and C = 5.26 cu.ft. Substituting the numbers in the equation: [47 - (.12 × 47)]/5.26] = 7.86 lb per cu.ft. dry density.

Field experience and testing have shown the density required in this section to be a minimum threshold for proper performance of strawbale walls in terms of insulation and structure and as a substrate for plaster. Most bales made by industrialized balers significantly exceed this density. See the commentary to Section

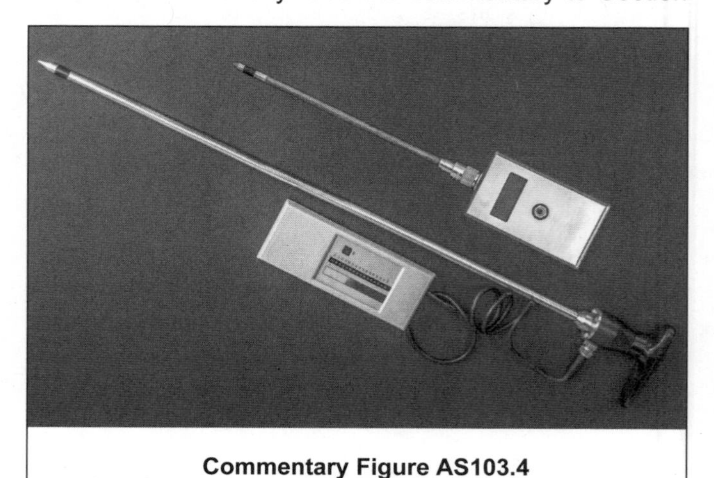

Commentary Figure AS103.4
EXAMPLE MOISTURE METERS FOR STRAW BALES

AS103.3 for information regarding the practice of intentionally cutting the ties of installed bales.

AS103.6 Partial bales. Partial *bales* made after original fabrication shall be retied with *ties* complying with Section AS103.3.

❖ Partial or custom-size bales, most commonly half bales, are typically made on site where sizes other than a full-size bale are required. Also see the commentary to Section AS103.3.

AS103.7 Types of straw. *Bales* shall be composed of *straw* from wheat, rice, rye, barley or oat.

❖ Bales made of straw from the five named cereal grains have been successfully used in modern strawbale buildings on six continents since 1990. Straw is an agricultural byproduct baled after nutrient grains have been harvested and is commonly used for livestock bedding and erosion control.

Baled grasses such as hay or alfalfa must not be used. They are cultivated and baled as livestock feed, are baled green, contain nutrients that support active decomposition by micro-organisms and are unacceptable for building.

AS103.8 Other baled material. The dry stems of other cereal grains shall be acceptable where *approved* by the building official.

❖ In addition to straw from the five cereal grains named in Section AS103.7, dry stalks from other plants such as flax have been successfully used for bales in strawbale buildings in North America. The building official can accept other baled dry plant material for construction where such material is demonstrated to be equivalent in performance to the types of straw allowed in Section AS103.7. Also see the commentary to Section AS103.7 for information regarding baled plants such as hay and alfalfa that are not acceptable for use in buildings.

SECTION AS104 FINISHES

AS104.1 General. Finishes applied to *strawbale* walls shall be any type permitted by this code, and shall comply with this section and with Chapters 3 and 7 unless stated otherwise in this section.

❖ Historically, plaster has been the most common finish on strawbale walls. Plaster requirements are covered in detail in Section AS104. However, other cladding materials permitted by the code have also been used and are not prohibited by this appendix. Non-plaster finishes must be installed in accordance with the requirements in the body of the code and must comply with Sections AS104.2 and AS104.3 of this appendix. As such, any water-resistive or air barrier used with a nonplaster finish must have a permeance rating of at least 3 perms. See the commentary to Sections AS104.2 and AS104.3.

AS104.2 Purpose, and where required. *Strawbale* walls shall be finished so as to provide mechanical protection, fire resistance and protection from weather and to restrict the passage of air through the *bales*, in accordance with this appendix and this code. Vertical *strawbale* wall surfaces shall receive a coat of *plaster* not less than $^3/_8$ inch (10 mm) thick, or greater where required elsewhere in this appendix, or shall fit tightly against a solid wall panel or dense-packed cellulose insulation with a density of not less than 3.5 pounds per cubic foot (56 kg/m^3) blown into an adjacent framed wall. The tops of *strawbale* walls shall receive a coat of *plaster* not less than $^3/_8$ inch (10 mm) thick where *straw* would otherwise be exposed.

Exception: *Truth windows* shall be permitted where a fire-resistance rating is not required. Weather-exposed *truth windows* shall be fitted with a weather-tight cover. Interior truth windows in Climate Zones 5, 6, 7, 8 and Marine 4 shall be fitted with an air-tight cover.

❖ The requirement that vertical surfaces of strawbale walls receive a coat of plaster or fit tightly against a wall panel or other abutting material, including dense-packed cellulose insulation in an adjacent framed wall, is intended to restrict air movement. Restriction of air movement across the surface of the bales, especially vertical movement, is important for inhibiting the potential spread of fire and for minimizing convective heat loss (see commentary, Section AS107). Restriction of air movement through the bales reduces the potential for convective heat transfer and controls migration of moist air through the wall assembly.

The requirement that exposed tops of strawbale walls receive a coat of plaster (where not covered tightly by materials such as wood framing, plywood, or gypsum board) is intended to protect the straw against fire and to inhibit air and moisture movement into the wall. Clay plaster has proven to be especially effective in managing moisture at the top of strawbale walls.

The exception to Section AS104.2 allows truth windows where a fire-resistance rating is not required. See the definition of "Truth window." The purpose of the air-tight cover requirement for interior truth windows in the indicated climate zones is to prevent condensation in the wall from warm, moist interior air bypassing the air barrier of the interior finish. Weather-tight and air-tight covers can be operable and can include a transparent material, such as glass or acrylic, to maintain the truth window's purpose, which is to display the straw in the wall. In the cover's closed position it must be weather-tight, or air-tight, as required. Also see Section AS105.6.2 and its commentary.

AS104.3 Vapor retarders. Class I and II vapor retarders shall not be used on a *strawbale wall*, nor shall any other material be used that has a vapor permeance rating of less than 3 perms, except as permitted or required elsewhere in this appendix.

❖ High vapor permeability of both interior and exterior finishes is desirable for strawbale walls to allow moisture migration through the wall and dispersion of any moisture that enters the wall. For this reason, Class I and Class II vapor retarders are not allowed on strawbale walls, except in extreme situations as required in this appendix, where the importance of keeping water vapor from entering the wall exceeds the importance of enabling moisture to exit the wall.

There are two sections in this appendix where such a situation is defined. In Section AS105.6.2, a Class I or Class II vapor retarder is required on strawbale walls enclosing showers or steam rooms. In Section AS105.6.5, a Class II vapor retarder is required to separate straw bales from adjacent concrete or masonry.

The minimum specified permeance rating of 3 perms is intended to be the "wet bulb" perm rating of the material because it is under wet conditions that it is most important for the materials to be vapor permeable. Some materials have identical wet bulb and dry bulb perm ratings, whereas others, such as plywood, have substantially different wet bulb and dry bulb perm ratings. Unless identical, the dry bulb perm rating of a material is always lower than the wet bulb rating, and can be used to comply with this section if the wet bulb rating is unknown because the dry bulb rating will be conservative.

AS104.4 Plaster. *Plaster* applied to *bales* shall be any type described in this section, and as required or limited in this appendix. *Plaster* thickness shall not exceed 2 inches (51 mm).

❖ Plaster has been the most common finish on strawbale walls and as such is given particular attention in this appendix. This appendix contains provisions for two plaster types that also appear in the body of the code —gypsum plaster and cement plaster—and contains provisions for four plaster types that do not appear in the body of the code: clay plaster, soil-cement plaster, lime plaster and cement-lime plaster. Plasters used on structural walls must comply with Section AS106.6.

All plaster types contained in this appendix have been successfully used in strawbale construction. Each has comparative advantages and disadvantages in terms of compressive strength, durability, vapor permeability, moisture management, availability, cost, ease of application, acoustic properties, fire resistance and aesthetic qualities. The choice of plaster should be carefully considered based on the context and requirements of the building project. Such considerations are typically influenced by climate and the required or desired structural performance of the strawbale walls. Similar consideration should be given when choosing between plaster and other finish materials or assemblies for strawbale walls.

The reason for limiting plaster thickness to 2 inches (51 mm) is that the additional weight of thicker plaster has potential structural consequences, especially in areas of high seismic risk. Thicker plaster may also unacceptably reduce vapor permeability depending on the plaster used; however, there are also potential benefits of thicker plaster, including improved thermal performance in some climates due to the increased mass. The building official may allow plaster thicker than 2 inches (51 mm) if deemed to present no significant additional risk. Also see the commentary to Section AS106.2.

AS104.4.1 Plaster and membranes. *Plaster* shall be applied directly to *strawbale* walls to facilitate transpiration of moisture from the *bales*, and to secure a mechanical bond between the *skin* and the *bales*, except where a membrane is allowed or required elsewhere in this appendix.

❖ Strawbale construction has historically been practiced with plaster applied directly to the stacked straw bales, without any membrane air barrier or water-resistive barrier between the plaster and the straw. This allows the plaster to mechanically bond with the straw, which is especially important for structural strawbale walls (see commentary, Section AS106.8). Also, under certain conditions, the presence of a membrane may impede the dispersion of moisture through the plaster to the interior or exterior.

It is important to note that while the code requires a water-resistive barrier for exterior cement plaster over wood-frame construction in accordance with Section R703.7.3, a water-resistive barrier is not required for cement plaster, or any exterior plaster allowed by this appendix, for plaster over strawbale walls. The moisture management characteristics of strawbale construction compared with wood-frame construction account for the differing requirements. In strawbale construction, the many tubular lengths of straw in the bale core give a strawbale wall considerably more capacity than a wood-frame wall to safely absorb, store and disperse moisture.

There are situations where a membrane is allowed or required by this appendix. For example, Section AS105.6.2 requires a Class I or Class II vapor retarder on strawbale walls enclosing showers or steam rooms. One means of achieving this is the installation of a membrane vapor retarder. See the commentary to Section AS105.6.1 for situations where a water-resistive barrier membrane is sometimes used between straw and plaster.

Installation of a membrane between the straw bales and plaster always requires mesh or lath that is adequately attached to the bales or through-tied to mesh or lath on the other side, including on nonstructural walls. This mesh or lath substitutes for the typical lathing bond between plaster and straw (see Section AS104.4.2) that is interrupted where a membrane is installed. Where mesh is required structurally, such attachment must be of an approved engineered design (see Section AS106.8).

AS104.4.2 Lath and mesh for plaster. The surface of the *straw bales* functions as lath, and other lath or *mesh* shall not be required, except as required for out-of-plane resistance by Table AS105.4 or for structural walls by Tables AS106.12 and AS106.13(1).

AS104.4.3 Clay plaster. *Clay plaster* shall comply with Sections AS104.4.3.1 through AS104.4.3.6.

AS104.4.3.1 General. *Clay plaster* shall be any plaster having a clay or *clay subsoil* binder. Such plaster shall contain sufficient clay to fully bind the plaster, sand or other inert granular material, and shall be permitted to contain reinforcing fibers. Acceptable reinforcing fibers include chopped straw, sisal and animal hair.

❖ The relative amounts of clay subsoil and added sand in the plaster mix depend on the amount of clay, sand and silt in the subsoil. Experimentation or experience is necessary to determine mixes that are workable and that yield a finished plaster with minimal or no cracking and sufficient compressive strength for the application.

See Section AS104.4.3.2 regarding methods used to determine the suitability of clay subsoil for clay plasters.

In addition to the acceptable reinforcing fibers listed, hemp fiber and agave fibers also have a history of successful use.

AS104.4.3.2 Clay subsoil requirements. The suitability of *clay subsoil* shall be determined in accordance with the Figure 2 Ribbon Test or the Figure 3 Ball Test in the appendix of ASTM E2392/E2392M.

❖ This section requires the use of one of two tests referenced in ASTM E2392-10 to determine the suitability of clay subsoil for clay plaster. These tests are commonly used by clay plaster practitioners. However this section is not meant to prohibit the use of other tests that demonstrate clay subsoil suitability, for example a laboratory soils analysis.

AS104.4.3.3 Thickness and coats. *Clay plaster* shall be not less than 1 inch (25 mm) thick, except where required to be thicker for structural walls as described elsewhere in this appendix, and shall be applied in not less than two coats.

❖ The primary purpose of requiring a minimum of two coats for clay plaster is to prevent or minimize through-cracks. Minor cracks sometimes occur in the first coat, and the second coat fills or bridges those cracks. The use of two coats also aids in drying clay plaster.

AS104.4.3.4 Rain-exposed. *Clay plaster,* where exposed to rain, shall be finished with lime wash, lime plaster, linseed oil or other *approved* erosion-resistant finish.

❖ Exception 2 in the definition of "Weather-exposed surfaces" in the IBC might be considered when determining whether a clay plaster should be considered "exposed to rain," subject to the evaluation of the building official, along with consideration of the local climate. That exception reads: "Walls or portions of walls beneath an unenclosed roof area, where located a horizontal distance from an open exterior opening equal to at least twice the height of the opening."

Lime plaster and lime wash over clay plaster should only be used with careful consideration of local climate, materials and experience, as spalling can occur under certain conditions. Lime in the bonding coat of the clay plaster has been shown to decrease the chance of spalling of these lime-based protective coats.

AS104.4.3.5 Prohibited finish coat. *Plaster* containing Portland cement shall not be permitted as a finish coat over *clay plasters.*

❖ Finish coats of Portland cement plaster over clay plasters have a history of spalling and, therefore, are prohibited. In addition, cement plaster has a significantly lower vapor permeability than clay plaster and, without sufficient lime, risks reducing the vapor permeability of the plaster below the minimum 3 perm requirement of Section AS104.3.

An often-desired attribute of clay plaster is its high vapor permeability (16-19 perms per inch). Applying cement plaster, or any finish that significantly reduces its vapor permeability, runs counter to this common and important intention of its use. Other finishes that protect clay plaster from erosion without significantly reducing its vapor permeability are available and commonly used (see Section AS104.4.3.4 and its commentary).

AS104.4.3.6 Plaster additives. Additives shall be permitted to increase *plaster* workability, durability, strength or water resistance.

❖ Additives commonly used in clay plasters that increase one or more of the qualities stated in this section include cactus juice, cooked flour paste, casein, animal manure, and linseed oil (finish coat only). Such additives should only be used with experience or sufficient trials.

AS104.4.4 Soil-cement plaster. Soil-cement plaster shall comply with Sections AS104.4.4.1 through AS104.4.4.3.

AS104.4.4.1 General. Soil-cement *plaster* shall be composed of *clay subsoil*, sand and not less than 10 percent and not more than 20 percent Portland cement by volume, and shall be permitted to contain reinforcing fibers.

❖ The relative amounts of clay subsoil and added sand in the plaster mix depend on the amount of clay, sand and silt in the subsoil. Experimentation or experience is necessary to determine mixes that are workable and that yield a finished plaster with minimal or no cracking and sufficient compressive strength for the application.

The water-to-cement ratio should be similar to the ratio used in mixing conventional cement plaster or concrete. That is, a ratio of approximately .5 water to 1 cement by weight. Lower ratios lead to higher strength and durability, but may make the mix difficult to work with. Higher ratios reduce the strength and durability.

AS104.4.4.2 Lath and mesh. Soil-cement *plaster* shall use any corrosion-resistant lath or *mesh* permitted by this code, or as required in Section AS106 where used on structural walls.

AS104.4.4.3 Thickness. Soil-cement *plaster* shall be not less than 1 inch (25 mm) thick.

AS104.4.5 Gypsum plaster. Gypsum *plaster* shall comply with Section R702.2.1. Gypsum *plaster* shall be limited to use on interior surfaces of *nonstructural* walls, and as an interior *finish* coat over a structural *plaster* that complies with this appendix.

❖ The subsection of Section R702 that pertains to gypsum plaster used on strawbale walls is Section R702.2.1. Gypsum plaster, like all other plasters described in this appendix, can be applied directly to strawbale walls in accordance with Section AS104.4.2.

AS104.4.6 Lime plaster. Lime *plaster* shall comply with Sections AS104.4.6.1 through AS104.4.6.3.

AS104.4.6.1 General. Lime *plaster* is any *plaster* with a binder that is composed of calcium hydroxide (CaOH) including Type N or S hydrated lime, hydraulic lime, natural hydraulic lime or quicklime. Hydrated lime shall comply with ASTM C206. Hydraulic lime shall comply with ASTM C1707. Natural hydraulic lime shall comply with ASTM C141 and EN 459. Quicklime shall comply with ASTM C5.

AS104.4.6.2 Thickness and coats. Lime *plaster* shall be not less than $^7/_8$ inch (22 mm) thick, and shall be applied in not less than three coats.

AS104.4.6.3 On structural walls. Lime *plaster* on *strawbale* structural walls in accordance with Table AS106.12 or Table AS106.13(1) shall use a binder of hydraulic or natural hydraulic lime.

❖ For structural strawbale walls using lime plaster, the binder is required to be hydraulic or natural hydraulic lime because those binders develop the minimum compressive strength required by Table AS106.6.1 more reliably than nonhydraulic limes. However, the building official may decide to allow Type N or S hydrated lime, or quicklime for lime plasters on structural walls where such plasters demonstrate the minimum required compressive strength through testing in accordance with Section AS106.6.1.

AS104.4.7 Cement-lime plaster. Cement-lime plaster shall be *plaster* mixes CL, F or FL, as described in ASTM C926.

❖ Cement-lime plasters use a dry mix of Portland cement and lime together as the binder in roughly equal proportions. The mix types listed in this section are shown in Tables 3 and 4 of ASTM C926 and are the only ones allowed on strawbale walls because only they contain sufficient lime to achieve a vapor permeability of at least 3 perms as required by Section AS104.3. The Portland cement and lime proportions for CL, F and FL plaster range from 1:0.75 to 1:2.

AS104.4.8 Cement plaster. Cement *plaster* shall conform to ASTM/C926 and shall comply with Sections R703.7.4 and R703.7.5, except that the amount of lime in plaster coats shall be not less than 1 part lime to 6 parts cement to allow a minimum acceptable vapor permeability. The combined thickness of *plaster* coats shall be not more than $1^1/_2$ inches (38 mm) thick.

❖ The subsections of Section R703.7 that pertain to cement plaster used on strawbale walls are Sections R703.7.4 and R703.7.5. Cement plaster is required to contain lime in the proportion stated in order to achieve a vapor permeability of at least 3 perms as required by Section AS104.3.

SECTION AS105 STRAWBALE WALLS—GENERAL

AS105.1 General. *Strawbale walls* shall be designed and constructed in accordance with this section and with Figures AS105.1(1) through AS105.1(4) or an *approved* alternative design. *Strawbale structural walls* shall be in accordance with the additional requirements of Section AS106.

❖ The provisions of this section apply to all strawbale walls, including nonstructural and structural walls (see the definitions of "Nonstructural wall" and "Structural wall"), except where a subsection states that the provision(s) apply only to nonstructural walls. In addition, structural strawbale walls must be designed and constructed in accordance with Section AS106.

Figures AS105.1(1) through AS105.1(4) illustrate acceptable strawbale wall systems. Each element in the figures references a section in this appendix or the code. Historically, many variations of strawbale construction have been practiced, and the illustrated systems are not meant to preclude viable variations. However, systems that vary substantially from those shown in Figures AS105.1(1) through AS105.1(4) must be approved by the building official.

Also see the commentary to Section AS101.

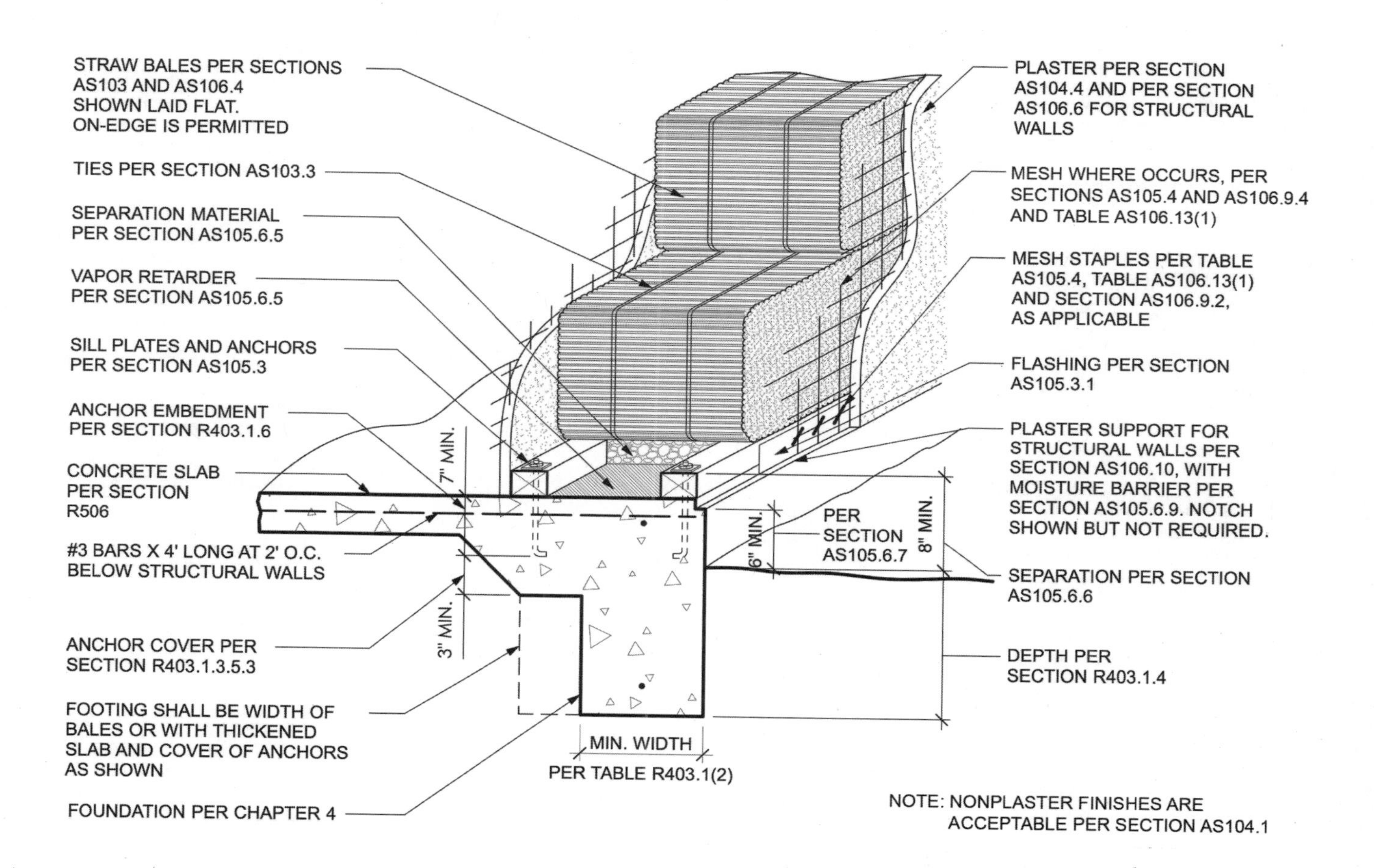

For SI: 1 inch = 25.4 mm.

FIGURE AS105.1(1)
TYPICAL BASE OF PLASTERED STRAWBALE WALL ON CONCRETE SLAB AND FOOTING

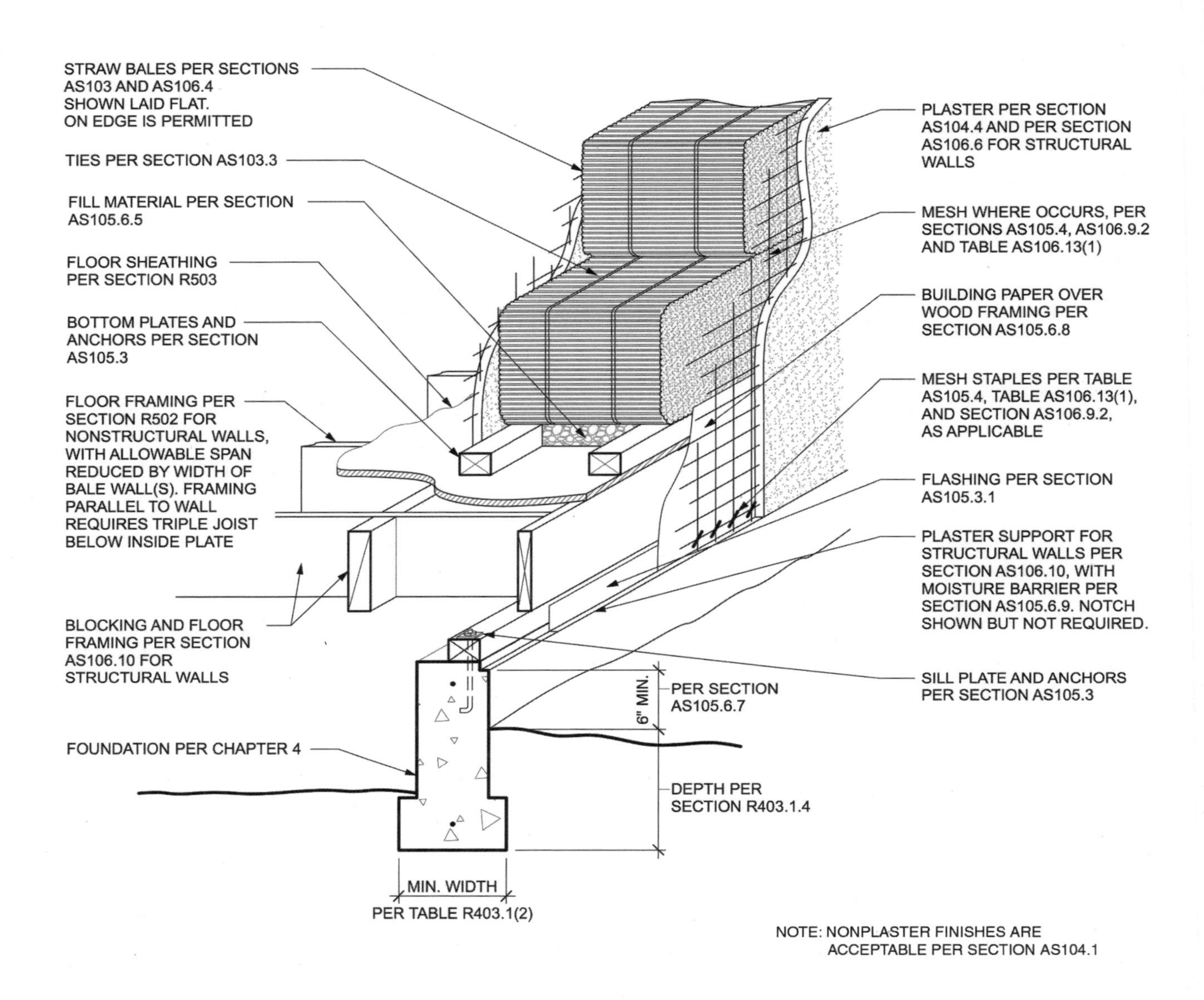

For SI: 1 inch = 25.4 mm.

FIGURE AS105.1(2)
TYPICAL BASE OF PLASTERED STRAWBALE WALL OVER RAISED FLOOR

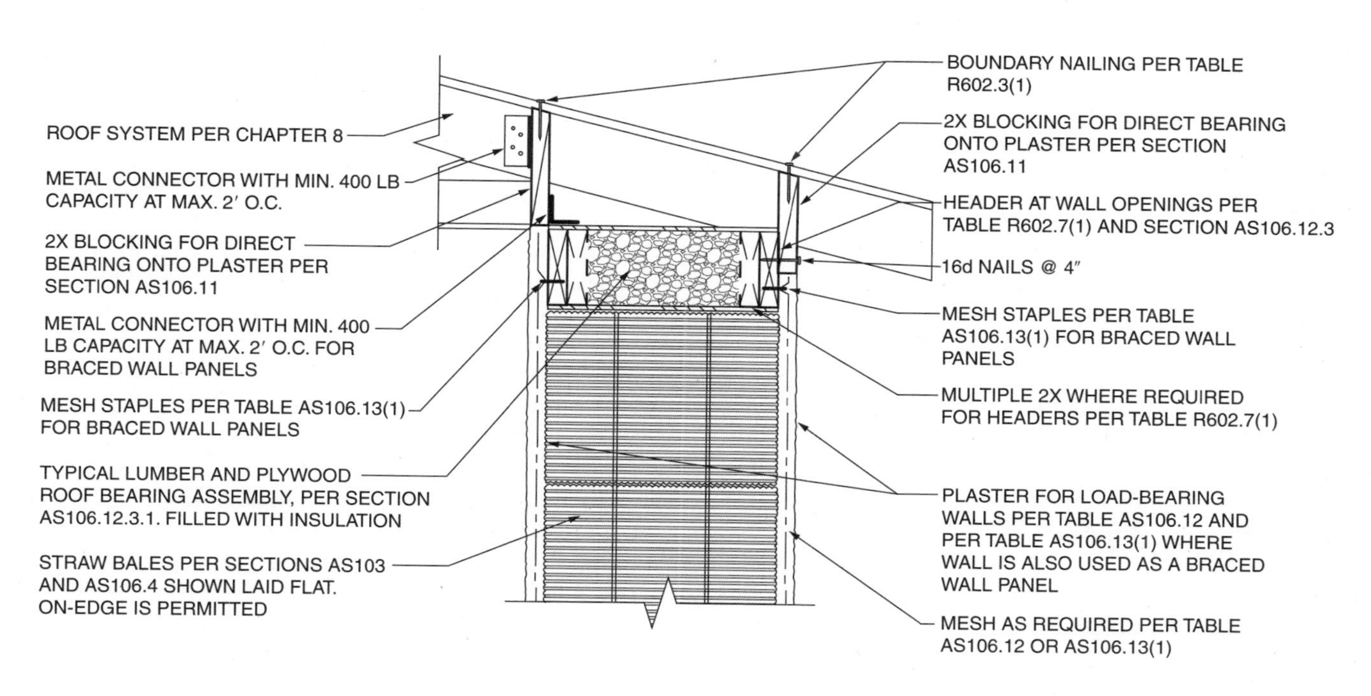

For SI: 1 inch = 25.4 mm, 1 pound = 2.2 kg.

FIGURE AS105.1(3)
TYPICAL TOP OF LOAD-BEARING STRAWBALE WALL

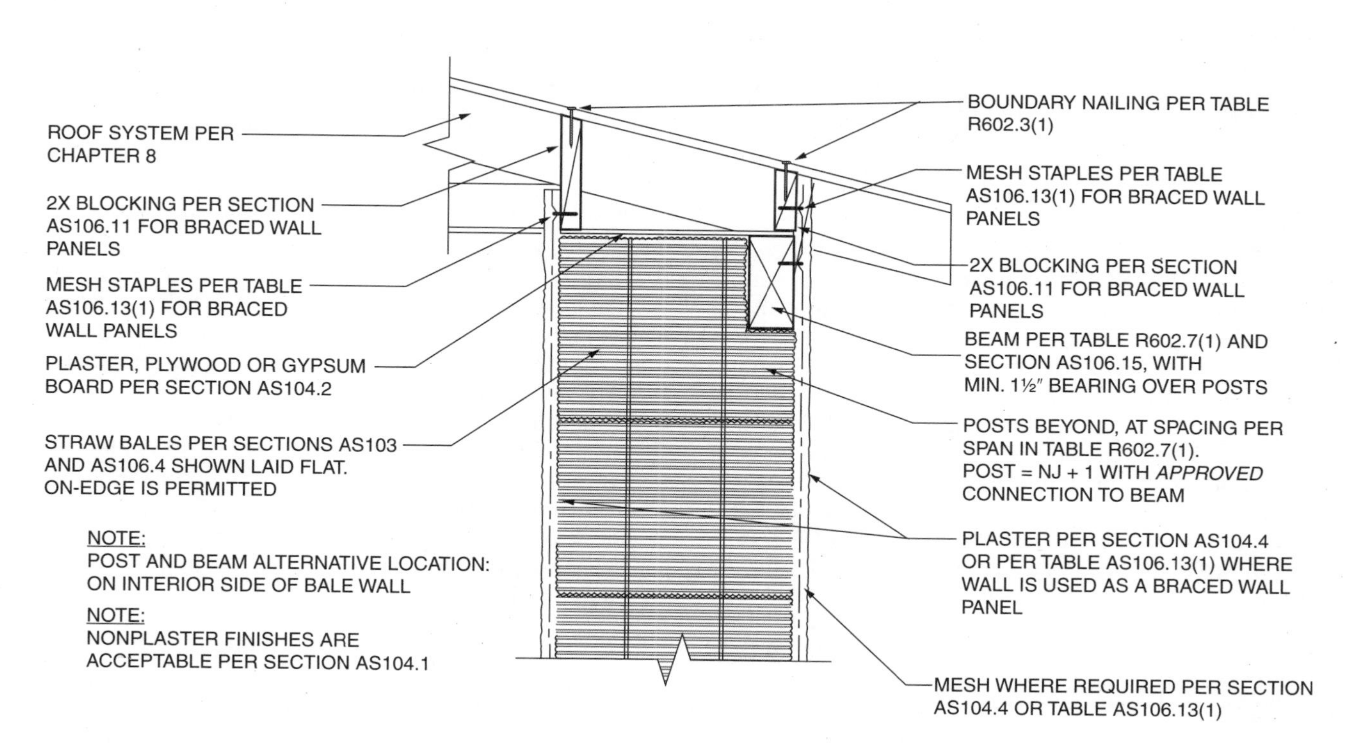

For SI: 1 inch = 25.4 mm.

FIGURE AS105.1(4)
TYPICAL TOP OF POST-AND-BEAM WALL WITH PLASTERED STRAWBALE INFILL

AS105.2 Building limitations and requirements for use of strawbale nonstructural walls. *Buildings* using *strawbale nonstructural walls* shall be subject to the following limitations and requirements:

1. Number of stories: not more than one, except that two stories shall be allowed with an *approved* engineered design.
2. *Building* height: not more than 25 feet (7620 mm), except that greater heights shall be allowed with an *approved* engineered design.
3. Wall height: in accordance with Table AS105.4.
4. Braced wall panel lengths: in accordance with Section R602.10.3, with the additional requirements that Table R602.10.3(3) shall apply to all *buildings* in Seismic Design Category C, and the minimum total length of braced wall panels in Table R602.10.3(3) shall be increased by 60 percent for *buildings* in Seismic Design Categories C, D_0, D_1 and D_2.

❖ **Item 4:** All buildings with nonstructural strawbale walls must employ braced wall panels that comply with Section R602.10.3. For such buildings in Seismic Design Categories C, D_0, D_1 and D_2, the minimum total length of braced wall panels in Table R602.10.3(3) must be increased by 60 percent. This is due to the additional weight of strawbale walls (especially where finished with plaster on both sides) compared to wood-framed walls. This weight imposes additional seismic load on the braced wall panels.

These increased minimum total lengths must be compared with the lengths based on wind speed in Table R602.10.2(1) with the applicable adjustment factors in Table R602.10.3(2). The larger values must be used, as required by Section R602.10.3, Item 4. The 60-percent increase does not apply to the use of Table R602.10.2(1) for wind, because wind loads are independent of building weight.

In Seismic Design Categories A and B, one simply uses Table R602.10.3(1) for wind with the applicable adjustment factors because wind governs the lateral force-resisting system in those SDCs

It is also possible for buildings with load-bearing strawbale walls to employ one of the conventional braced wall panels described in Section R602.10.3. In Seismic Design Categories C, D_0, D_1 and D_2, the same increase in minimum total length applies. In Seismic Design Categories A and B, the wind tables are used without the 60-percent increase as explained above.

Buildings employing strawbale braced wall panels as their lateral force-resisting system must comply with Section AS106.13. Where used in Seismic Design Categories C, D_0, D_1 and D_2, the required minimum total lengths of the strawbale braced wall panels in Table AS106.13(3) already account for the seismic load due to the weight of the plastered strawbale walls; therefore, no increase is necessary.

The 60-percent increase in Item 4 assumes a wall dead load of up to 60 psf. This is consistent with the wall dead load limit stipulated in Section AS106.3. Smaller wall dead loads, especially for strawbale walls with a nonplaster finish, could justify a smaller percentage increase if demonstrated in accordance with accepted engineering practice.

AS105.3 Sill plates. Sill plates shall be installed in accordance with Figure AS105.1(1) or AS105.1(2). Sill plates shall support and be flush with each face of the *straw bales* above and shall be of naturally durable or preservative-treated wood where required by this code. Sill plates shall be not less than nominal 2 inches by 4 inches (51 mm by 102 mm) with anchoring complying with Section R403.1.6 and the additional requirements of Tables AS105.4 and AS106.6(1), where applicable.

AS105.3.1 Exterior sill plate flashing. Exterior sill plates shall receive flashing across the plate to slab or foundation joints.

❖ Continuous flashing must be installed that covers the joint between the bottom of the sill plate and its supporting material (e.g., concrete slab, foundation or raised floor). Such flashing should drain to the exterior, and can be combined with flashing used to satisfy Section AS105.6.9. Examples of this flashing are shown in Figures AS105.1(1) and AS105.1(2).

The purpose of this flashing is to protect against intrusion of moisture that might accumulate near the bottom of the exterior plaster. If a water-resistive barrier is installed over the exterior face of the bales (e.g., under a finish other than plaster, between wood framing and plaster, or between the straw and plaster in some applications), it should lap over the flashing.

Also see Section AS105.6.9 and its commentary.

AS105.4 Out-of-plane resistance methods and unrestrained wall dimension limits. *Strawbale* walls shall employ a method of out-of-plane load resistance in accordance with Table AS105.4, and comply with its associated limits and requirements.

❖ Every strawbale wall must employ an acceptable method of out-of-plane resistance in accordance with Table AS105.4. The maximum unrestrained height of any strawbale wall, both nonstructural and structural, is a function of its method of out-of-plane resistance, as well as other parameters included in Table AS105.4. The overall height of a strawbale wall can exceed the unrestrained height limits if it employs an approved horizontal restraint (see Table AS105.4, Note b) at an intermediate height between the sill plates and the top plate or roof bearing assembly. Also see commentary to Table AS105.4.

TABLE AS105.4. See page S-15.

❖ See the commentary to Sections AS105.4 and AS105.4.2.

Where employing studs as the means of out-of-plane resistance (typically with an unreinforced plaster or nonplaster wall finish), Note f requires an approved attachment method of the bales to the studs. This is commonly achieved with wood stakes or plywood gussets, with sufficient connection to the studs and bales, and sufficient quantity and distribution for the design wind and seismic loads for the building and its location. This method should be used only with an engineered design, unless a local history of acceptable experience exists, and must be approved by the building official.

AS105.4.1 Determination of out-of-plane loading. Out-of-plane loading for the use of Table AS105.4 shall be in terms of the ultimate design wind speed and seismic design category as determined in accordance with Sections R301.2.1 and R301.2.2.

AS105.4.2 Pins. *Pins* used for out-of-plane resistance shall comply with the following or shall be in accordance with an *approved* engineered design. *Pins* shall be external, internal or a combination of the two.

1. *Pins* shall be $^1/_2$-inch-diameter (12.7 mm) steel, $^3/_4$-inch-diameter (19.1 mm) wood or $^1/_2$-inch-diameter (12.7 mm) bamboo.
2. External *pins* shall be installed vertically on both sides of the wall at a spacing of not more than 24 inches (610 mm) on center. External *pins* shall have full lateral bearing on the sill plate and the top plate or roof-bearing element, and shall be tightly tied through the wall to an opposing *pin* with *ties* spaced not more than 32 inches (813 mm) apart and not more than 8 inches (203 mm) from each end of the *pins*.
3. Internal *pins* shall be installed vertically within the center third of the *bales*, at spacing of not more than 24 inches (610 mm) and shall extend from top course to bottom course. The bottom course shall be connected to its support and the top course shall be connected to the roof- or floor-bearing member above with *pins* or other *approved* means. Internal *pins* shall be continuous or shall overlap through not less than one *bale* course.

❖ The term "pin" (see definition of "Pin") comes from the once-common use of internal pins, typically made of steel reinforcing bar, in modern stawbale construction.

TABLE AS105.4
OUT-OF-PLANE RESISTANCE METHODS AND UNRESTRAINED WALL DIMENSION LIMITS

METHOD OF OUT-OF-PLANE LOAD RESISTANCE[a]	FOR ULTIMATE DESIGN WIND SPEEDS (mph)	FOR SEISMIC DESIGN CATEGORIES	UNRESTRAINED WALL DIMENSIONS, H[b]		MESH STAPLE SPACING AT BOUNDARY RESTRAINTS
			Absolute limit in feet	Limit based on bale thickness T[c] in feet (mm)	
Nonplaster finish or unreinforced plaster	≤ 130	A, B, C, D_0	$H \le 8$	$H \le 5T$	None required
Pins per **Section** AS105.4.2	≤ 130	A, B, C, D_0	$H \le 12$	$H \le 8T$	None required
Pins per **Section** AS105.4.2	≤ 140	A, B, C, D_0, D_1, D_2	$H \le 10$	$H \le 7T$	None required
Reinforced[d] clay plaster	≤ 140	A, B, C, D_0, D_1, D_2	$H \le 10$	$H \le 8T^{0.5}$ ($H \le 140T^{0.5}$)	≤ 6 inches
Reinforced[d] clay plaster	≤ 140	A, B, C, D_0, D_1, D_2	$10 < H \le 12$	$H \le 8T^{0.5}$ ($H \le 140T^{0.5}$)	≤ 4 inches[e]
Reinforced[d] cement, cement-lime, lime or soil-cement plaster	≤ 140	A, B, C, D_0, D_1, D_2	$H \le 10$	$H \le 9T^{0.5}$ ($H \le 157T^{0.5}$)	≤ 6 inches
Reinforced[d] cement, cement-lime, lime or soil-cement plaster	≤ 155	A, B, C, D_0, D_1, D_2	$H \le 12$	$H \le 9T^{0.5}$ ($H \le 157T^{0.5}$)	≤ 4 inches[e]
2x6 load-bearing studs[f] at max. 6′ o.c.	≤ 140	A, B, C, D_0, D_1, D_2	$H^g \le 9$	N/A	None required
2x6 load-bearing studs[f] at max. 4′ o.c.	≤ 140	A, B, C, D_0, D_1, D_2	$H^g \le 10$	N/A	None required
2x6 load-bearing studs[f] at max. 2′ o.c.	≤ 140	A, B, C, D_0, D_1, D_2	$H^g \le 12$	N/A	None required
2x4 load-bearing studs[f] at max. 2′ o.c.	≤ 140	A, B, C, D_0, D_1, D_2	$H^g \le 10$	N/A	None required
2x6 nonload-bearing studs[f] at max. 6′ o.c.	≤ 140	A, B, C, D_0, D_1, D_2	$H^g \le 12$	N/A	None required

For SI: 1 inch = 25.4 mm, 1 foot = 304.8 mm, 1 mile per hour = 0.447 m/s.

N/A = Not Applicable

a. Finishes applied to both sides of stacked bales. Where different finishes are used on opposite sides of a wall, the more restrictive requirements shall apply.

b. *H* = Stacked bale height in feet (mm) between sill plate and top plate or other approved horizontal restraint, or the horizontal distance in feet (mm) between *approved* vertical restraints. For load-bearing walls, *H* refers to vertical height only.

c. *T* = Bale thickness in feet (mm).

d. Plaster reinforcement shall be any mesh allowed in **Table** AS106.16 for the matching plaster type, and with staple spacing in accordance with this table. Mesh shall be installed in accordance with **Section** AS106.9.

e. Sill plate attachment shall be with $^5/_8$-inch anchor bolts or approved equivalent at not more than 48 inches on center where staple spacing is required to be ≤ 4 inches.

f. Bales shall be attached to the studs by an approved method. Horizontal framing and attachment at top and bottom of studs shall be in accordance with Section R602 or an *approved* alternative. Table R602.7(1) shall be used to determine the top framing member where load-bearing stud spacing exceeds 24 inches o.c.

g. H is vertical height only.

In this method, the bales are skewered with steel bar as a means of stabilizing the walls during construction and to provide resistance to out-of-plane wind and seismic forces. The practice of internal pinning (with steel or other materials) has fallen into disuse by most practitioners of strawbale construction due to its greater difficulty and cost compared with other means of achieving the same or better wall stability. There is also debate among strawbale building practitioners regarding whether steel pins may be a location of potential condensation in some conditions.

The term "external pin" might be considered a misnomer relative to general use or understanding of the word "pin" (typically being inside another material), but it is commonly used terminology in strawbale construction.

AS105.5 Connection of light-framed walls to strawbale walls. *Light-framed* walls perpendicular to, or at an angle to a *strawbale* wall assembly, shall be fastened to the bottom and top wood members of the *strawbale* wall in accordance with requirements for wood or cold-formed steel *light-framed* walls in this code, or the abutting stud shall be connected to alternating *strawbale* courses with a $^{1}/_{2}$-inch diameter (12.7 mm) steel, $^{3}/_{4}$-inch-diameter (19.1 mm) wood or $^{5}/_{8}$-inch-diameter (15.9 mm) bamboo dowel, with not less than 8-inch (203 mm) penetration.

❖ The connection methods stated in this section are not meant to preclude the use of other common and effective methods, such as an all-thread rod inserted through the abutting stud and bale wall, with a plywood "washer" on the opposite side of the bale and a steel nut and washer on both ends of the threaded rod.

AS105.6 Moisture control. *Strawbale* walls shall be protected from moisture intrusion and damage in accordance with Sections AS105.6.1 through AS105.6.9.

❖ Preventing intrusion of moisture into strawbale walls is important for maximizing service life and for maintaining the insulating value of the straw.

There is a colloquial metaphor commonly used by practitioners of strawbale construction that summarizes the basic principles for keeping a wall dry: "Good boots, a good hat, and a coat that breathes." This translates into keeping the bottom of the wall protected from ground and weather-related moisture, providing ample roof overhangs (especially in wet climates) to shield the wall and its openings from rain and providing a protective wall finish that is vapor permeable. This dictum is also considered to be wise advice for buildings constructed of wood frame and other materials.

Straw, like wood, is composed primarily of cellulose, hemicellulose and lignin. Straw and wood can last indefinitely as long as there is insufficient free moisture (moisture not bound in the material) to cause deterioration. If free moisture becomes available (generally where the moisture content of the material exceeds 20 percent) and other necessary environmental conditions exist—including temperatures above 50°F (10°C) and the presence of oxygen—then microbial degradation of the straw can occur. In buildings, the only practically controllable condition is the availability of free moisture.

Though straw and wood have similar material makeup, straw is more susceptible to damage from moisture over a shorter period of time. This is due to the greater surface area of the many lengths of tubular straw, compared with only the perimeter surface area of wood framing members. However, the greater surface area of straw also has the comparative advantage of potentially absorbing and storing more moisture before it becomes free moisture and causes degradation. This, combined with the much larger storage capacity of a continuous core of straw compared to the spaced framing members in a wood-framed wall, gives a strawbale wall significantly greater moisture management capacity.

Subsections AS105.6.1 through AS105.6.9 contain requirements to minimize the possibility of moisture intrusion into strawbale walls from common sources such as rain or snow and condensation (from uncontrolled flow of relatively warm, moist air into the wall). See the additional commentary after each subsection.

Potential sources that are not addressed in these subsections are leaks and condensation from plumbing pipes in the walls. While there is no requirement regarding plumbing pipes in this appendix, it is recommended and common practice to minimize the installation of supply and waste pipes in strawbale walls, or to install them at the bottom of the wall (between sill plates) or in a continuous sleeve that drains to the exterior.

In addition to the importance of preventing moisture from entering a strawbale wall, moisture must also be allowed to readily exit the wall. Thus, the finish on each side should be as vapor permeable as possible within the requirements of this appendix and the code. Sections AS105.6.1, AS105.6.2 and AS104.3 include minimum vapor permeability requirements for finishes and any associated vapor retarder or water-resistive barrier.

AS105.6.1 Water-resistant barriers and vapor permeance ratings. *Plastered bale* walls shall be constructed without any membrane barrier between *straw* and *plaster* to facilitate transpiration of moisture from the *bales*, and to secure a structural bond between *straw* and *plaster*, except as permitted or required elsewhere in this appendix. Where a water-resis**tant** barrier is placed behind an exterior finish, it shall have a vapor permeance rating of not less than 5 perms, except as permitted or required elsewhere in this appendix.

❖ This section is intended to both facilitate transpiration of moisture out of the strawbale core through its finish and to allow a mechanical bond between plaster finishes and the straw. Some strawbale designs incorporate water-resistive barriers. This typically includes water-resistive barriers on the exterior face of bales below window sills, on the first course(s) of bales that are particularly exposed to weather-related moisture, behind nonplaster finishes, over sheathing, or as part of a rain screen. Where a water-resistive barrier is used, its vapor permeance rating must be at least 5 perms (which is the definition of vapor permeable in the code).

Where a water-resistive barrier is used behind a plaster finish on a structural wall, its mesh must be through-tied to the plaster skin on the other side by means of an approved engineered design (see Section AS106.8). Though not addressed in this appendix, where a water-resistive barrier is used behind plaster on a nonstructural wall, it should contain mesh or other lath that is adequately attached to the bale core or to the finish on the other side of the wall to ensure that the plaster will remain in place in normal service.

AS105.6.2 Vapor retarders. Wall *finishes* shall have an equivalent vapor permeance rating of a Class III vapor retarder on the interior side of exterior *strawbale walls* in Climate Zones 5, 6, 7, 8 and Marine 4, as defined in Chapter 11. *Bales* in walls enclosing showers or steam rooms shall be protected on the interior side by a Class I or Class II vapor retarder.

❖ The interior wall finish of exterior strawbale walls requires an equivalent vapor permeance rating of a Class III vapor retarder (> 1 perm and ≤ 10 perms) in the stated climate zones. This requirement is intended to prevent condensation in the wall in cold winter climates by retarding the passage of water vapor into the wall from interior air. However, sufficient vapor permeability of the interior finish is also important to allow migration of moisture from the wall to the interior under some conditions. This includes conditions that occur in many locations in the stated climate zones where interiors are mechanically cooled during warm, humid seasons.

Because of potentially conflicting seasonal demands in the stated climate zones, a balance between the vapor-retarding ability and the vapor permeability of the interior finish of exterior strawbale walls is desirable. Field experience of strawbale practitioners, coupled with principles of building science regarding moisture migration, dictate a perm rating in the upper half of a Class III vapor retarder (between 5 and 10 perms) as optimal for the interior finish of exterior strawbale walls in these climate zones.

In cold climates, the comparative vapor permeance of the interior and exterior finishes are at least as important as the vapor permeability of the interior finish alone. Although not a requirement in this appendix, it is recommended that the vapor permeance of the exterior finish of strawbale walls be greater than that of the interior finish. This will allow water vapor that is driven from the relatively warm and moist interior air toward the colder and drier exterior to continue out of the wall before accumulating and condensing.

The requirement for a Class I or Class II vapor retarder on walls enclosing showers or steam rooms applies to all climate zones and interior and exterior strawbale walls. The term "shower" means a shower or a combination tub/shower. Walls enclosing tubs without showers are not intended to be subject to this requirement.

Also see the commentary to Section AS104.3.

AS105.6.3 Penetrations in exterior strawbale walls. Penetrations in exterior *strawbale* walls shall be sealed with an *approved* sealant or gasket on the exterior side of the wall in all climate zones, and on the interior side of the wall in Climate Zones 5, 6, 7, 8 and Marine 4, as defined in Chapter 11.

❖ In all climate zones, sealing penetrations (e.g., by piping, conduit, electrical boxes or ventilation caps) in the exterior finish of strawbale walls helps minimize or prevent the intrusion of weather-related moisture into the straw core of the wall. In climate zones with cold winters, and in marine climate zones where the air is especially laden with moisture, sealing penetrations on the interior side of exterior walls is important to prevent warm, moist interior air from bypassing the air barrier of the interior finish of the wall, which could condense in the bale core.

Window and door openings are considered large penetrations in exterior walls and are subject to the requirements of this section. The interfaces between window units and door units and exterior strawbale walls must be sealed on the exterior to protect against weather-related moisture intrusion. They must also be sealed on the interior in the listed climate zones in order to protect against intrusion of moisture from interior air. Window sills are subject to the requirements of Section AS105.6.4.

Windows and doors require particular attention to minimize moisture intrusion into the wall. Appropriate flashing measures should be taken and detailed according to the exterior finish (plaster or other cladding), the profile of any window or door trim and the location of the unit within the thickness of the strawbale wall. Common practice is to seal the plaster to the rough opening, which is in turn sealed to the window or door unit. Conventional "peel-and-stick" window flashings are often used to bridge from the window or door unit onto the surrounding bales at the jambs, head and sill.

Penetrations in strawbale walls that are not weather-exposed are not intended to require sealing on the exterior. However, penetrations in such walls must be sealed on the interior in Climate Zones 5, 6, 7, 8 and Marine 4. Exception 2 in the definition of "Weather-exposed surfaces" in the IBC might be used to determine whether a penetration in a strawbale wall is considered weather exposed, subject to the evaluation of the local building official, with additional consideration for the local climate. That exception reads, "Walls or portions of walls beneath an unenclosed roof area, where located a horizontal distance from an open exterior opening equal to at least twice the height of the opening."

AS105.6.4 Horizontal surfaces. *Bale* walls and other *bale* elements shall be provided with a water-resistant barrier at weather-exposed horizontal surfaces. The water-resistant barrier shall be of a material and installation that will prevent water from entering the wall system. Horizontal surfaces shall include exterior window sills, sills at exterior niches and buttresses. Horizontal surfaces shall be sloped not less than 1 unit vertical in 12 units horizontal (8-percent slope) and shall drain away from *bale* walls and elements. Where the water-resistant barrier is below the finish material, it shall be sloped not less than 1 unit vertical in 12 units horizontal (8-percent slope) and shall drain to the outside surface of the *bale* wall's vertical finish.

❖ As with wood-frame walls, horizontal (or nearly horizontal) surfaces in or on strawbale walls can be vulnerable to weather-related moisture entering the wall. Window sills for windows recessed into the wall are especially vulnerable and should be carefully detailed and constructed to meet the prescriptive and performance criteria of this section. It is not uncommon for the water-resistive barrier under window sills to lap over a vapor permeable water-resistive barrier that continues down the exterior face of the bales below in order to provide additional protection for those potentially vulnerable bales.

Horizontal surfaces in or on strawbale walls that are not weather exposed do not require a water-resistive barrier. Exception 2 in the definition of "Weather-exposed surfaces" in the IBC might be used to determine whether a horizontal surface in a strawbale wall is considered weather exposed, subject to the evaluation of the local building official, with additional consideration of the local climate. That exception reads, "Walls or portions of walls beneath an unenclosed roof area, where located a horizontal distance from an open exterior opening equal to at least twice the height of the opening."

Ample roof overhangs and wrap-around porch roofs are often employed by strawbale building practitioners, especially in wet climates and heavy snow fall areas, to protect walls and their openings from weather-related moisture intrusion.

AS105.6.5 Separation of bales and concrete. A sheet or liquid-applied Class II *vapor retarder* shall be installed between bales and supporting concrete or masonry. The bales shall be separated from the vapor retarder by not less than $^3/_4$ inch (19.1 mm), and that space shall be filled with an insulating material such as wood or rigid insulation, or a material that allows vapor dispersion such as gravel, or other approved insulating or vapor dispersion material. Sill plates shall be installed at this interface in accordance with Section AS105.3. Where bales abut a concrete or masonry wall that retains earth, a Class II vapor retarder shall be provided between such wall and the bales.

❖ A Class II vapor retarder is required to separate bales from supporting concrete or masonry to prevent "rising damp" from reaching the bales and potentially causing mold or mildew on the straw. Further, a minimum $^3/_4$-inch space is required between the vapor retarder and the straw and insulating or vapor dispersion material. Insulating materials help keep the underside of the first course of bales above the dew point to avoid condensation on the straw. Vapor dispersion materials also allow safe dispersion of any moisture that might condense or otherwise reach the area at the base of the wall.

See Figure AS105.1(1) for a typical base-of-wall detail for slab-on-grade conditions.

AS105.6.6 Separation of bales and earth. *Bales* shall be separated from earth by not less than 8 inches (203 mm).

❖ Straw, like wood, is subject to decay with prolonged exposure to excessive free moisture; therefore, straw bales are required to be separated from potentially moisture-laden earth by at least 8 inches (203 mm). This separation also reduces exposure of straw at the bottom of the wall to weather-related moisture, such as rain splash-back and snow melt against the wall's exterior plaster. Even larger separation is recommended in situations where excessive moisture exposure is expected due to local climate conditions.

AS105.6.7 Separation of exterior plaster and earth. Exterior plaster applied to *straw bales* shall be located not less than 6 inches (102 mm) above earth or 3 inches (51 mm) above paved areas.

❖ Section R703.7.2.1 requires the bottom of weep screeds for exterior cement plaster to be not less than 4 inches above the earth and 2 inches (51 mm) above paved areas. For exterior plaster applied to straw bales, Section AS105.6.7 requires the plaster to be not less than 6 inches (152 mm) above earth and 3 inches (51 mm) above paved areas. The additional comparative separation is to give greater assurance that exterior moisture at grade will not reach the straw. This separation requirement is for all exterior plasters allowed in this appendix, not only for cement plaster as referenced in Section R703.7.2.1.

Even larger separation is recommended where excessive, weather-related moisture exposure, including rain splash-back, is expected due to local climatic conditions, and for weather-exposed clay plasters to minimize erosion, even where protected by an erosion-resistant finish as required by Section AS104.4.3.4.

AS105.6.8 Separation of wood and plaster. Where wood framing or wood sheathing occurs at the exterior face of *strawbale* walls, such wood surfaces shall be separated from exterior plaster with two layers of Grade D paper, No. 15 asphalt felt or other *approved* material in accordance with Section R703.7.3.

Exceptions:

1. Where the wood is preservative treated or *naturally durable* and is not greater than $1^1/_2$ inches (38 mm) in width.
2. Clay plaster shall not be required to be separated from untreated wood that is not greater than $1^1/_2$ inches (38 mm) in width.

❖ A water-resistive barrier such as two layers of Grade D paper or No. 15 asphalt felt is required to separate exterior plaster from wood framing or sheathing due to the possibility of moisture penetrating the plaster and reaching the framing or sheathing. The water-resistive barrier protects the framing or sheathing from potential decay.

A water-resistive barrier is not required under the conditions noted in Exception 1 because such materials are not susceptible to decay in the presence of moisture.

In Exception 2, a water-resistive barrier is not required between clay plaster and untreated wood because clay is hydrophilic (i.e., water-attracting, drawing moisture away from adjacent materials) and blocks passage of water due to its strong surface bonding with available moisture. Also, clay has considerable capacity to capture and store water due to the large surface area of its microscopic layers. This combination of hydrophilic properties, water-blocking properties and water storage capacity protects adjacent materials that might otherwise be harmed by prolonged exposure to moisture.

The $1^1/_2$-inch (38 mm) maximum width in Exceptions 1 and 2 relates to the prevention of plaster cracking. Plasters (all types identified in this appendix) on strawbale walls have demonstrated the ability to span across $1^1/_2$-inch-wide (38 mm) wood members without significant cracking, even where unreinforced and without the "slip sheet" effect of a water-resistive barrier between the plaster and the wood member. For wood members greater than $1^1/_2$ inches (38 mm) wide, a two-layered water-resistive barrier, along with an appropriate reinforcing mesh that extends at least 1 inch (25 mm) beyond the edges of the wood, has been shown to greatly reduce or eliminate plaster cracking in these locations.

Also see Sections AS104.4.1 and AS106.8 and their commentary for information regarding where a water-resistive barrier between plaster and straw bales is not required.

AS105.6.9 Separation of exterior plaster and foundation. Exterior plaster shall be separated from the building foundation with a moisture barrier.

❖ A moisture barrier is required to separate the exterior plaster on a strawbale wall from the foundation so that moisture contained in the foundation (as a result of contact with damp ground or exposure to rain) cannot be wicked into the plaster and then delivered to the straw bales, potentially causing degradation of the straw. Commonly used moisture barriers for this purpose include metal flashing and liquid-applied products made for sealing concrete. Also see Section AS105.3.1 and its commentary.

AS105.7 Inspections. The *building official* shall inspect the following aspects of *strawbale* construction in accordance with Section R109.1:

1. Sill plate anchors, as part of and in accordance with Section R109.1.1.
2. Mesh placement and attachment, where mesh is required by this appendix.
3. *Pins*, where required by and in accordance with Section AS105.4.

❖ Items 2 and 3, regarding inspections for mesh and pins, are necessary only where their use is required by this appendix. Mesh or pins are considered required where used as a means of compliance with Section AS105.4 for out-of-plane resistance and for mesh in structural applications in accordance with Section AS106. Where mesh or pins are installed voluntarily, no inspection of their installation is required.

AS105.8 Voids and stuffing. Voids between *bales* and between *bales* and framing members shall not exceed 4 inches (102 mm) in width, and such voids shall be tightly stuffed with *flakes*, loose straw or *straw-clay* before application of finish.

❖ Stuffing of voids between bales and between bales and framing members is considered necessary for the thermal performance of all exterior strawbale walls, and for the structural performance of strawbale structural walls. The density of straw or straw-clay that fills voids should achieve the minimum required density for bales (see Section AS103.5) to the degree practicable. Also see the commentary to Section AS108.

SECTION AS106 STRAWBALE WALLS—STRUCTURAL

AS106.1 General. Plastered *strawbale* walls shall be permitted to be used as structural walls in accordance with the prescriptive provisions of this section.

❖ The provisions of Section AS106 apply to structural strawbale walls (load-bearing and/or shear walls). Some provisions in this section are also considered good practice for nonstructural walls and are identified as such in this commentary.

Structural strawbale walls require a plaster finish (as prescribed in this section), as the plaster skins together with the strawbale core provide the structural capacity of the composite wall system. The plaster must be reinforced if the wall is used as a shear wall (braced wall panel). Strawbale braced wall panels may be utilized in load-bearing or nonload-bearing applications. In both strawbale load-bearing walls and shear walls the relatively stiff plaster skins carry most of the loads while the strawbale core braces the relatively thin plaster skin against buckling. The strawbale core also serves as a backup load-carrying system should the plaster skins degrade.

AS106.2 Building limitations and requirements for use of strawbale structural walls. *Buildings* using strawbale *structural* walls shall be subject to the following limitations and requirements:

1. Number of stories: Not more than one.
2. *Building* height: Not more than 25 feet (7620 mm).
3. Wall height: In accordance with Table AS105.4, AS106.13(2) or AS106.13(3) as applicable, whichever is most restrictive.
4. Braced wall panel lengths: The greater of the values determined in accordance with Tables AS106.13(2) and AS106.13(3) for *buildings* using strawbale braced wall panels, or in accordance with Item 4 of Section AS105.2 for *buildings* with *load-bearing strawbale walls* that do not use *strawbale* braced wall panels.

❖ **Item 1:** Use of strawbale structural walls is limited to one-story buildings. However, there are many examples of two-story buildings constructed with structural strawbale walls in the U.S and other countries. This section is not meant to prohibit such structures, but they are outside of the prescriptive and performance structural requirements of this appendix and, as such, should be accompanied by an approved engineered design. Note that Section AS105.2, Item 1 allows nonstructural strawbale walls in two-story buildings with an approved engineered design.

Item 2: The 25-foot (7620 mm) height limit is for the building height as defined in the body of the code. Greater heights may be possible with an approved engineered design. Note that Section AS105.2, Item 2 allows buildings with nonstructural strawbale walls to exceed 25 feet (7620 mm) in height with an approved engineered design.

Also see Section AS105.2 and its commentary.

AS106.3 Loads and other limitations. Live and dead loads and other limitations shall be in accordance with Section R301. *Strawbale* wall dead loads shall not exceed 60 psf (2872 N/m^2) per face area of wall.

❖ The amount of bracing required for seismic loads is directly related to the weight of the structure. The dead load limit of 60 psf for strawbale walls is used to avoid potential overburden of the lateral load-resisting system in Seismic Design Categories C, D_0, D_1 and D_2. A larger wall dead load could be acceptable with an approved engineered design, but may require total braced wall panel lengths that are greater than the 60 percent stipulated in Section AS105.2, Item 4 for conventional braced wall panels.

The weight "per face area of wall" means the weight of all materials within a horizontally projected area from one face of the wall to the other. This includes finishes on both sides and the straw bale core.

For example, the weight per square foot of a strawbale wall with two-string bales laid flat and 1-inch-thick lime plaster on both sides would be determined as follows: First, determine the weight per cubic foot of a typical bale for the project (total bale weight / total bale volume). For this example, say: 47 lb/(1.16 feet × 1.5 feet × 3 feet) = 8.5 pcf. For this example assume the lime plaster weighs 130 pcf. Therefore, the weight per square foot of the strawbale wall is: (1 foot × 1 foot × 1.5 foot × 8.5 pcf) + 2(1 foot × 1 foot × $^1/_{12}$ foot × 130 pcf) = 12.75 lb + 21.67 lb = 34.42 psf.

AS106.4 Foundations. Foundations for plastered *strawbale* walls shall be in accordance with Chapter 4, Figure AS105.1(1) or Figure AS105.1(2).

❖ Foundations that satisfy the requirements of Chapter 4 are acceptable for buildings with strawbale walls.

Three tables in the code are used to determine footing width and thickness, depending on the type of wall construction. Strawbale walls, with any finish allowed in this appendix, are closest to the weight of light-frame walls with brick veneer in Table R403.1(2); therefore, that table should be used.

See Figure AS105.1(1) for typical concrete slab-on-grade construction with monolithic foundation supporting a strawbale wall. The figure shows a strawbale structural wall with its required anchor bolts for exterior and interior sill plates and a thickened slab below the interior sill plate. This is necessary to provide the 7-inch (178 mm) embedment for the anchor bolts as required by Section R403.1.6 and the minimum 3-inch (76 mm) cover for reinforcement as required by Section R403.1.3.5.3. Nonstructrual strawbale walls may also require anchor bolts for their sill plates, and therefore also may require a similar thickened slab.

For structural strawbale walls on a raised floor, Section AS106.10 requires an approved engineered design. See the commentary to that section for recommendations for raised floors supporting structural strawbale walls. Also, Figure AS105.1(2) illustrates the typical elements and configuration for a strawbale wall on a raised floor.

AS106.5 Configuration of bales. *Bales* in *strawbale* structural walls shall be laid flat or on-edge and in a running bond or stack bond, except that bales in structural walls with unreinforced plasters shall be laid in a running bond only.

AS106.6 Plaster on structural walls. Plaster on *load-bearing* walls shall be in accordance with Table AS106.12. Plaster on shear walls shall be in accordance with Table AS106.13(1).

❖ Requirements for plaster on strawbale structural walls are more stringent than for nonstructural applications. The term "shear wall" is used interchangeably with the term "braced wall panel" in this appendix.

AS106.6.1 Compressive strength. For plaster on *strawbale* structural walls, the *building official* is authorized to require a 2-inch (51mm) cube test conforming **to** ASTM C109 to demonstrate a minimum compressive strength in accordance with Table AS106.6.1.

❖ Minimum compressive strengths in Table AS106.6.1 are provided to clarify minimum expectations for builders and building officials. These minimum values relate (with appropriate factors of safety) to the allowable bearing capacities for walls with plaster types indicated in Table AS106.12, and to the minimum braced wall panel lengths for wall types indicated in Tables AS106.13(2) and AS106.13(3).

The minimum compressive strengths in Table AS106.6.1 are on the low end of the range of compressive strengths typical for each plaster type. As such, a building official may expect these values to be achieved without testing for a plaster made with good-quality materials and workmanship. However, the building official is authorized to require an ASTM C109 2-inch cube test to confirm a plaster's minimum compressive strength for structural strawbale walls.

Since ASTM C109 applies to the compressive strength testing of hydraulic cement mortars, the following modifications are needed for the testing of soil-cement and clay plasters:

1. Soil-cement plaster samples must be cured in a moist environment in accordance with Section 10.5 of ASTM C109, and must not be immersed in lime-saturated water in accordance with the same section.

2. Clay plaster samples must be dried to the approximate ambient moisture conditions of the

project site, and must not be not cured in the moist or wet curing environments described in Section 10.5 of ASTM C109.

The rationale behind those modifications is that hydraulic cement mortar and the cement, cement-lime, lime and soil-cement plasters described in this appendix all contain sand and hydraulic cement and/or hydraulic lime. These mixes develop their strength by hydration (a water-based chemical reaction, sometimes called curing) over time. Clay plaster develops its strength by drying, not by moist curing. Soil-cement plaster contains both hydraulic cement that must be moist cured and clay that must dry to an acceptable moisture content. As such it should be moist cured, but not immersed in water.

TABLE AS106.6.1
MINIMUM COMPRESSIVE STRENGTH FOR PLASTERS ON STRUCTURAL WALLS

PLASTER TYPE	MINIMUM COMPRESSIVE STRENGTH (psi)
Clay	100
Soil-cement	1000
Lime	600
Cement-lime	1000
Cement	1400

For SI: 1 pound per square inch = 6894.76 N/m².

AS106.7 Straightness of plaster. Plaster on *strawbale* structural walls shall be straight, as a function of the bale wall surfaces they are applied to, in accordance with all of the following:

1. As measured across the face of a *bale*, *straw* bulges shall not protrude more than $^3/_4$ inch (19.1 mm) across 2 feet (610 mm) of its height or length.
2. As measured across the face of a *bale* wall, *straw* bulges shall not protrude from the vertical plane of a *bale* wall more than 2 inches (51 mm) over 8 feet (2438 mm).
3. The vertical faces of adjacent *bales* shall not be offset more than $^3/_8$ inch (9.5 mm).

❖ The plaster on structural strawbale walls carries in-plane forces. Significant deviation from straightness may compromise a wall's strength and structural performance. The requirements of Section AS106.7 are intended to ensure a minimum degree of straightness for the expected structural performance of the wall.

Note that some practitioners have constructed large-radius strawbale walls by curving bales laid flat. Such a wall can be considered to comply with this section and be used as a load-bearing wall if the wall's plaster is straight vertically, and if it meets all other requirements in this appendix for load-bearing strawbale walls.

AS106.8 Plaster and membranes. *Strawbale* structural walls shall not have a membrane between straw and plaster, or shall have attachment through the *bale* wall from one plaster skin to the other in accordance with an *approved* engineered design.

❖ The application of plaster with no membrane between the straw and plaster allows the plaster to bond directly to the straw, thus allowing composite structural performance of the wall assembly. Specifically it allows: the plaster to be well braced by the strawbale core; the plaster to resist delamination from the straw core under load; and the strawbale core to transfer shear forces from one plaster skin to the other. In circumstances where a membrane is required or desired, the mesh in one plaster skin must be through-tied to the mesh in the opposite plaster skin in accordance with an approved engineered design.

See the commentary to Section AS104.4.1 for information regarding membranes and issues of moisture management and the support of plaster on nonstructural walls.

AS106.9 Mesh. Mesh in plasters on *strawbale* structural walls, and where required by Table AS105.4, shall be installed in accordance with Sections AS106.9.1 through AS106.9.4.

❖ Mesh required on structural strawbale walls and mesh used as a method of out-of-plane resistance (in accordance with Table AS105.4) must be installed in accordance with all subsections of Section AS106.9. Where mesh is installed but not required, these provisions are recommended as good practice.

AS106.9.1 Mesh laps. Mesh required by Table AS105.4 or AS106.12 shall be installed with not less than 4-inch (102 mm) laps. Mesh required by Table AS106.13(1) or in walls designed to resist wind uplift of more than 100 plf (1459 N/m), shall run continuous vertically from sill plate to the top plate or roof-bearing element, or shall lap not less than 8 inches (203 mm). Horizontal laps in such mesh shall be not less than 4 inches (102 mm).

❖ Mesh "required by Table AS106.13(1)" refers to the wall types described in that table where used as braced wall panels to resist wind or seismic forces in accordance with Table AS106.13(2) or AS106.13(3). Mesh "in walls designed to resist uplift of more than 100 plf" refers to the use of mesh in accordance with Section AS106.14.

AS106.9.2 Mesh attachment. Mesh shall be attached with staples to top plates or roof-bearing elements and to sill plates in accordance with all of the following:

1. **Staples.** Staples shall be pneumatically driven, stainless steel or electro-galvanized, 16 gage with $1^1/_2$-inch (38 mm) legs, $^7/_{16}$-inch (11.1 mm) crown; or manually driven, galvanized, 15 gage with 1-inch (25 mm) legs. Other staples shall be as designed by a registered design professional. Staples into preservative-treated wood shall be stainless steel.
2. **Staple orientation.** Staples shall be firmly driven diagonally across mesh intersections at the required spacing.

3. **Staple spacing.** Staples shall be spaced not more than 4 inches (102 mm) on center, except where a lesser spacing is required by Table AS106.13(1) or Section AS106.14, as applicable.

❖ Mesh is required to be attached to a wall's horizontal boundary elements, such as roof-bearing elements and sill plates. Attachment to posts or studs (where present) is not required, but is recommended, particularly at vertical boundaries, such as wall corners or where a wall is interrupted by a window or door opening.

Staples are required to be stainless steel or galvanized steel to minimize or prevent corrosion of staples in the event that significant moisture is present. Staples are required to be stainless steel where driven into preservative-treated wood.

Installing staples diagonally across mesh intersections is important for strawbale braced wall panels (shear walls) in order to minimize slip of the mesh during load transfer between the mesh and the top and bottom plates, regardless of the direction of movement. This can also be important for transfer of loads for mesh used on load-bearing walls, or to resist out-of-plane loads or wind uplift loads. Diagonal stapling also helps limit splitting of the wood member, thus helping to maintain the strength of the connection. Diagonally placed staples are required for structural strawbale walls and are recommended good practice wherever mesh is used on strawbale walls.

AS106.9.3 Steel mesh. Steel mesh shall be galvanized, and shall be separated from preservative-treated wood by Grade D paper, No. 15 roofing felt or other *approved* barrier.

❖ Where steel mesh is used in structural strawbale plasters, it is required to be galvanized to reduce the potential for corrosion. Where galvanized, Section R317.3.1 requires that the fasteners be hot-dipped, zinc coated galvanized steel. The required separation of steel mesh from preservative-treated wood in this section is intended to prevent corrosion in the mesh.

AS106.9.4 Mesh in plaster. Required mesh shall be embedded in the plaster except where staples fasten the mesh to horizontal boundary elements.

❖ Required mesh must be embedded in the plaster so that they work together as a structural composite to resist in-plane forces. The plaster provides resistance to compressive forces and the mesh provides resistance to tensile and shear forces. The plaster helps resist deformation of the embedded mesh under load, while resolving diagonal tensile forces into alignment with the principal directions of the mesh.

Ideally, mesh is embedded in the middle third of the overall thickness of the plaster. In practice, deviation from ideal embedment can occur due to the irregular surface of the bales and normal construction practices. Various spacers have been used to hold mesh away from the bales, and U-shaped wire "staples" or wire "through-ties" have been used to bring mesh closer to the bales and to help locate the mesh in the middle third of the plaster application.

Where the mesh is stapled to the top-of-wall member or assembly and to the sill plates, the mesh cannot be fully embedded in the plaster. Tested strawbale wall assemblies used as the basis for the structural provisions of this appendix were constructed using direct fastening of the mesh to their horizontal boundary elements with not more than reasonable effort to locate the mesh in the middle third of their plasters.

AS106.10 Support of plaster skins. Plaster *skins* on *strawbale* structural walls shall be continuously supported along their bottom edge. Acceptable supports include: a concrete or masonry stem wall, a concrete slab-on-grade, a wood-framed floor in accordance with Figure AS105.1(2) and an *approved* engineered design or a steel angle anchored with an *approved* engineered design. A weep screed as described in Section R702.7.2.1 is not an acceptable support.

❖ Plaster skins are much stiffer than the strawbale core of a strawbale wall and thus carry most of the applied load. The strawbale core provides lateral bracing to the relatively thin plaster skins.

Both vertical and lateral loads are delivered through the plaster skins to the skins' bottom edges. It is vital that the bottom edge of the plaster skins is supported in a way that is capable of delivering these loads to the foundation. Section AS106.10 describes acceptable ways of supporting plaster skins for structural strawbale walls. See Figure AS105.1(1) for typical plaster support on a concrete slab-on-grade with monolithic foundation, and Figure AS105.1(2) for typical plaster support for a raised floor condition.

Where structural strawbale walls are supported by a wood-framed floor, an approved engineered design is required. The requirement that these floors be blocked refers to blocking between joists that supports the interior plaster where floor joists run perpendicular to the wall. Where the joists run parallel to the wall, a joist directly below the interior plaster provides that support. Such blocking or joists should also support the interior sill plate of the wall, and all sill plates should have adequate attachment to the floor framing, so that the system maintains a continuous load path to the foundation. Where strawbale braced wall panels are used, the engineered design should also address any associated uplift in the floor system.

Note that the commonly used weep screed for cement plaster over wood sheathing, as described in Section R703.7.2.1, is not an acceptable structural support for strawbale construction. Conventional weep screeds are not designed to support structural loads; however, such a weep screed is acceptable for non-structural strawbale walls.

AS106.11 Transfer of loads to and from plaster skins. Where plastered *strawbale* walls are used to support superimposed vertical loads, such loads shall be transferred to the plaster *skins* by continuous direct bearing in accordance with Figure AS105.1(3) or by an *approved* engineered design. Where plastered *strawbale* walls are used to resist in-plane lateral loads, such loads shall be transferred to the reinforcing mesh from the structural member or assembly above in accordance with Figure AS105.1(3) or AS105.1(4) and to the sill plate in accordance with Figure AS105.1(1) or AS105.1(2) and with Table AS106.13(1).

❖ Section AS106.11 contains criteria for the transfer of loads to the top of load-bearing strawbale walls, to or from the top-of-wall structural members and to the sill plates of strawbale shear walls (braced wall panels). Prescriptive requirements for the sill plates are contained in Section AS105.3. There are no prescriptive requirements for the top-of-wall members or assemblies for structural or nonstructural strawbale walls.

Roof loads imposed on load-bearing strawbale walls require continuous direct bearing on the top edges of the plaster, or an engineered means of transferring the loads to the plaster skins. Figure AS105.1(3) shows an example of direct bearing onto the top edge of the plaster. Other configurations are also possible. Roof loads should be transferred approximately equally to both plaster skins, or only half the allowable bearing capacity should be used. See Section AS106.12.2 for information regarding concentrated loads.

One example of an engineered means of load transfer without direct bearing is from the roof bearing assembly through attached mesh into a plaster-mesh matrix, with approved calculations or test results demonstrating that the allowable capacity of the load-transfer system is greater than the design loads. Any system other than direct bearing should demonstrate a load path with sufficient capacity.

For strawbale walls used to resist in-plane lateral loads (wind or seismic), the load transferred from the structural member or assembly at the top of the wall through the required mesh and staples and into the required sill plate shall be as stipulated in Table AS106.13(1).

Historically, load-bearing strawbale walls have used some variation of a wooden box beam as a roof- or floor-bearing assembly at the top of the wall, though other members or assemblies have been used. Figure AS105.1(3) shows a generic box beam at the top of a load-bearing strawbale wall. Box beams and other roof-bearing assemblies can also serve as headers over wall openings, as indicated in Figure AS105.1(3). The figure also shows an example of locations for mesh attachment to the box beam where the strawbale wall is used as a braced wall panel (shear wall).

For nonstructural strawbale walls, the members at the top of the walls are typically the beams of a post-and-beam frame, where the top course of straw bales is infilled tightly against the underside of the beams, or notched to accommodate them. Figure AS105.1(4) shows a generic configuration of the top of a post-and-beam wall with strawbale infill. The figure also shows example locations for mesh attachment to the beams and roof assembly where the strawbale wall is used as a braced wall panel (shear wall).

AS106.12 Load-bearing walls. Bearing capacities for plastered *strawbale* walls used as load-bearing walls in one-*story buildings* to support vertical loads imposed in accordance with Section R301 shall be in accordance with Table AS106.12.

❖ See Figure AS105.1(3) for the top of a typical load-bearing wall. See Figures AS105.1(1) and AS105.1(2) for the base of a typical load-bearing wall over a slab-on-grade and for a raised wood-framed floor, respectively.

TABLE AS106.12. See below.

❖ The allowable bearing capacities in Table AS106.12 are for strawbale walls with plaster on both sides. Plaster can be installed on one side only; however, the allowable bearing capacity is then half of that shown in the table.

AS106.12.1 Precompression of load-bearing strawbale walls. Prior to application of plaster, walls designed to be load-bearing shall be precompressed by a uniform load of not less than 100 plf (1459 N/m).

❖ The purpose of requiring precompression of load-bearing strawbale walls is to accelerate settling of the stacked straw bales before plaster is applied so that potential cracking of the plaster is minimized. The required 100 plf (1459 N/m) precompression is intended to be permanent and can be achieved by a variety of means, including vertically tensioned steel mesh fastened to sill plates and top plates, or tensioned packaging straps, or other tensile material looped under sill plates and over top plates. In many cases, the dead load of the roof on the wall (prior to plaster) may meet the 100 plf (1459 N/m) precompression requirement.

TABLE AS106.12
ALLOWABLE SUPERIMPOSED VERTICAL LOADS (LBS/FOOT) FOR PLASTERED LOAD-BEARING STRAWBALE WALLS

WALL DESIGNATION	PLASTER[a] (both sides) Minimum thickness in inches each side	MESH[b]	STAPLES[c]	ALLOWABLE BEARING CAPACITY[d] (plf)
A	Clay $1^1/_2$	None required	None required	400
B	Soil-cement 1	Required	Required	800
C	Lime $^7/_8$	Required	Required	500
D	Cement-lime $^7/_8$	Required	Required	800
E	Cement $^7/_8$	Required	Required	800

For SI: 1 inch = 25.4mm, 1 pound per foot = 14.5939 N/m.

a. Plasters shall conform to Sections AS104.4.3 through AS104.4.8, AS106.7 and AS106.10.

b. Any metal mesh allowed by this appendix and installed in accordance with Section AS106.9.

c. In accordance with Section AS106.9.2, except as required to transfer roof loads to the plaster skins in accordance with Section AS106.11.

d. For walls with a different plaster on each side, the lower value shall be used.

AS106.12.2 Concentrated loads. Concentrated loads shall be distributed by structural elements capable of distributing the loads to the bearing wall within the allowable bearing capacity listed in Table AS106.12 for the plaster type used.

❖ Roof-bearing members or assemblies (see Section AS106.12.3) must be used to distribute concentrated loads to avoid exceeding the allowable bearing capacity listed in Table AS106.12 for the plaster type used. The load transfer to the plaster skins must be by continuous direct bearing or with an approved engineered design, as required in Section AS106.11.

Concentrated loads commonly occur at both sides of a window or door opening. For openings not exceeding 4 feet, the header (typically as part of a continuous roof-bearing assembly) may be used to transfer the load over the first 2 feet on each side of the opening (see Section AS106.12.3.1). For larger openings, studs or posts on each side of the opening are recommended. Headers must comply with Tables R602.7(1) and R602.7(2) and may be part of a continuous roof-bearing assembly.

Concentrated loads from a beam or post resting on the roof-bearing assembly can be considered to be carried by a 4-foot length of wall (2 feet on each side of the load) where the wall is unbroken by an opening.

Any uniform loads carried by a wall where a concentrated load is supported must be taken into account when determining whether the allowable bearing capacity of the wall is exceeded. Concentrated loads should be transferred approximately equally to both plaster skins, or only half the allowable bearing capacity should be used.

AS106.12.3 Roof-bearing assembly. Roof-bearing assemblies shall be of nominal 2-inch by 6-inch (51 mm by 152 mm) lumber with $^{15}/_{32}$-inch (12 mm) plywood or OSB panels fastened with 8d nails at 6 inches (152 mm) on center in accordance with Figure AS105.1(3) and Items 1 through 6, or be of an *approved* engineered design.

1. Assembly shall be a box assembly on the top course of *bales*, with the panels horizontal.
2. Assembly shall be the width of the *strawbale* wall and shall comply with Section AS106.11.
3. Discontinuous lumber shall be spliced with a metal strap with not less than a 500-pound (2224 N) allowable wind or seismic load tension capacity. Where the wall line includes a braced wall panel the strap shall have not less than a 2,000-pound (8896 N) capacity.
4. Panel joints shall be blocked.
5. Roof and ceiling framing shall be attached to the roof-bearing assembly in accordance with Table R602.3(1), Items 2 and 6.
6. Where the roof-bearing assembly spans wall openings, it shall comply with Section AS106.12.3.1

❖ See Section AS106.12.2 and its commentary.

AS106.12.3.1 Roof-bearing assembly spanning openings. Roof-bearing assemblies that span openings in *strawbale* walls shall comply with the following at each opening:

1. Lumber on each side of the assembly shall be of the dimensions and quantity required to span each opening in accordance with Table R602.7(1).
2. The required lumber in the assembly shall be supported at each side of the opening by the number of jack studs required by Table R602.7(1), or shall shall extend beyond the opening on both sides a distance, D, using the following formula:

$$D = S \times R/2 \,/\, (1-R) \qquad \textbf{(Equation AS-1)}$$

where:

D = Minimum distance (in feet) for required spanning lumber to extend beyond the opening

S = Span in feet

R = B_L / B_C

B_L = Design load on the wall (in pounds per lineal foot) in accordance with Sections R301.4 and R301.6

B_C = Allowable bearing capacity of the wall in accordance with Table AS106.12

❖ See Section AS106.12.2 and its commentary.

AS106.13 Braced wall panels. Plastered *strawbale* walls used as braced wall panels for one-story *buildings* shall be in accordance with Section R602.10 and Tables AS106.13(1), AS106.13(2) and AS106.13(3). Wind design criteria shall be in accordance with Section R301.2.1. Seismic design criteria shall be in accordance with Section R301.2.2.

❖ Plastered strawbale walls can be used as braced wall panels (shear walls) to resist in-plane wind and/or seismic loads where they are constructed in accordance with Table AS106.13(1), and where meeting the minimum total lengths stipulated in Table AS106.13(2) for wind loads and Table AS106.13(3) for seismic loads. The greater of the lengths in Tables AS106.13(2) and AS106.13(3) must be used, as required by Section AS106.2, Item 4.

Testing of plastered strawbale walls has demonstrated their ability to absorb and dissipate energy under extreme in-plane loading. The reinforced plaster provides an in-plane lateral load-resisting element with stiffness comparable to wood panels on light wood framing. The strawbale core braces the relatively thin plaster skins and provides significant backup capacity, including the ability to support superimposed gravity loads.

Conventional hold-downs are not required on strawbale braced wall panels to resist overturning from seismic or wind forces. The required reinforcing mesh and the weight of the roof and wall assemblies provide adequate resistance to overturning forces from seismic and wind loads.

TABLE AS106.13(1). See below.

❖ The strawbale braced wall panel types in Table AS106.13(1) are with plaster on both sides. Plaster is permitted to be installed on one side only if the minimum total braced wall panel lengths shown in Tables AS106.13(2) and AS106.13(3) are doubled.

For strawbale braced wall panels used to resist wind forces in accordance with Table AS106.13(2) or seismic forces in accordance with Table AS106.13(3), it is recommended that the mesh required in Table AS106.13(1) be supplemented by a 1-foot wide band of mesh along the sill plate with the same fastening as for the required mesh. This improves the transfer of forces to the sill plate.

Although strawbale braced wall panel type A1 appears in Table AS106.13(1), it is not listed as a braced wall panel type in Table AS106.13(2) for resisting wind loads or in Table AS106.13(3) for resisting seismic loads. Only mesh-reinforced plasters are used to resist in-plane lateral loads in this appendix. However, wall panel type A1 can be used as a nonload-bearing or load-bearing wall in conjunction with conventional braced wall panels in accordance with Tables R602.10.3(1) and R602.10.3(3). The lengths in Table R602.10.3(3) must increased by 60 percent for buildings in Seismic Design Categories C, D_0, D_1 and D_2 as required by Section AS105.2, Item 4. See the commentary to Section AS105.2.

TABLE AS106.13(2). See page S-26.

❖ The wind speed used in this table is the ultimate design wind speed in Table R301.2(1) as determined from Figure R301.2(5)A in accordance with Section R301.2.1.

Note c clarifies the minimum total lengths of braced wall panels along each braced wall line in Table AS106.13(2). Where the required length is satisfied with a single braced wall panel, the aspect ratio (H:L) of the braced wall panel must not exceed 2:1. The required length may also be satisfied using multiple braced wall panels. If each of those braced wall panels has an aspect ratio not exceeding 1:1, the sum of their lengths must simply satisfy the required total length in the table. If any of those braced wall panels has an aspect ratio exceeding 1:1, the required total length must be increased by multiplying it by the largest aspect ratio (H:L) of the braced wall panels in that line.

Where a strawbale braced wall panel ends at a corner, the lengths of its plaster skins on opposite sides of the wall may differ. In such cases the braced wall panel length should be taken as the average of the two lengths.

TABLE AS106.13(3). See page S-27.

❖ For buildings located in Seismic Design Categories A and B, the minimum total length of braced wall panels is governed by the basic wind speed in accordance with Table AS106.13(2).

Buildings located in Seismic Design Categories C, D_0, D_1, and D_2 are also subject to wind loads. The required length of their bracing along each braced wall line is the greater value determined from Table AS106.13(2) or AS106.13(3), with applicable adjustment factors.

Soil Class D is assumed in these tables and in Table R602.10.3(3), which applies to wood-frame construction. For site-specific soil classes other than D, adjustments can be made to the minimum total length of strawbale braced wall panels, as is allowed for braced wall panels in wood-frame construction in accordance with Note b of Table R602.10.3(3).

Note c clarifies the minimum required total lengths of braced wall panels along each braced wall line in Table

TABLE AS106.13(1)
PLASTERED STRAWBALE BRACED WALL PANEL TYPES

WALL DESIGNATION	PLASTER[a] (both sides)		SILL PLATES[b] (nominal size in inches)	ANCHOR BOLT[c] SPACING (inches on center)	MESH[d] (inches)	STAPLE SPACING[e] (inches on center)
	Type	Thickness (minimum in inches each side)				
A1	Clay	1.5	2 × 4	32	None	None
A2	Clay	1.5	2 × 4	32	2 × 2 high-density polypropylene	2
A3	Clay	1.5	2 × 4	32	2 × 2 × 14 gage	4
B	Soil-cement	1	4 × 4	24	2 × 2 × 14 gage	2
C1	Lime	$^7/_8$	2 × 4	32	17-gage woven wire	3
C2	Lime	$^7/_8$	4 × 4	24	2 × 2 × 14 gage	2
D1	Cement-lime	$^7/_8$	4 × 4	32	17 gage woven wire	2
D2	Cement-lime	$^7/_8$	4 × 4	24	2 × 2 × 14 gage	2
E1	Cement	$^7/_8$	4 × 4	32	2 × 2 × 14 gage	2
E2	Cement	1.5	4 × 4	24	2 × 2 × 14 gage	2

SI: 1 inch = 25.4 mm.

a. Plasters shall comply with Sections AS104.4.3 through AS104.4.8, AS106.7, AS106.8 and AS106.12.

b. Sill plates shall be Douglas fir-larch or southern pine and shall be preservative treated where required by the *International Residential Code*.

c. Anchor bolts shall be in accordance with **Section** AS106.13.3 at the spacing shown in this table.

d. Installed in accordance with **Section** AS106.9.

e. Staples shall be in accordance with **Section** AS106.9.2 at the spacing shown in this table.

AS106.13(3). Where the required length is satisfied with a single braced wall panel, the aspect ratio (H:L) of the braced wall panel must not exceed 2:1. The required length may also be satisfied using multiple braced wall panels. If each of those braced wall panels has an aspect ratio not exceeding 1:1, the sum of their lengths must simply satisfy the required total length in the table. If any of those braced wall panels has an aspect ratio exceeding 1:1, the required total length must be increased by multiplying it by the largest aspect ratio (H:L) of the braced wall panels in that line.

Where a strawbale braced wall panel ends at a corner, the lengths of its plaster skins on opposite sides of the wall may differ. In such cases, the braced wall panel length should be taken as the average of the two lengths.

Braced wall panel designation A3 is not intended to be excluded from use in resisting seismic forces in accordance with Table AS106.13(3). If this panel type is used, the minimum total lengths should be 25 percent greater than the lengths required for braced wall panel designation A2.

TABLE AS106.13(2)
BRACING REQUIREMENTS FOR STRAWBALE-BRACED WALL PANELS BASED ON WIND SPEED

• EXPOSURE CATEGORY B[d] • 25-FOOT MEAN ROOF HEIGHT • 10-FOOT EAVE-TO-RIDGE HEIGHT[d] • 10-FOOT WALL HEIGHT[d] • 2 BRACED WALL LINES[d]			MINIMUM TOTAL LENGTH (FEET) OF STRAWBALE BRACED WALL PANELS REQUIRED ALONG EACH BRACED WALL LINE[a, b, c, d]		
Ultimate design wind speed (mph)	Story location	Braced wall line spacing (feet)	Strawbale-braced wall panel[e] A2, A3	Strawbale-braced wall panel[e] C1, C2, D1	Strawbale-braced wall panel[e] B, D2, E1, E2
≤ 110	One-story building	10	6.4	3.8	3.0
		20	8.5	5.1	4.0
		30	10.2	6.1	4.8
		40	13.3	6.9	5.5
		50	16.3	7.7	6.1
		60	19.4	8.3	6.6
≤ 115	One-story building	10	6.4	3.8	3.0
		20	8.5	5.1	4.0
		30	11.2	6.4	5.1
		40	14.3	7.2	5.7
		50	18.4	8.1	6.5
		60	21.4	8.8	7.0
≤ 120	One-story building	10	7.1	4.3	3.4
		20	9.0	5.4	4.3
		30	12.2	6.6	5.3
		40	16.3	7.7	6.1
		50	19.4	8.3	6.6
		60	23.5	9.2	7.3
≤ 130	One-story building	10	7.1	4.3	3.4
		20	10.2	6.1	4.8
		30	14.3	7.2	5.7
		40	18.4	8.1	6.5
		50	22.4	9.0	7.1
		60	26.5	9.8	7.8
≤ 140	One-story building	10	7.8	4.7	3.7
		20	11.2	6.4	5.1
		30	16.3	7.7	6.1
		40	21.4	8.8	7.0
		50	26.5	9.8	7.8
		60	30.6	11.0	8.3

For SI: 1 inch = 25.4 mm, 1 foot = 305 mm, 1 mile per hour = 0.447 m/s.

a. Linear interpolation shall be permitted.

b. All braced wall panels shall be without openings and shall have an aspect ratio (H:L) ≤ 2:1.

c. Tabulated minimum total lengths are for braced wall lines using single-braced wall panels with an aspect ratio (H:L) ≤ 2:1, or using multiple braced wall panels with aspect ratios (H:L) ≤ 1:1. For braced wall lines using two or more braced wall panels with an aspect ratio (H:L) > 1:1, the minimum total length shall be multiplied by the largest aspect ratio (H:L) of braced wall panels in that line.

d. Subject to applicable wind adjustment factors associated with "All methods" in Table R602.10.3(2)

e. Strawbale braced panel types indicated shall comply with Sections AS106.13.1 through AS106.13.3 and with Table AS106.13(1).

AS106.13.1 Bale wall thickness. The thickness of *strawbale* braced wall panels without their plaster shall be not less than 15 inches (381 mm).

❖ The thickness of a strawbale wall contributes to its stability under strong ground shaking because: the width of the bale ensures overlying bales are well supported as gaps open during in-plane rocking; the width of the bale enhances out-of-plane strength and stiffness; and lateral restoring forces are developed as walls rock out-of-plane. This section is not meant to restrict the use of two-string bales on edge [which typically results in a bale wall thickness of 14 inches (356 mm)] where a history of good experience exists and where wall-height-to-thickness requirements in Table AS105.4 are satisfied, particularly in Seismic Design Categories A and B. Walls with 12-inch-wide (305 mm) bales for use in small residential buildings have performed well in the field and in full-scale structural tests (Donovan, 2014, see the bibliography).

AS106.13.2 Sill plates. Sill plates shall be in accordance with Table AS106.13(1).

❖ See Section AS105.3 and its commentary for general sill plate requirements

AS106.13.3 Sill plate fasteners. Sill plates shall be fastened with not less than $^5/_8$-inch-diameter (15.9 mm) steel anchor bolts with 3-inch by 3-inch by $^3/_{16}$-inch (76.2 mm by 76.2 mm by 4.8 mm) steel washers, with not less than 7-inch (177.8 mm) embedment in a concrete or masonry foundation, or shall be an *approved* equivalent, with the spacing shown in Table AS106.13(1). Anchor bolts or other fasteners into framed floors shall be of an *approved* engineered design.

AS106.14 Resistance to wind uplift forces. Plaster mesh in *skins* of *strawbale walls* that resist uplift forces from the roof assembly, as determined in accordance with Section R802.11, shall be in accordance with all of the following:

1. Plaster shall be any type and thickness allowed in Section AS104.
2. Mesh shall be any type allowed in Table AS106.13(1), and shall be attached to top plates or roof-bearing elements and to sill plates in accordance with Section AS106.9.2.
3. Sill plates shall be not less than nominal 2-inch by 4-inch (51 mm by 102 mm) with anchoring complying with Section R403.1.6.

TABLE AS106.13(3)
BRACING REQUIREMENTS FOR STRAWBALE-BRACED WALL PANELS BASED ON SEISMIC DESIGN CATEGORY

• SOIL CLASS D[f] • WALL HEIGHT = 10 FEET[d] • 15 PSF ROOF-CEILING DEAD LOAD[d] • BRACED WALL LINE SPACING ≤ 25 FEET[d]			MINIMUM TOTAL LENGTH (FEET) OF STRAWBALE-BRACED WALL PANELS REQUIRED ALONG EACH BRACED WALL LINE[a, b, c, d]	
Seismic Design Category	Story location	Braced wall line length (feet)	Strawbale-braced wall panel[e] A2, C1, C2, D1	Strawbale-braced wall panel[e] B, D2, E1, E2
C	One-story building	10	5.7	4.6
		20	8.0	6.5
		30	9.8	7.9
		40	12.9	9.1
		50	16.1	10.4
D_0	One-story building	10	6.0	4.8
		20	8.5	6.8
		30	10.9	8.4
		40	14.5	9.7
		50	18.1	11.7
D_1	One-story building	10	6.3	5.1
		20	9.0	7.2
		30	12.1	8.8
		40	16.1	10.4
		50	20.1	13.0
D_2	One-story building	10	7.1	5.7
		20	10.1	8.1
		30	15.1	9.9
		40	20.1	13.0
		50	25.1	16.3

For SI: 1 inch = 25.4 mm, 1 foot = 305 mm, 1 pound per square foot = 0.0479 kPa.

a. Linear interpolation shall be permitted.

b. Braced wall panels shall be without openings and shall have an aspect ratio (H:L) ≤ 2:1.

c. Tabulated minimum total lengths are for braced wall lines using single braced wall panels with an aspect ratio (H:L) ≤ 2:1, or using multiple braced wall panels with aspect ratios (H:L) ≤ 1:1. For braced wall lines using two or more braced wall panels with an aspect ratio (H:L) > 1:1, the minimum total length shall be multiplied by the largest aspect ratio (H:L) of braced wall panels in that line.

d. Subject to applicable seismic adjustment factors associated with "All methods" in Table R602.10.3(4), except "Wall dead load."

e. Strawbale braced wall panel types indicated shall comply with Sections AS106.13.1 through AS106.13.3 and Table AS106.13(1).

f. Wall bracing lengths are based on a soil site class "D." Interpolation of bracing lengths between S_{ds} values associated with the seismic design categories is allowable where a site-specific S_{ds} value is determined in accordance with Section 1613.3 of the *International Building Code*.

4. Mesh attached with staples at 4 inches (51 mm) on center shall be considered to be capable of resisting uplift forces of 100 plf (1459 N/m) for each plaster skin.
5. Mesh attached with staples at 2 inches (51 mm) on center shall be considered to be capable of resisting uplift forces of 200 plf (2918 N/m) for each plaster skin.

❖ Section R802.11 gives wind-uplift-resistance requirements for roof assemblies. Mesh used to satisfy plaster reinforcement requirements for strawbale load-bearing walls or braced wall panels may also be used to resist wind uplift forces. Mesh may also be provided in the plaster of structural or nonstructural walls for this purpose alone.

Mesh must be adequately anchored as part of a complete load path in accordance with this section. Also see Section AS106.9.1, which indicates that mesh used in walls designed to resist wind uplift of more than 100 plf must run continuous vertically from sill plate to the top plate or roof-bearing element, or shall lap not less than 8 inches.

AS106.15 Post-and-beam with strawbale infill. Post-and-beam with *strawbale* infill systems shall be in accordance with Figure AS105.1(4) and Items 1 through 6, or be of an *approved* engineered design.

1. Beams shall be of the dimensions and number of members in accordance with Table R602.7(1), where the space between posts equals the span in the table.
2. Beam ends shall bear over posts not less than $1^1/_2$ inches (38 mm) or be supported by a framing anchor in accordance with Table R602.7(1).
3. Discontinuous beam ends shall be spliced with a metal strap with not less than 1,000-pound (454 kg) wind or seismic load tension capacity. Where the wall line includes a braced wall pane,l the strap shall have not less than a 4,000-pound (1814 kg) capacity.
4. Each post shall equal NJ + 1 in accordance with Table R602.7(1), where the space between posts equals the span in the table.
5. Posts shall be connected to the beam by an *approved* means.
6. Roof and ceiling framing shall be attached to the beam in accordance with Table R602.3(1), Items 2 and 6.

❖ Strawbale buildings are commonly designed and constructed with a post-and-beam structural system with straw bales "infilled" between and around the posts and beams. Typically the posts are flush with the exterior or interior face of the stacked bales. However, the posts can be in the middle of the bales or are sometimes located completely to the inside or outside of the stacked bales. Where the posts are within the width of the bales, the bales are typically notched to fit around the posts (unless the bale vertical joints align with the posts). Bales stacked on-edge cannot be notched because notching them requires cutting the ties.

Figure AS105.1(4) illustrates an acceptable post-and-beam strawbale wall system (with the alternative location of the post and beam on the interior side of the wall stated and allowed). Historically, many variations of post-and-beam strawbale wall systems have been practiced, and the illustrated system is not meant to preclude viable variations. However, systems that vary substantially from those shown in Figure AS105.1(4) must be approved by the building official.

Although most post-and-beam systems for strawbale walls are engineered, this section gives a prescriptive option using the wood wall framing provisions in Section R602 of the code. Those provisions describe conventional wood stud construction, typically with single bottom plates and double top plates, with headers installed at door and window openings supported by jack studs. This section utilizes those provisions along with its own requirements to allow prescriptive design of a post-and-beam system.

Item 1: Table R602.7(1) gives header sizes in terms of single or multiple 2x members. Those sizes can be used for a post-and-beam system, as long as they have the required bearing over the posts described in Item 2, and a connection to the posts as described in Item 5. Beams of multiple 2x members should be nailed together in accordance with Table R602.3(1), Item 10. A 4x or 6x member can be substituted for the required 2x members in the table, as long as its width and height are at least those of the cumulative 2x members.

Item 3: Where a beam consists of multiple 2x members, staggering the joints can substitute for the metal strap requirement for discontinuous beam ends, but the length and fastening of such splices should be determined by a registered design professional, consistent with the 1000 pound (2224 N) and 4000 pound (17,783 N) requirements in this item.

Item 4: The post size is determined by taking NJ (number of jack studs) from the appropriate place in Table R602.7(1) and adding 1. For example, if the loading conditions and clear space between posts (equivalent to the span in the table) dictates an NJ of 2, then the post size will be 3 (NJ +1). If the post consists of multiple 2x members (they can be 2x4 or 2x6) they should be nailed together in accordance with Table R602.3(1), Item 9. A 4x or 6x post can be substituted for the required 2x members, as long as its width is at least that of the cumulative 2x members.

Item 5: The connection of the posts to the beams can be in accordance with Table R602.3(1), Item 16, or with a metal post cap connector, or by other means, depending on the structural conditions, subject to the approval of the building official.

Although there are no requirements in this section regarding the connection at the bottom of the posts, it is recommended that the posts be supported by the sill plate of the bale wall (required in Section AS105.3 or AS106.13.2) with fastening in accordance with Table R602.3(1), Item 16, or with a metal post base connector. Where posts are not located at a sill plate, they should be supported and fastened with a design by a registered design professional

SECTION AS107 FIRE RESISTANCE

AS107.1 Fire-resistance rating. *Strawbale* walls shall not be considered to exhibit a fire-resistance rating, except for walls constructed in accordance with Section AS107.1.1 or AS107.1.2. Alternately, fire-resistance ratings of strawbale walls shall be determined in accordance with Section R302.

❖ Many variations of strawbale wall assemblies have been used in practice, but only two assemblies, tested in accordance with one of the two standards required for determining fire-resistance ratings of walls, are included in the appendix. These two assemblies, as described in Sections AS107.1.1 and AS107.1.2, can be used as fire-resistance-rated walls with their stated ratings, where a fire-resistance-rated wall is required by the code. All other strawbale wall assemblies, whether finished with plaster or another acceptable finish, are considered nonrated unless proven otherwise. Although the minimum requirement for the density of bales in accordance with Section A103.5 is 6.5 pcf, the minimum density for bales used in plastered fire-resistance-rated walls is required to be 7.5 pcf. This increase in density is based on the average density of the bale specimens that were utilized in the physical fire testing.

This section also refers to Sections R302 and 302.1 and Tables R302.1(1) and R302.1(2), which give minimum fire-resistance ratings for exterior walls in certain circumstances and state that such fire-resistance-rated walls must be tested in accordance with ASTM E119 or UL 263. The rated wall assemblies described in Sections AS107.1.1 and AS107.1.2 were tested in accordance with ASTM E119 at a certified testing laboratory.

Although only two strawbale wall assemblies are given fire-resistance ratings in this appendix (based on specific ASTM E119 tests), two decades of field experience with and testing of other assemblies, including European tests to German Institute for Standardization (DIN) standards, have more broadly shown plastered strawbale walls to be highly resistant to fire.

As with wood-frame construction or other assemblies that are flammable prior to being finished, strawbale walls are vulnerable to fire while the straw is exposed. However, because of the density of the bales and the limited oxygen within the compressed straw, even unplastered strawbale wall assemblies have shown good resistance to fire. Their burning characteristics have been compared to large wood timbers where charring occurs at the surface but the lack of oxygen and the insulating effect of the charred material greatly slows combustion beyond the charred surface.

An important fire-related issue that is not addressed in this section, but is addressed in other sections of this appendix, is that strawbale walls must not contain vertical spaces that could act as a chimney to spread fire across unprotected straw where an ignition source is present. This is mostly a matter of preventing fire from spreading to wood roof or floor framing above the wall because, although fire can spread quickly across the face of bales, the compressed bales generally do not contain sufficient oxygen to sustain a flame. Vertical spaces in a strawbale wall that must be eliminated include: space between the face of the stacked bales and their finish; space between bales (especially in a stack bond wall); and spaces between vertical framing members and the straw where straw bales are used as infill in a wood frame wall.

Section AS104.2 requires either a plaster finish on strawbale walls or that the bales "fit tightly against a solid wall panel" (such as plywood, fiberboard, or gypsum board). This is intended to address the first condition described in the preceding paragraph. Section AS106.5 requires stuffing of voids for structural walls between bales, but this is important for all walls as a matter of fire blocking to address the second and third conditions described in the preceding paragraph, both between bales and between bales and any vertical framing. Bales should be notched around framing members wherever possible, or the spaces should be stuffed with straw or straw-clay.

Section R302.10.1 requires that the flame spread index and smoke-developed index of insulation not exceed 25 and 450 respectively. Straw bales in exterior strawbale walls typically function as insulation (among other functions) and thus are subject to this section. Straw bales have been shown through testing to meet these requirements. See Section AS108.2 and its commentary.

AS107.1.1 One-hour-rated clay-plastered wall. One-hour fire-resistance-rated nonload-bearing clay plastered *strawbale* walls shall comply with all of the following:

1. *Bales* shall be laid flat or on-edge in a running bond.
2. *Bales* shall maintain thickness of not less than 18 inches (457 mm).
3. *Bales* shall have a minimum density of 7.5 pounds per cubic foot (120 kg/m^3).
4. Gaps shall be stuffed with *straw-clay*.
5. *Clay* plaster on each side of the wall shall be not less than 1 inch (25 mm) thick and shall be composed of a mixture of 3 parts clay, 2 parts chopped straw and 6 parts sand, or an alternative approved *clay* plaster.
6. Plaster application shall be in accordance with Section AS104.4.3.3 for the number and thickness of coats.

❖ See the commentary to Section AS107.1.

AS107.1.2 Two-hour-rated cement-plastered wall. Two-hour fire-resistance-rated nonload-bearing cement-plastered strawbale walls shall comply with all of the following:

1. Bales shall be laid flat or on-edge in a running bond.
2. Bales shall maintain a thickness of not less than 14 inches (356 mm).
3. *Bales* shall have a minimum density of 7.5 pounds per cubic foot (120 kg/m^3).
4. Gaps shall be stuffed with *straw-clay*.

5. A single section of $^1/_2$-inch (38 mm) by 17-gage galvanized woven wire mesh shall be attached to wood members with $1^1/_2$-inch (38 mm) staples at 6 inches (152 mm) on center. 9 gage U-pins with not less than 8-inch (203 mm) legs shall be installed at 18 inches (457 mm) on center to fasten the mesh to the *bales*.
6. Cement plaster on each side of the wall shall be not less than 1 inch (25 mm) thick.
7. Plaster application shall be in accordance with Section AS104.4.8 for the number and thickness of coats.

❖ See the commentary to Section AS107.1.

AS107.2 Openings in rated walls. Openings and penetrations in *bale* walls required to have a fire-resistance rating shall satisfy the same requirements for openings and penetrations as prescribed in this code.

AS107.3 Clearance to fireplaces and chimneys. *Strawbale* surfaces adjacent to fireplaces or chimneys shall be finished with not less than $^3/_8$-inch-thick (10 mm) plaster of any type permitted by this appendix. Clearance from the face of such plaster to fireplaces and chimneys shall be maintained as required from fireplaces and chimneys to combustibles in Chapter 10, or as required by manufacturer's instructions, whichever is more restrictive.

❖ Section R1001.11 addresses the required clearances from masonry fireplaces to combustible materials, Section R1002.5 addresses the required clearances from masonry heaters to combustible materials and Section R1003.18 addresses the required clearances from masonry chimneys to combustible materials.

Strawbale walls must maintain those clearances to the face of the strawbale wall's required plaster. For factory-built fireplaces and chimneys, strawbale walls must maintain the clearances to combustible materials required by the manufacturer's installation instructions, from the fireplace or chimney to the face of the strawbale wall's required plaster.

These clearance requirements are more conservative for strawbale construction than for wood-frame construction, adding both a finish requirement and the resulting increased distance to the combustible material in the assembly. Whereas wood framing is allowed to be directly exposed in these locations, this appendix requires the straw to be finished with plaster, and the distances required are to the face of the plaster, thereby increasing the distance to combustible material by the thickness of the plaster.

The requirement for a plaster finish is not intended to preclude the use of materials that afford similar protection, such as gypsum board, wood framing, or plywood, where those materials are tight against the straw.

SECTION AS108 THERMAL INSULATION

AS108.1 R-value. The unit *R*-value of a *strawbale* wall with bales laid flat is R-1.55 for each inch of *bale* thickness. The unit *R*-value of a *strawbale* wall with *bales* on-edge is R-1.85 for each inch of *bale* thickness.

❖ The unit *R*-values in this section are used for exterior strawbale walls to determine compliance with Section N1102.1.1 and Table N1102.1.1. Following the principles of Section N1102.1.2, only the *R*-value of the straw (as the cavity insulation) and any insulating sheathing should be used when computing the *R*-value, not finishes such as plaster or siding. The computed *R*-value for the strawbale wall must be at least the *R*-value in Table N1102.1.1 in the column "Wood Frame Wall *R*-value," for the climate zone where the structure is located.

The unit *R*-values in this section reflect data and analysis based on thermal resistance tests conducted in Denmark (2004) and the United Kingdom (2012), along with tests at the Oak Ridge National Laboratory (ORNL) in Tennessee (1998). The ORNL tests were conducted in accordance with the guarded hot box protocol of ASTM C236, and the Danish and UK tests in accordance with its ISO equivalent, ISO 8990. (See the bibliography at the end of this appendix for *Thermal Performance of Straw Bale Wall Systems - II.*)

Strawbale walls have been shown to have different unit *R*-values depending on the orientation of the bales (i.e., laid flat or on-edge). This is because industrialized baling machines typically create bales where the strands of straw are predominantly oriented across the "width" of the bale (across its strings). Straw conducts heat more readily along than across its length (where more air pockets interrupt the flow of heat). Thus, the thermal resistance per inch of a bale on-edge is typically greater than that of a bale laid flat.

Other factors that affect the thermal performance of a strawbale wall include bale density, adhesion of the facing materials (e.g., plaster) to the straw bales and homogeneity of the wall system. Although a minimum dry density of 6.5 lbs/cu.ft. is required, higher-density bales generally provide higher thermal performance. Additionally, good adhesion of a plaster finish to the straw bales reduces air infiltration, which helps maintain or increase the *R*-value of the assembly.

Finally, any gaps between bales, or between a bale and other wall elements such as framing members, should be tightly filled with straw or straw-clay. Otherwise a significant negative impact on the overall thermal performance of the wall assembly can occur. See Section AS105.8 for information regarding voids and stuffing.

Straw bales have been and may be used as insulation in nonstructural strawbale walls in any orientation, including laid flat, on-edge, and on-end (see Figure AS102.1). In nonstructural strawbale walls, elements in the same wall other than the bales and their plaster, such as wood or steel framing or structural panels, may carry structural loads.

In addition to their function as thermal insulation in walls, straw bales have been used as insulation in raised floors and ceilings. However, issues of structure and of protection from fire must be addressed where straw bales are used in these locations. The floor or ceiling structure must be adequate for carrying the additional weight of the bales, for both static and seismic loading. A complete load path to the foundation must be provided for these loads. For fire protection the bales must be tightly confined on all sides (similar to their required confinement in a wall system) by materials such as plaster, wood framing, plywood or gypsum board.

AS108.2 Compliance with Section R302.10.1. *Straw bales* meet the requirements for insulation materials in Section R302.10.1 for flame spread index and smoke-developed index as tested in accordance with ASTM E84.

❖ Section R302.10.1 requires that the flame spread index and smoke-developed index of insulation not exceed 25 and 450 respectively, as tested in accordance with ASTM E84 or UL 723. Straw bales in exterior strawbale walls typically function as insulation (among other functions) and thus are subject to this requirement. An ASTM E84 test of unplastered straw bales conducted in an accredited testing laboratory in the year 2000 indicated a flame spread index of 10 and a smoke-developed index of 350, thus satisfying the requirements of Section R302.10.1.

Compliance is stated in Section AS108.2 to assist building officials and design and building professionals who may be unaware of this ASTM E84 test that demonstrates that straw bales meet these requirements. The test report is posted at: http://ecobuildnetwork.org/projects/straw-bale-construction-supporting-documents.

SECTION AS109 REFERENCED STANDARDS

ASTM C5—10	Standard Specification for Quicklime for Structural Purposes	AS104.4.6.1
ASTM C109/C 109M—2015el	Standard Test Method for Compressive Strength of Hydraulic Cement Mortars	AS106.6.1
ASTM C141/C 141M—14	Standard Specification for Hydrated Hydraulic Lime for Structural Purposes	AS104.4.6.1
ASTM C206—14	Standard Specification for Finishing Hydrated Lime	AS104.4.6.1
ASTM C926—15B	Standard Specification for Application of Portland Cement Based Plaster	AS104.4.7 AS104.4.8
ASTM C1707—11	Standard Specification for Pozzolanic Hydraulic Lime for Structural Purposes	AS104.4.6.1
ASTM E2392/ ASTM E2392M —10	Standard Guide for Design of Earthen Wall Building Systems	AS104.4.3.2
ASTM BS1 ASTM BS EN 459 —2015	Part 1: Building Lime. Definitions, Specifications and Conformity Criteria; Part 2: Test Methods	AS104.4.6.1

Bibliography

The following resource materials were used in the preparation of the commentary for this appendix of the code:

ASCE 7–10, *Minimum Design Loads for Buildings and Other Structures*. Reston, VA: American Society of Civil Engineers, 2010.

Aschheim, Mark, S. Jalali, C. Ash, K. Donahue, and M. Hammer. "Allowable Shears for Plastered Straw-Bale Walls." *Journal of Structural Engineering*. New York, NY: American Society of Civil Engineers, July 1, 2014.

Aschheim, Mark, Kevin Donahue, and Martin Hammer. *Strawbale Construction—Parts 1 & 2*, Structure Magazine. Chicago, IL: National Council of Structural Engineers Associations, September & October 2012.

ASTM C109–12, *Standard Test Method for Compressive Strength of Hydraulic Cement Mortars*. West Conshohocken, PA: ASTM International, 2012.

CMHC Technical Series 08-107, *Effect of Mesh and Bale Orientation on Strength of Straw Bale Walls*. Ottowa, Ontario: Canada Mortgage and Housing Corporation, 2008.

DOE-G010094–01, *House of Straw*. Washington DC: U.S. Department of Energy, Office of Energy Efficiency and Renewable Energy, 1995.

Donovan, Darcey. *Seismic Performance of Innovative Straw Bale Wall Systems*, NEES Research Report 2009-0666. Network for Earthquake Engineering Simulation, 2014. https://nees.org/warehouse/project/666

Eisenberg, David, and Martin Hammer. "Strawbale Construction and Its Evolution in Building Codes." *Building Safety Journal*. Washington, DC: International Code Council, Inc., February 2014.

Hammer, Martin. "The Status of Straw-bale Codes and Permitting Worldwide," *The Last Straw Journal*. Lincoln, NB: Green Prairie Foundation, Issue #54, April 2006.

King, Bruce, et al. *Design of Straw Bale Buildings*. San Rafael, CA: Green Building Press, 2006.

King, Bruce. "Straw-Bale Construction: A Review of Testing and Lessons Learned to Date." *Building Safety Journal*. Washington, DC: International Code Council, Inc., May-June 2004.

King, Bruce. "Straw-Bale Construction: What Have We Learned?" *The Last Straw Journal*. Lincoln, NB: Green Prairie Foundation, Issue #53, Spring 2006.

Listiburek, Joseph. *Builder's Guide to Mixed Climates*. Newtown, CT: Taunton Press, Inc., 2000.

Lstiburek, Joseph. *Insulations, Sheathings and Vapor Retarders*, Research Report 0412. Westford, MA: Building Science Press, 2004.

Magwood, Chris. "Nylon Strapping for Pre-tensioning Bale Walls." *The Last Straw Journal*. Lincoln, NB: Green Prairie Foundation, Issue #53, Spring 2006.

Parker, Andrew, M. Aschheim, D. Mar, M. Hammer, and B. King. "Recommended Mesh Anchorage Details For Straw Bale Walls." *Journal of Green Building*. Glen Allen, VA: College Publishing, Vol.1, No.4, 2004.

Racusin, Jacob and Ace McArleton, *The Natural Building Companion*. White River Junction, VT: Chelsea Green Publishing, 2012.

Stone, Nehemiah. *Thermal Performance of Straw Bale Wall Systems*. San Rafael, CA: Ecological Building Network, 2003.

Stone, Nehemiah. *Thermal Performance of Straw Bale Wall Systems - II*, 2016, posted at: http://ecobuildnetwork.org/projects/straw-bale-construction-supporting-documents

Straube, John. "Moisture Basics and Straw-Bale Moisture Basics." *The Last Straw Journal*. Lincoln, NB: Green Prairie Foundation, Issue #53, Spring 2006.

The Straw Bale Alternative Solutions Resource. The Province of British Columbia, Canada: Alternative Solutions Resource Initiative, 2013.

Appendix 1
Fire and Straw Bale Walls

(See also Chapter 7, IRC Appendix S and Commentary, Section AS107 Fire Resistance)

Straw bale structures are fire-resistant because dense straw bales covered with thick plasters resist both ignition and heat transference. A straw bale wall with one inch of exterior cement-lime plaster has earned a two-hour fire rating, and with earthen plasters, a one-hour fire rating. This equals or exceeds many conventional wood-frame walls that offer a one-hour or 20-minute fire rating. However, plasters can transmit heat to the straw behind them, just as sheetrock can transmit heat to flammable insulation for framing. Over time—sometimes years—the straw could ignite if high heat sources are too close to the wall. There haven't been many documented instances of straw bale house fires, but reports suggest that fire in a plastered straw bale wall will "smolder" and "creep around." Building occupants may smell smoke or see smoke seeping from cracks in the plaster. Parts of the wall may feel warm to the touch. Place wood-burning stoves and other heat sources like flues the required clearances from plastered straw bale walls.

Fire Safety During Construction

Plumbers and other subcontractors often work with halogen lamps, torches, welding equipment, angle grinders, and other heat, flame, and spark sources. Loose straw and unplastered bale walls readily ignite; if fire spreads to wood framing, it could engulf the entire building. During the bale stack, keep loose straw swept and stored away from the building.

Remind contractors daily to take extra care to keep sparks and flames away from straw. If flame-producing activity must occur near the wall, wet it before the work and then carefully check for embers or smoldering when the work is completed. Keep working hoses and fire extinguishers onsite.

Post signs that remind workers of the potential hazard. Provide a designated smoking area away from the building. Check the jobsite after work to shut down compressors, lights, power cords, and anything else that could spark a fire.

Instructions for the Fire Department

There haven't been enough straw bale structure fires to develop an established protocol for fighting them. Much depends on the fire's stage at the time a fire crew responds to a structure fire. When the fire is in the straw bale wall and hasn't spread to the attic (see Figures 2-21, 2-22 showing various fire barriers at the top of the wall to slow this spread), fire departments have reported removing the plaster to get at the source of smoke. This, of course, adds oxygen, allowing a smoldering fire to burst into flames. Water has been used to douse flames, but Class A foam retardant is much more effective in suppressing the fire without soaking adjacent bales. Fire departments may pull out all of the bales in the vicinity of a fire and return several times to ensure the fire is completely out.

Appendix 2
Managing Successful and Effective Work Parties

Two defining characteristics of straw bale buildings—thick, secure walls and natural plasters—lend themselves to a community effort reminiscent of "barn raisings." Many people who eventually built straw bale homes first learned about straw building by helping at a neighbor's or friend's bale raising or plaster party.

Work parties can be successful and effective; accomplishing the goals of raising the bale walls or applying a plaster coat, and participants can enjoy and learn from the experience. But unprepared and disorganized work parties usually result in work that needs to be redone, and participants who leave feeling their time wasn't well spent, holding a negative view of straw bale building, or natural building in general.

Not every project is suited to a work party involving enthusiastic volunteers—many of whom have no experience working on a construction site, let alone with bales or plasters. Although usually avoidable, accidents can happen. If a builder's liability insurance doesn't cover work parties, make sure that the homeowner's liability insurance does. Work sites may be remote or have few amenities to support a work party; the project may be technical and challenging; and members of the building team may be averse to having "the public" wandering around the construction site for much of a day or weekend. Many builders or owners are more comfortable hiring experienced crews for bale stacking and plastering—and for them that's the right choice.

Others take the opportunity to enlist interested family, friends, and neighbors for some hands-on exposure to building with straw, and save on labor costs in the process. People with experience running successful and effective work parties always:

- **Prepare as though a paid crew was arriving**, with everything in place and ready for everyone to work efficiently. For bale raisings, this means enough hand trucks and carts to move bales, baling needles and saws for retying, notching, and fitting bales to posts, and ladders or scaffolding to safely work the upper courses. Enough bales and baling twine must be available to complete the project. For plaster parties, it means having enough buckets, tubs, ladders, hawks, and trowels, and, of course, enough plaster ingredients so plastering proceeds without delays. Work slows or stops when the work party has too few tools to work effectively or runs out of key materials.
- **Make sure several work party members already have straw bale stacking or plastering experience** and are comfortable leading the work party. They need to instruct, manage, and monitor volunteers to ensure the bale walls are to be stacked and plastered properly.

Work Parties and Workshops differ in significant ways. Work party volunteers work in exchange for experience or to help a neighbor or friend. Workshop participants pay an instructor to teach them how to plaster or build with straw bales. The instructor may spend time demonstrating a variety of alternative methods that may not be part of the current building project. While much the same ground is covered, the work party emphasis is getting the job done, while the workshop focuses on learning.

Reach out to local or regional natural building organizations to learn of experienced people in your area. Join with a straw building association and participate in other work parties. Many do this to gain experience, learn how to plan and manage a work party, and recruit experienced workers. Consider hiring a small crew of experienced professionals to manage volunteer bale stacking or plastering activities. You can also propose your project as a workshop, where a paid instructor teaches participants about building with straw or plastering.

- **Have someone in charge of the overall project.** The work at successful, effective work parties progresses steadily, with everyone engaged in a seamless, coordinated process. People with experience working together sometimes achieve this automatically, but more often it's consciously designed into the work party. Someone needs to keep the overall project in mind, monitoring material flow, team progress, orienting and plugging newly arrived volunteers into the work, and looking ahead to avert potential bottlenecks. Many work parties divide into teams led by one or more experienced people who focus on a smaller picture—stacking the bales and stuffing between them, keeping the walls straight, keeping the team supplied with plaster, or applying it at a uniform thickness. Telltale indications that people need more direction include standing around wondering what to do or how to do it, or uncoordinated efforts.
- **Keep things fun and involve everyone.** Take short "learning" breaks: show the group correct bale alignment, gap stuffing, or efficiently applying plaster—things everyone must know. More experienced workers can teach interested volunteers new skills. Work parties often involve people of all ages and abilities—someone needs to stomp bales into place, sweep up loose straw, scrape up clods of fallen plaster for remixing... there's a job for everyone! Hold a "friendly competition" between teams to see who builds the best wall (per project goals) as judged by the work party leader.
- **Anticipate avoidable problems.** Construction sites can be hazardous—sharp tools, dust, trip hazards, tools falling from ladders, etc. People need to arrive properly clothed for the work. Most people benefit from wearing gloves, especially when working with sharp tools or materials. A hat with visor is handy when applying plasters overhead, and rubber gloves, safety glasses, and long sleeves are a must when working with lime plasters. Review jobsite safety before the work party starts. This may include a two-person bale lift demonstration. Make sure that experienced people using electric chainsaws operate them in a "safe zone" away from passersby, and that others know to steer clear of chainsaw operator. Those using cutting tools like chainsaws must wear safety glasses and dust masks.
- **Provide amenities.** It's nice to rest in the shade, wash hands before a meal, and, if it's hot, refresh or relax with a cold drink; or, if the weather is cold, it's good to take a break in a warm place with a hot drink. Don't forget about toilets and cleanup stations, too. Work parties lasting longer than a day should offer a place to clean up and overnight lodging for those traveling a long distance.
- **Reward volunteers in some way.** For some, the opportunity to learn about straw building and spend time with friends and neighbors is reward enough. But still, thank participants—after all, their effort may contribute to lowering the project cost by thousands of dollars. Nothing says "THANKS" better than a well-prepared feast! Depending on the situation, a leisurely, satisfying lunch or an all-out dinner with music and dancing may be called for.

Glossary

ABS: Acrylonitrile butadiene styrene. A plastic used for plumbing waste and vent pipes.

Beam: In a post-and-beam system, a structural member spanning from post to post and carrying the roof or floor load above. Usually 4x lumber.

Bearing Wall: A wall that supports gravity loads (weights of building materials, occupants, moveable contents, and accumulated snow).

Blocking: Short pieces of dimensional lumber used to fill, reinforce, or provide attachment.

Boundary: The perimeter or edge of structural *element.*

Box Beam: A roof or floor bearing assembly at the top of a straw bale wall usually fabricated with 2x lumber and plywood top and bottom panels.

Buck: A rough wood frame for windows or doors.

Clay Slip: A suspension of clay particles in water used in making straw-clay or as a fire-retardant and adhesion coat prior to plastering straw bale walls.

Cob: Clay subsoil mixed with straw and sand applied wet in free-form masses to create a monolithic wall.

Cold Joint: A joint in plaster, concrete, or other wet-applied material where one application that has partly or fully dried or cured meets a subsequent application.

Composite: Two or more structural materials interacting as a system. In a plastered straw bale wall, the interacting materials include:

- The core of stacked straw bales
- Plaster on both sides of the wall
- The plaster's reinforcing mesh

Diaphragm: The horizontal structural element (usually a floor or roof) used to distribute lateral seismic or wind forces to vertical elements (shear walls, diagonal braces, or moment frames) of the lateral force-resisting system.

Drift: In single-story buildings, the lateral displacement of the roof relative to the foundation. For a wall, the lateral displacement of the top of the wall relative to the base.

Drywall: (aka sheetrock, wallboard, plasterboard, gypsum board) A gypsum panel sandwiched between paper, and used to cover ceilings and interior walls.

Ductility: The property of a material that allows it to sustain *inelastic* deformations (such as cracks, leaning, or other visible permanent change) without breaking. A ductile structure is able to sustain its load-carrying capacity and dissipate energy when subjected to cyclic inelastic displacements during an earthquake.

Eaves: The roof edges overhanging the exterior wall surface.

Elastic: The ability of a material or structural system to deform due to external force, and return to its original shape after the force is removed.

Element: An assembly of structural materials or components that act together in carrying or resisting forces, including structural frames, floor and roof *diaphragms*, and load-bearing walls or shear walls.

Flashing: Material used to cover and protect roof, wall, window and door joints from water intrusion by safely shedding water.

Fresco: The technique of applying pigment to wet plaster; the pigment penetrates into the plaster and becomes fixed as the plaster cures.

Frieze Board: Trim covering the gap between the wall top and soffit.

Head: Upper horizontal member of a window.

Header: Framing member that spans openings and transfers loads to vertical framing.

HVAC: Heating, Ventilation, Air Conditioning.

Hydrophilic: water-attracting, drawing moisture away from other materials, (e.g., clay).

Lintel: Horizontal support above a wall opening.

Inelastic: Material or structural system behavior where deformation occurs due to external force, and remains permanent after the force is removed.

In-Plane Load: A force parallel to the strongest or primary plane of a structural element.

Lateral Force-Resisting System: The combination of shear walls, braced frames, moment frames, and horizontal diaphragms in a building designed to resist wind and seismic lateral loads.

Ledger: In straw bale walls, horizontal lumber installed to support a porch roof, floor, or cabinets.

Load-Bearing: A structural wall or frame used to carry gravity loads. Plastered straw bale walls used to carry gravity loads are classified as load-bearing.

Load Path: A path of elements and connections, through which forces in a building travel to its foundation.

Load-Resisting or (Force-Resisting): A structural wall or other element used to resist any structural load, including static (gravity) and dynamic (seismic and wind) loads in any direction. A load-resisting wall can resist gravity loads, lateral loads, or both.

LVL: Laminated Veneer Lumber

Niche: Shallow recess carved into a wall. Sometimes referred to as a "nicho," especially in the Southwest.

Out-of-Plane Load: A force perpendicular to the strongest or primary plane of a structural element.

Pins: Vertical elements, the full height of a straw bale wall, employed to stiffen walls during construction and/or to resist out-of-plane loads, especially for walls with a large height-to-thickness ratio. Steel, wood, and bamboo pins have been used, either driven down through the courses of bales after the first course is pinned to the foundation or as opposing external pins on the face of the bales that are through-tied.

PEX: Cross-linked polyethylene. A plastic used for water-supply tubing in plumbing.

Post-and-Beam: A structural frame used to carry gravity loads. Straw bale walls containing a post-and-beam structural system are classified as non-load-bearing, where the straw bales are infilled between the frame, and function as insulation, enclosure, and a substrate for plaster. Such a plastered straw bale wall may also function as a *shear wall* if designed and constructed for that purpose.

PV: Photovoltaic.

PVC: Polyvinyl chloride. A plastic used for electrical conduit and for water-supply pipes in plumbing.

Redundancy: The provision of more than one *load path* to carry or resist loads in a structure.

Regime: Plaster application method or procedure.

Robert Pins: "U"-shaped galvanized metal staples approximately 6" long made from ⅛" diameter wire, used to secure lath and mesh to bale walls, whimsically named for the much smaller "Bobby pins," which they resemble.

Shear Wall: A wall element that transfers in-plane lateral forces (shear and overturning) from the roof or raised floor to the foundation.

Sill: Exterior horizontal shelf at the bottom of a window directs water away from wall.

Sill Plate: Continuous horizontal member at the bottom of a wall that connects it to the floor. When it connects the wall to a concrete slab, it is often called a sole plate.
Soffit: The underside of eaves from the exterior wall top to the roof's outer edge. Also the horizontal surface at the head of a deep-set window or doorway.
Stiffness: The rigidity of a material, assembly, or system. The extent to which it resists deformation in response to an applied force.
Stool: Interior horizontal shelf at the bottom of a window. See *sill.*
Straw-clay: Loose straw mixed and coated with clay slip.
Strength (or Ultimate Strength): The maximum axial, shear, or moment force that can be resisted by a structural component, element, or assembly.
Stud: 2× vertical wood framing.
Thermal Mass: A material's ability to absorb and store heat energy.
Top Plate: A continuous horizontal framing member at the top of a wall.
Vapor Permeance: A measure of the amount of vapor passing through a unit thickness under a unit vapor pressure difference. Measured in perms.
Vapor Permeability: A material's ability to allow water vapor to pass through it.
Yielding: *Inelastic* deformation of a structural element. In non-*ductile* materials, it is usually visible as cracking at the surface. Yielding always reduces the *stiffness* of the element, but not necessarily its *strength* (see also *Ductility*).

Principal Contributor Biographies

Janet Armstrong Johnston is a licensed architect and contractor. She graduated from the University of Cincinnati architecture program and headed to California to design and build, working with John Swearingen and Skillful Means for over ten years. She is a founding member of CASBA; she served on the advisory board and developed CASBA's first Pro-Courses. She moved to Joshua Tree to build the Harrison straw bale vault residence; it was there that she met her husband, George Armstrong. They settled in Joshua Tree to raise a family, build their own straw bale home, contribute to the desert and community, and run StrongArm Construction.

Cole Butler is an award-winning and published architect and ambassador of socially significant architecture that uses collaborative leadership to emphasize, analyze, and debate the role of the architect/citizen as cultural communicator and builder responsive to societal, cultural, and environmental challenges. She believes architecture is not about things, it's about people and life—economic feasibility and ecological responsibility are two halves of holistic sustainable design. Understanding how we can restore the environment while creating places for the soul may be this generation's great work. Cole has served on several city council environmental commissions and with various non-profit organizations.

Lesley Christiana first encountered CASBA in 1996. She is a long-standing advisory board member, and deeply committed to the use of straw as a building material. She has organized conferences, Green Building show booths, and has assisted with straw bale home tours. She also oversees CASBA's website and event registrations, and does general research. Lesley holds a British BA (Hons) in Fine Art.

Anthony Dente is a licensed engineer and principal of Verdant Structural Engineers in Berkeley, California. For over a decade, Anthony has designed natural building systems under and alongside his business partner, natural building veteran engineer Kevin Donahue, SE. Anthony is on the Cob Research Institute board of directors and has acted as field engineer and project manager for university testing programs of natural building systems. He is passionate about effective material use and supports the natural and green building community by developing design procedures to increase the number of safe and effective options available to future engineers.

Martin Hammer is an architect in California and co-director of Builders Without Borders. Throughout a 30-year career, he has emphasized sustainable building design, with particular focus on the design, testing, and construction of straw bale buildings. Martin has written and lectured widely on the subject. Since 2001, he has worked to include sustainable building systems in building codes. He is lead author of the International Residential Code's Appendix S: Strawbale Construction, and co-author of Appendix R: Light Straw-Clay Construction. Since 2006, Martin has worked extensively to introduce straw bale construction to earthquake-affected Pakistan, Haiti, and Nepal.

Devin Kinney works as an architect, designer, and illustrator in the Bay Area. For the past eight years, he has been a project manager with Arkin Tilt Architects, a firm that focuses on sustainable design utilizing natural materials like straw. He has helped with the design, permitting, and construction of 14 straw bale homes and has provided illustrations for multiple issues of *Home Energy Magazine* and the second edition of *No Regrets Remodeling*. Devin has a BA in Architecture from UC Berkeley and has been a CASBA member since 2013.

John Koester came to CASBA interested in becoming an owner-builder. When he learned they were working on a book about details (and looking for volunteers), he thought he could assist the organization while learning more about straw bale construction. John leveraged his background as a graphic artist to bring to life many of the details illustrated in this book. He and his wife Kirsten Sjoquist, former CASBA web master and advisory board member, plan to begin construction on their straw bale house this year.

Dennis LaGrande is a third-generation farmer growing rice, wheat, and hay in the Northern Sacramento Valley. When new state regulations restricted post-harvest residue burning in the mid-1990s, Dennis developed techniques to produce two-string rice straw bales for the emerging straw bale building market. An early CASBA member, Dennis supplied bales for many Northern California and Southern Oregon building projects, CASBA workshops, and wall assembly tests.

Darcey Messner became involved in residential design, engineering, and construction in 1994. In 2000, she established EcoEngineering, a structural engineering consulting practice that specialized in designing sustainable residences, including 25 straw bale buildings. In 2006, Darcey co-founded the Pakistan Straw Bale and Appropriate Building (PAKSBAB) organization, which built 40 straw bale houses and trained 70 local builders in Pakistan. She was the principal investigator for PAKSBAB's seismic research project at the University of Nevada-Reno, which culminated in successful shake table tests of a full-scale straw bale house. She currently works as Plans Examiner for the town of Truckee, California.

Celine Pinet has a PhD in Architecture and is a lifelong educator, serving as Dean of Academic Affairs at Hartnell College. Through her career, she has taught research, life cycle analysis, and green design concepts. Her most recent efforts secured funding for several artistic performances on water and environmental conservation from the perspective of social activism.

Jim Reiland is a general contractor and owner of Many Hands Builders in Southern Oregon, where he lives in a straw bale home with his wife, Joy Rogalla. He started out as an owner-builder, then left a long career in sales and marketing to build full-time. He has worked on over 50 straw bale structures in Northern California and Oregon, in all phases of straw bale construction and remodeling, including renovation after fire and water damage. He joined CASBA in 2004 and has been on the advisory board since 2006. He is active in CASBA's workshop program.

Tim Rudolph is a California professional engineer and long-time CASBA member. He graduated from UC Irvine with a BS in Engineering and BA in Social Ecology. He has provided engineering on residential straw bale buildings since 2001. He was a peer reviewer for the straw bale building code proposals and was on a FEMA P695 committee related to straw bale shear walls. In 2003, he and wife, Dadre, designed and built the third straw bale structure in Mono County, California. They live in northern San Diego county.

Dan Smith is the principal of dsa architects in Berkeley, California, and has several decades of sustainable design experience. The projects of his award-winning firm include net-zero energy homes, LEED platinum and gold designs, as well as AIA sustainability awards. He is a founding member of CASBA and has completed over 40 straw bale building projects. Dan contributed to *Alternative Construction: Contemporary Natural Building Methods*, by Lynn Elizabeth and Cassandra Adams, eds., and *The Design of Straw Bale Buildings*, Bruce King, ed.

Eric Spletzer is a professional engineer, building contractor, and a CASBA advisory board member. He holds master's degrees from UC Berkeley in both Architecture and Engineering. Since 2002, his company, Crafted Earth, Inc., has designed or built nearly a dozen straw bale structures in California, including small informal spaces, off- and on-grid homes, a family winery, and a sprawling industrial agricultural facility. He lives in Marin County with his wife and fellow eco-champion, Kirstin Weeks, and their two sidekick daughters, Hana and Luna.

Nehemiah Stone oversaw some of the earliest (1996–98) tests of straw bale wall thermal performance and has written about it in several contexts, including: The California Energy Commission's report on SB R-Values, ACEEE's Summer Study proceedings, Bruce King's *Design of Straw Bale Buildings*, as background for the 2015 and 2018 Appendix—S Strawbale Construction in the IRC, and as a sidebar in several books and articles about straw bale construction. An economist by training, Nehemiah has worked in construction and as a building inspector. He is a past co-director and current CASBA advisory board member.

John Swearingen has been designing and building homes and community and religious architecture throughout the United States for over 40 years. An early pioneer in solar design and the use of natural materials, his firm, Skillful Means, built the first permitted straw bale building in California. John has designed, built, and consulted on over 75 projects since then. John's experience with a wide range of materials and techniques has been applied to developing efficient and sound straw bale wall system detailing.

Rebecca Tasker is a general contractor and co-owner of Simple Construct, a design-build company specializing in carbon-positive building that has been involved in the construction of numerous straw bale buildings in San Diego, including the Deer Park Monastery complex. Simple Construct designed and built the Fallgren Home, the first residence in Southern California to achieve Net Zero Energy Building certification through the Living Building Challenge and is a recipient of a 2018 Excellence in Energy Leadership award. Rebecca is a published author of articles on straw bale building and frequently presents on topics related to high-performance natural building.

Bob Theis, architect, had been looking for a building method that created thick, informal walls when straw bale construction walked into his life in 1992. His first straw bale building—the second one permitted in California—followed soon thereafter, and after the 12-week slog with the building department, he was sure this was not going anywhere quickly. Within three years, it was a majority of his work. He has designed over 100 bale buildings since, and contributed to several books on the subject. Nowadays, he spends spare moments creating and testing straw bale details suitable for beginners.

Anni Tilt, AIA, is a principal of Arkin Tilt Architects, a Berkeley-based firm widely recognized for excellence in sustainability and design. A founding CASBA member, Anni has been designing with bales for over 20 years, including *Fine Homebuilding*'s 2012 "Home of the Year." Previously, she worked as a designer with internationally acclaimed Fernau + Hartman Architects, and DEGW, and as a construction coordinator for a major building contractor in Seattle. With a B.S.E. in Civil Engineering from Princeton University, Anni explored ecological use of wood in construction and taught structures and design while pursuing a master's in Architecture at UC Berkeley.

Tracy Vogel Thieriot has worked as a contractor, teacher, and consultant specializing in natural plasters and fine finishes for over 18 years. Her career has spanned local vernacular building and large-scale straw bale projects. Material specialties include earth, clay, lime, and pigments used to create thick wall plasters, fine finishes, tadelakt, fresco, moldings, showers, sinks, tubs, countertops, floors, stairs, lighting sconces, and more. Tracy has been a part of more than 30 straw bale projects and many earthen structures. She resides in Ukiah, California, where she is raising three children and consulting on natural materials and plastering.

Index

ABOUT NEW SOCIETY PUBLISHERS

New Society Publishers is an activist, solutions-oriented publisher focused on publishing books for a world of change. Our books offer tips, tools, and insights from leading experts in sustainable building, homesteading, climate change, environment, conscientious commerce, renewable energy, and more—positive solutions for troubled times.

DON'T EAT THIS BOOK *(but you could)*

We're proud to hold to the highest environmental and social standards of any publisher in North America. This is why some of our books might cost a little more. We think it's worth it!

- We print all our books in North America, never overseas
- All our books are printed on **100% post-consumer recycled paper**, processed chlorine-free, with low-VOC vegetable-based inks (since 2002)
- Our corporate structure is an innovative employee shareholder agreement, so we're one-third employee-owned (since 2015)
- We're carbon-neutral (since 2006)
- We're certified as a B Corporation (since 2016)

At New Society Publishers, we care deeply about *what* we publish—but also about *how* we do business.

Download our catalog at https://newsociety.com/Our-Catalog or for a printed copy please email info@newsocietypub.com or call 1-800-567-6772 ext 111.

New Society Publishers
ENVIRONMENTAL BENEFITS STATEMENT

For every 5,000 books printed, New Society saves the following resources:[1]

46	Trees
4,168	Pounds of Solid Waste
4,586	Gallons of Water
5,982	Kilowatt Hours of Electricity
7,577	Pounds of Greenhouse Gases
33	Pounds of HAPs, VOCs, and AOX Combined
12	Cubic Yards of Landfill Space

[1] Environmental benefits are calculated based on research done by the Environmental Defense Fund and other members of the Paper Task Force who study the environmental impacts of the paper industry.